MW00837914

Environmental Management towards Sustainability

Environmental Management towards Sustainability

Dr. Prasad Modak

CRC Press is an imprint of the
Taylor & Francis Group, an **informa** business

CRC Press
Taylor & Francis Group
6000 Broken Sound Parkway NW, Suite 300
Boca Raton, FL 33487-2742

Printed on acid-free paper

International Standard Book Number-13: 978-1-4987-9624-8 (Hardback)

Visit the Taylor & Francis Web site at
http://www.taylorandfrancis.com

and the CRC Press Web site at
http://www.crcpress.com

Contents

Foreword

Achieving sustainability in the present era is necessary for this planet's survival. Mere economic development is not the solution and gross domestic product (GDP) alone is not the true indicator of growth. Despite the economic development, we witness poverty, disparities and inequalities, and serious adverse impacts like land degradation, air pollution, water scarcity, health concerns, loss of biodiversity, etc. Overpopulation, urban–rural polarization, skews in the global material flows, poor resource use efficiency, and unchecked pollution have been some of the influencing factors. Threats to resource security due to climate change have added a new dimension to the complexity of the problem. Impacts need to be understood today, recognizing the nexus. A piecemeal or isolated approach is not going to work. The Sustainable Development Goals (SDGs), Paris Agreement on Climate Change, and a focus on sustainability provide important guiding principles for sustainable development.

To move towards sustainability requires an innovative and rounded management strategy involving all the key stakeholders with efforts taken on multiple levels. In his book, Prasad Modak introduces such a need by illustrating key global issues with top end statistics, emphasizing the nexus and relevance of the SDGs.

Modak then introduces four key groups of stakeholders, viz., the government, financing institutions, business, and the communities. These stakeholders need to work at multiple levels, viz., local, regional, national, and global. This book devotes separate chapters to introduce the role played by each stakeholder with numerous case studies drawn from across the world. This structure of this book provides the reader the stakeholder-specific focus while also enabling appreciation of the interconnections. I would say that this is the uniqueness or distinguishing feature of this book.

Chapter 2 of this book focuses on governments and illustrates the responses by the national governments that include constitutional provisions, policies, plans, establishment of legislative framework, and institutional mechanisms at various levels to monitor implementation. A range of topics such as Environmental Impact Assessment, Environmental Tribunals, Common Environment Infrastructure, and Green Public Procurement are introduced with examples. An overview of government-supported market-based instruments, economic instruments, and information-based instruments is also provided. This chapter is thus a good read to understand the multi-level and multifarious efforts taken by the national governments on environmental management across the world.

Chapter 3 is devoted to the role played by the Development Financing Institutions (DFIs) and the Private Sector Financing Institutions (PSFIs). This chapter is very well presented with information that is not generally found in the standard text or reference books. The evolution of Environmental and Social Safeguards at DFIs and PSFIs is illustrated with time stamps, and sections are built on frameworks followed at the World Bank, Asian Development Bank, and the International Finance Corporation. Topics such as Community Driven Development (CDD), Equator Principles and FIs following an investment focus such as Renewable Energy and Biodiversity are also covered. This chapter also covers various Financial Intermediary Funds, Climate Finance, Conservation Finance, Green Bonds, Sustainable Stock Exchanges, and Impact Investing. Many examples provided on each of the above makes this chapter an interesting read.

Chapters 4 and 5 are devoted to the role played by business. Several illustrations are provided about how business looks at sustainability to achieve compliance and competitiveness and leverage on branding and innovation. In this context, concepts and tools such as Life Cycle Thinking, Ecolabels, Environmental Management systems ISO 14001, Sustainable Supply Chains, and Corporate Social Responsibility (CSR) are introduced with case studies. Having described the generic strategies, Chapter 5 presents the efforts taken by business in specific sectors. The sectors covered are Transport, Information and Communication Technology (ICT), Pulp and Paper, Cement, Waste Management, Cosmetics, Mining, Social Development, etc. This chapter ends with collective industry responses and examples of industry leadership.

Finally, Chapter 6 focuses on a community's role, a community being the common binder of the key three stakeholders – government, financing institutions, and businesses. It also puts forth various innovative mechanisms devised for involving and drawing support from the communities. Examples, such as participatory urban planning and participatory environmental monitoring and the role played by judicial activism, are provided. While intersecting the community role with business and FIs, this chapter discusses various micro-finance initiatives, eco-entrepreneurship models, and social enterprises. This chapter introduces concepts such as Traditional (Ecological) Knowledge, Shared Resources, and Collaborative Consumption, emphasizing a need for behavioral change. The author wants to state, and rightly so, that the only way to move toward sustainability is to change our lifestyles and adopt sustainable consumption.

We often talk a lot about sustainability, but examples of sustainability practices across the world that are followed by the key stakeholders are rarely compiled. I therefore like the approach Prasad has used in building this book through case studies. The examples chosen are recent, relevant, and inspirational. This book is essentially a cleverly woven set of examples to introduce key concepts and strategies (sometimes done implicitly) that are geographically diverse and represent different economies. This book does not teach you fundamentals. Instead it encourages you to read more through well-cited resources. And this is typically what a professional or a policy maker or an investment banker would like to see.

Each concept, when introduced, is supported by an evidence *story* ending with discussion questions. These discussion questions are rather stimulating and can be used by professors/students to generate discussions and assignments. This structure of this book makes learning both interesting and effective. This book can be thus used as a textbook for students to get to know in a very lucid form what is happening in the field of environmental management and sustainability across the world and to get a rounded perspective.

I would sum up that Prasad Modak has put an excellent compilation structured across key stakeholders on practicing sustainability. This book will certainly meet the interests of government officials, academia, and professionals. This book could serve as a textbook at universities as well as serve as a resource in continuing education programs related to environmental management and sustainability.

I first met Prasad at the Asian Institute of Technology (AIT), Bangkok, between 1981 and 1984 as my doctoral student, certainly a very bright and committed individual. Even then during his time at AIT, he was like a professional colleague, and I also learned from him. Since then we have been in touch as professional colleagues, as well as close friends, and I have seen his career grow. He is today a leading environmental management expert, sought after across the world.

I am delighted that Prasad has put years of his experience in this book to meet the gap of a resource that blends the concepts and experience on sustainability. I would like to congratulate Prasad for all the efforts he put in to bring out such a comprehensive and much needed compilation.

Dr. Lohani is ex-Vice-President of the Asian Development Bank (ADB) for Knowledge Management and Sustainable Development. In his almost 30 years at the ADB, he held several positions, including Chief Compliance Officer; Special Advisor to the President on Clean Energy, Climate Change and Environment; the Secretary of the Bank; and Deputy Director General of the Infrastructure Department and Office of Environment and Social Department. Before joining the ADB, he worked for the Government of Nepal (relating to the infrastructure departments) and was the Associate Professor and Division Chairman of the Environmental Engineering Program at the Asian Institute of Technology (AIT), Bangkok.

Dr. Bindu N. Lohani

Preface

In the last four decades, we have seen that the population of the world has been growing rapidly. We expect that in another 20 years, nearly 10 billion people will be living on this planet. With an increase in the population and rising per capita income, we can foresee a huge increase in the consumption leading to unsustainable extraction of our finite resources.

Unfortunately, we often consume the resources at low resource use efficiency. There are also wastages of the resources during transportation and storage. As the conversion of resources to products is not perfect, wastes and emissions get released to the environment. Release of these wastes and emissions, if uncontrolled, results in contamination of the resources and deterioration of the environment. In some cases, the contamination of resources is so severe that it becomes extremely difficult, both technologically as well as financially, to restore or remediate these resources. We thus face the challenge of rapid depletion of the scarce resources and at the same time run the risk of resource degradation because of poorly managed waste streams.

As a result, we have witnessed impacts like land degradation, air pollution, water contamination, loss of biodiversity, etc. These impacts are not limited to local scale but have escalated to regional and national levels and to the global scale. Depletion of the ozone layer, increased global warming leading to climate change and threats to biodiversity, deforestation, land degradation and desertification are examples. Today, there are no geographical boundaries to environmental issues. We now know that progress cannot be understood only by indicators like the Gross Domestic Product (GDP) but we have to look at more realistic indicators like the Gross Ecosystem Product (GEP).

It is also important that we understand the issues in the context of nexus and complexity. It is not correct to address issues in isolation or in silos, and we must understand the interrelationships between issues such as agriculture, land, water, food, and energy. Unless we look at all the contributing factors and the *connections*, the solutions we will arrive at will only be piecemeal and may not result in any effective outcomes and may even worsen the situation. The nexus is now getting more and more compounded with uncertainty because of climate change. Changes happening in precipitation or rainfall patterns, melting of the glaciers, and the likelihood of sea level rise are examples. The extremes in the weather we see due to climate change are leading to more vulnerability.

To respond to these challenges, countries started developing policies and regulations and set up institutions, giving them the responsibility of enforcement of related acts and rules. The institutional capacities were accordingly built. Delivery mechanisms were developed to put the policy principles in practice following precautionary, preventative as well as mitigative approaches. Efforts were made to set standards, procedures for permitting and requirements of documentation. Today many national governments have set up environmental, social and resource management frameworks that help minimize consumption of resources, especially of fossil fuels, promote use of renewable resources to the extent possible, and manage release of wastes and emissions within prescribed standards.

In addition to the problem of pollution, the world is facing two critical challenges. The first challenge is polarization of population that is happening due to intense urbanization. We are going to see at least 50% of the population residing in cities in the next 25 years.

This implies that there is going to be a concentration of a large number of the population in relatively compact areas. These growing urban centers are going to demand huge resources like water, infrastructure, energy, and food leading to large ecological footprints. The process of urban metabolism is going to lead to the generation of complex waste streams, wastewater, and air emissions leading to an environment of poor quality and increasing, thereby, health ailments. We will need to address this concern by reducing the in-migration to the cities and making cities livable through better planning and providing green infrastructure.

The second challenge is the skew in the global material flows especially between the low income high resource countries to high income industrialized nations. Today, the materials are extracted and moved from one country to another, taking advantage of the relatively low prices, resource abundance, and, in some cases, poor or lax environmental governance. Unless efforts are taken to improve the Resource Efficiency (RE) and Secondary Material Recycling (SMR) to reduce Domestic Material Consumption (DMC), the uneven and intense extraction of material flows will continue to be unsustainable. Here the national efforts alone are not going to be sufficient and countries will need to come together and forge partnerships. Moving towards a Circular Economy (CE) should be the agenda and a common goal.

The United Nations Environment Program (UNEP) has played an important role in placing a global agenda of issues at the apex level and introducing the Sustainable Development Goals (SDG). Several key Multilateral Environmental Agreements (MEAs) have been ratified by countries to address issues such as ozone layer depletion, loss of biodiversity, and global warming.

Development Financing Institutions (DFIs) play a very important role in fostering development. The DFIs have supported needed infrastructure through investments and helped build the capacities of the institutions. DFIs such as the World Bank also have influenced national policies towards sustainable development. The Private Sector Financing Institutions (PSFIs) have engaged with business and provided them capital with caveats on the environmental and social obligations. Many market based instruments have been evolved in partnership with governments to simulate sustainability. Both the DFIs and PSFIs have demonstrated a business case for sustainability by increasing the development effectiveness, improving the economic returns and widening the social impacts. But several businesses still need to be convinced to transform their operations. Addressing the Small and Medium Enterprises (SMEs) has always been the challenge. Reaching the SMEs by the greening of the supply chains has emerged as an effective strategy. The manufacturing and service sectors have reformed their operations by introducing green products through product redesign and by opening up waste recycling services and creating green jobs. Communities are playing a proactive role by demanding green products and waste recycling at the local level. All these efforts have led to innovations.

Awareness and education are important when it comes to behavioral change, especially on taming consumption patterns. Behavioral change is perhaps the most important element of the pyramid of paradigm shift towards a Circular Economy. Concepts such as sustainability literacy have emerged with steep targets to meet. Programs have been evolved such as Green Schools, and Tool kits have been developed for greening of campuses and curricula at universities. UNESCO has promoted the idea of Learning Cities by setting Lighthouses for learning. These experiments, when scaled up and replicated, are going to lead to behavioral change, which is the only solution finally towards putting sustainability in practice.

Governance at the national level must become more resource (not limited to management of residues) and sustainability centric. Both DFIs and PSFIs should lend or take stakes in a more responsible manner during engagements and investments ensuring that the business recognizes sustainability, not just as an element of risk, but also as an opportunity. Business should invest in the process of innovations and not only to expand the market or get higher economic return but also guide the consumers in making better choices. Innovations and eco-entrepreneurship towards sustainability are going to be the multipliers. In all of the above, working in partnership is going to be the key.

This book has been developed with the above background. It is targeted to both graduate students at universities undergoing a course and practitioners who would like to learn about several facets of sustainability to take a rounded approach. The practitioners include policy makers and regulators, senior management at financing institutions, heads of businesses and leaders at the community-based organizations and environmental Non-Governmental Organizations (NGOs).

This book has been structured with six chapters. The first chapter introduces the critical issues the world is facing today with relevant statistics, underscoring the importance of recognizing the nexus. The key concerns discussed are the polarization of the population due to urbanization and the skew in global material flows. The Sustainable Development Goals (SDGs) are outlined, highlighting the importance of Resource Efficiency (RE) and Secondary Material Recycling (SMR). This chapter ends by introducing the concept of Circular Economy (CE).

Chapter 2 introduces the stakeholders to sustainability such as government, financing institutions, and communities. It lists the key governing principles that need to be put into practice. Responses from the national governments at the policy level are then described – introducing examples of constitutional provisions. Next, the planning related interventions are illustrated with case studies such as zoning, eco-cities, and eco-industrial parks or eco-towns. Regulatory frameworks are then dealt with, citing examples of standards and their evolution in the life cycle perspective.

Chapter 3 is devoted mainly to Development Financial Institutions (DFIs) and Private Sector Financial Institutions (PSFIs). Examples are cited of the operations at the World Bank and Asian Development Bank, followed by PSFIs that have adopted Equator Principles. This chapter also introduces various financial instruments, trends, and opportunities such as Sustainable Stock Exchanges, Adaptation Funds, and Green Bonds.

Sustainability in business organizations (both manufacturing and services) is discussed in Chapters 4 and 5. Chapter 4 introduces the strategies practiced by businesses across sectors and Chapter 5 presents more sectoral experience with numerous case studies.

In Chapter 6, the focus is on the role played by communities. Communities play an extremely key role when it comes to achieving results on the ground. Sometimes the community plays a role as a watchdog, sometimes as a facilitator, and sometimes takes leadership, which we often call Community Driven Development (CDD). Chapter 6 also underscores the importance of awareness, education, training, and innovation. Topics such as Traditional Ecological Knowledge (TEK) and eco-entrepreneurship are also introduced in this context. This Chapter presents several case studies where social enterprises have been set up for the benefit of the community at large. The importance of linkages between sustainability and innovation is brought out when we show how businesses and governments along with the local community are moving on the path of innovation. This chapter presents case studies that emphasize the importance of partnerships across communities, government, financing institutions, and business.

This book uses around 140 examples/case studies in the form of boxes. In each box, there are discussion questions and references for further reading for the interest of the students and the faculty.

This book may be used as a textbook or a principal reference to design and conduct a 36-lecture course at the graduate level on Environmental Management and Sustainability. The instructor could intersperse these lectures with practicums, discussion sessions, and brainstorming events. For each of the case studies, this book has the lead references given. So, the student is encouraged to go through the original references from where the case studies have been drawn and then discuss the case studies in much more detail.

A recommended plan to use this book may be as below:

Chapter 1 (4 lectures); Chapter 2 (6 lectures); Chapter 3 (6–8 lecturers); Chapters 4 and 5 (10–12 lectures); and Chapter 6 (6 lectures), totaling 32–36 lectures.

One could use this book selectively depending on the audience. The contents of Chapters 4 and 5, which are a little more focused on, for example, business, could be used in combination with Chapter 1 (as an introduction) to conduct short-term training programs. A similar approach could be used to train officers of the government, financing institutions, and communities.

This book does not delve too deeply into details of specific tools. However, concepts of Life Cycle Assessment (LCA), practices in Environmental Impact Assessment (EIA), ISO 14001 Environmental Management System (EMS), and frameworks such as Sustainability Assessment of Technologies (SAT) are introduced. This book cites a number of references to give directions to the reader on how to deepen knowledge on some of these tools. The students and practitioners are encouraged to follow the references. The instructor can even consider exposing the students further by setting up reading assignments and preparing notes. For example, students may be encouraged to do a study on the application of LCA on some of the interesting products such as washing machines, or plastic bags, and then present these assignments in group work and share with each other the methodologies used.

It was difficult to pick case studies given the numerous and exemplary initiatives happening across the world in the sustainability space. I have picked the case studies mainly to illustrate examples, draw learning and provide inspiration. The examples are not my endorsements.

This book uses the most recent Web-based references while building the boxes of case studies as well as for recommending further reading. These Web links have been checked and verified; however, it is possible that some of the links moved elsewhere or became inactive. My apologies in case you find difficulties or surprises. Do reach me at my email, prasad.modak@emcentre.com, should you encounter such situations.

This book makes an attempt to present an interesting and useful compilation of experiences put in the perspectives of key stakeholders such as government financing institutions, business, and communities. I do hope that you find this resource useful and helpful to put sustainability in practice.

Acknowledgments

This book would not have been possible without the support I received from the team working with me at the Environmental Management Centre LLP (EMC) and Ekonnect Knowledge Foundation. Many of the assignments we did at EMC over the last 20 years helped me in structuring this book and sharing the experience. EMC's motto has been *Practicing Sustainability to the Advantage of All* and this book is in some sense a reflection.

Specifically, I would like to thank the extensive research inputs provided by Vibhuti Agarwal on the case studies that constitute a major portion of this book. Pooja Joshi and Mayuri Raichur supported the referencing work, editing, and formatting as well as tracking of the various versions of the drafts and assisted in the required permissions. Without their constant Pooja's constant support over the last year, this book would not have been possible. Other colleagues who supported me include Lucille Andrade, Sonal Alvares, and Mayuri Raichur who contributed to the research, and Bhushan Bhaud who worked on the images.

My gratitude to Dr. Gangandeep Singh, who managed publication of this book on behalf of CRC Press. His tolerance to my numerous requests of extensions and guidance is really appreciated.

Finally, writing of this book took away a lot of time from my personal life. I would like to thank my wife, Kiran, and children Devika and Pranav for their patience, encouragement, and understanding.

Author

Dr. Prasad Modak is an alumnus of Indian Institute of Technology (IIT) Bombay and holds a Doctor of Environmental Engineering degree from the Asian Institute of Technology, Bangkok. Dr. Modak was Professor at Centre for Environmental Science and Engineering at IIT Bombay between 1984 and 1995. Currently, he is back again with his alma mater as Professor (Adjunct) at the Centre for Technology Alternatives in Rural Areas.

Dr. Modak runs a strategic consulting company—Environmental Management Centre LLP—that focuses on sustainability. He is also Director of Ekonnect Knowledge Foundation, a Section 8 and not-for-profit company with a mandate to impart education and training in the arena of environmental management. Dr. Modak was Chief Sustainability Officer at Infrastructure Leasing & Financial Services Limited (IL&FS), and the first Dean of IL&FS Academy for Applied Development as a consultant.

Dr. Modak has worked with almost all key UN, multi-lateral and bi-lateral developmental institutions in the world. Apart from the government of India and various state governments, Dr. Modak's advice is sought by the governments of Bangladesh, Egypt, Indonesia, Mauritius, Thailand, and Vietnam. Dr. Modak has been invited by several international agencies, public bodies, professional journals and governments to serve as a jury member and as a reviewer. He has authored and executed more than 500 consulting reports and trained more than 8,000 professionals across the world.

Dr. Modak has published books with the UN University, Oxford University Press, UNEP, and Centre for Environmental Education in India. He served as Hon. Editor of the *Journal of Indian Water Works Association* between 1998 and 2004. Recently, he contributed chapters on Waste Management & Recycling in UNEP's *Green Economy Report* and *Global Waste Management Outlook*. Dr. Modak was the Chief Editor and contributor to the *Asia Waste Management Outlook* for UNEP.

Dr. Modak has regularly posted blogs since 2014 at https://prasadmodakblog.wordpress.com/. These posts have resulted in two self-published books titled *Sixty Shades of Green and Musings on Sustainability* (2016) and *Blue, Green – and Everything in Between* (2017). Dr. Modak has been the Editor of *Sustainability Quotient* – a quarterly newsletter with the Bombay Chamber of Commerce and Industries since 2013.

Dr. Modak has worked on a number of assignments concerning institutional assessment and capacity building in the field of environmental management. His blend of experience with academic and research institutions, private sector and financing institutions, bi-lateral and multilateral development agencies as well as governments has provided him an advantage in taking a rounded approach. Very recently, in 2015, Dr. Modak was invited to join the prestigious Indian Resources Panel by the Ministry of Environment, Forests & Climate Change.

Dr. Modak has received a number of awards and recognitions, and his name has been listed in distinguished personalities on environmental management. He has been a recipient of the Distinguished Alumni Award of AITAA in 2010 for Significant Contribution to International Affairs. In 2011, Dr. Modak was elected by the American Association of Environmental Engineers as a Board Certified Environmental Engineering Member for his work in research, teaching, and professional practice.

Visit www.linkedin.com/in/prasadmodak for Dr. Modak's detailed profile.

1

The Challenges We Face Today

1.1 Challenges We Face Today: Population, Consumption and Limited Resources

In 2016, the world's population had reached 7.4 billion people. Assuming that the average fertility levels will continue to decline, the global population is still expected to reach between 8.4 and 8.6 billion by 2030, 9.4 and 10 billion by 2050, and 10 and 12.5 billion by 2100.[1] This rampant population growth has been and will remain the main driver for our unsustainable present and the future as population drives consumption. Figure 1.1 shows the global material extraction in billion tons and global GDP in trillion $ (2005 prices) from 1970 to 2015.

Figure 1.2 shows the trend in the GDP across regions of the world. It may be observed that the trend in the GDP growth is higher in the South Asia, East Asia, and Pacific regions where the population and its growth is also high.

The Human Development Index (HDI) is a composite statistic of life expectancy, education, and per capita income indicators, that are used to rank countries into four tiers of human development. A country scores higher HDI when the lifespan is higher, the education level is higher, and the GDP per capita is higher.[2]

Figure 1.3 shows the relation between per capita Material Footprint (MF) with HDI between 1990 and 2010. MF is the attribution of global material extraction to domestic final demand of a country. It is calculated as the raw material equivalent of imports plus domestic extraction minus raw material equivalents of exports. Box 1.1 explains the concept of MF.

It may be observed that the MF of countries with High HDI is nearly 4–5 times higher as compared to low HDI countries. There is significant unevenness in the distribution of MF across the world.

1.2 Indicators of Development

The Gross Domestic Product (GDP) is often used as an indicator of the economic progress of a country, which is measured through the consumption of products and services. It is assumed that if a population's consumption is high, then the overall well-being and welfare of that country is also high. The GDP, however, is not the indicator of social welfare and quality of life. Further, the GDP does not account for the negative externalities associated

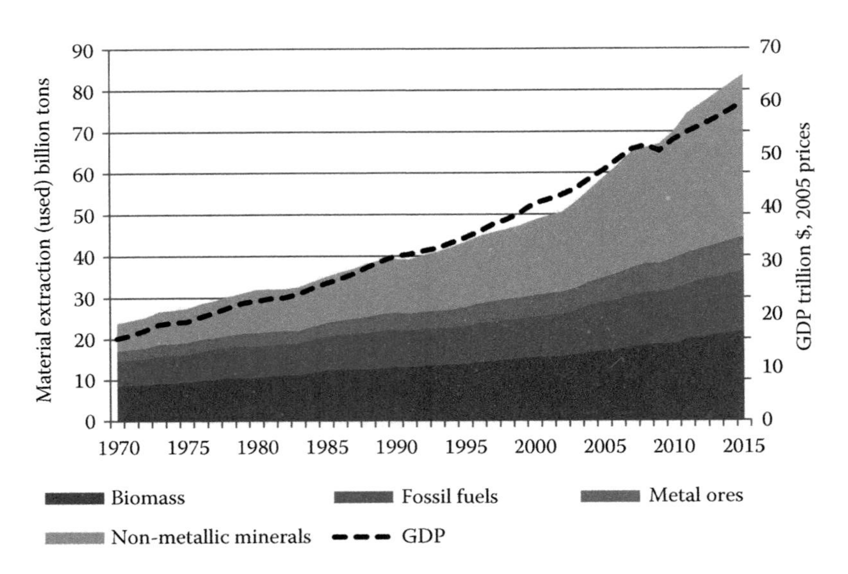

FIGURE 1.1
Global material extraction in billion tons and global GDP in trillion $ (2005 prices) from 1970 to 2015. (From Ekins, P. and Hughes, N., *Resource Efficiency: Potential and Economic Implications*, International Resource Panel, UN Environment, UNEP, p. 13, 2016. With permission, online source: https://www.env.go.jp/press/files/jp/102839.pdf.)

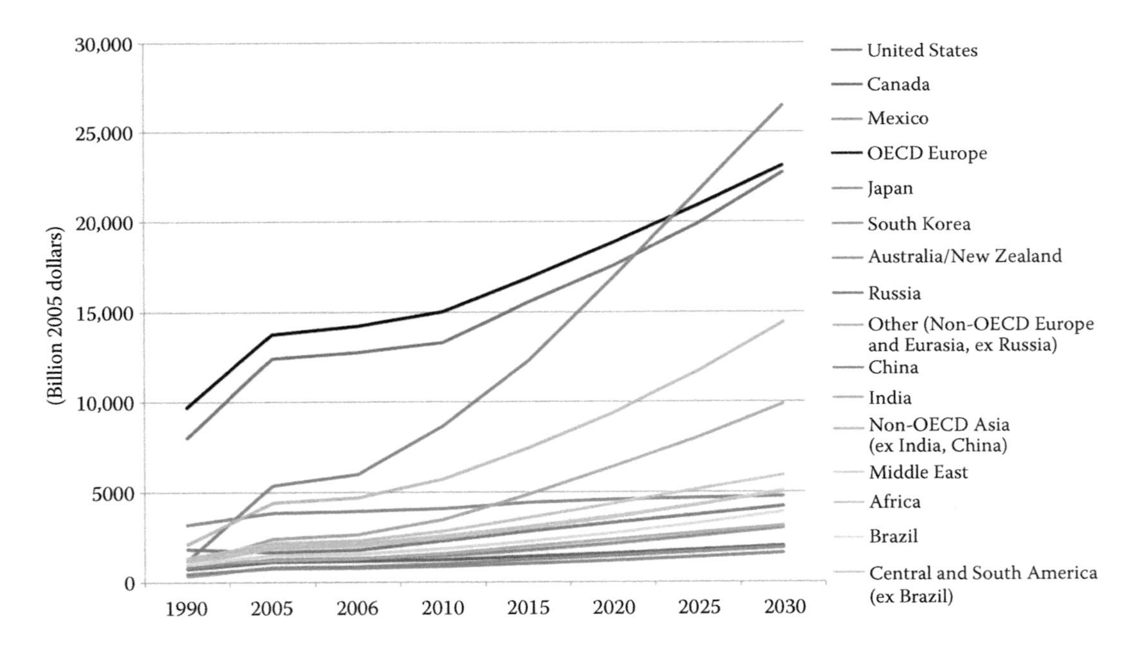

FIGURE 1.2
Trends in GDP growth across regions. (From Projected GDP, 1990–2030, Mongabay, 2009. With permission, online source: http://rainforests.mongabay.com/energy/gdp.html.)

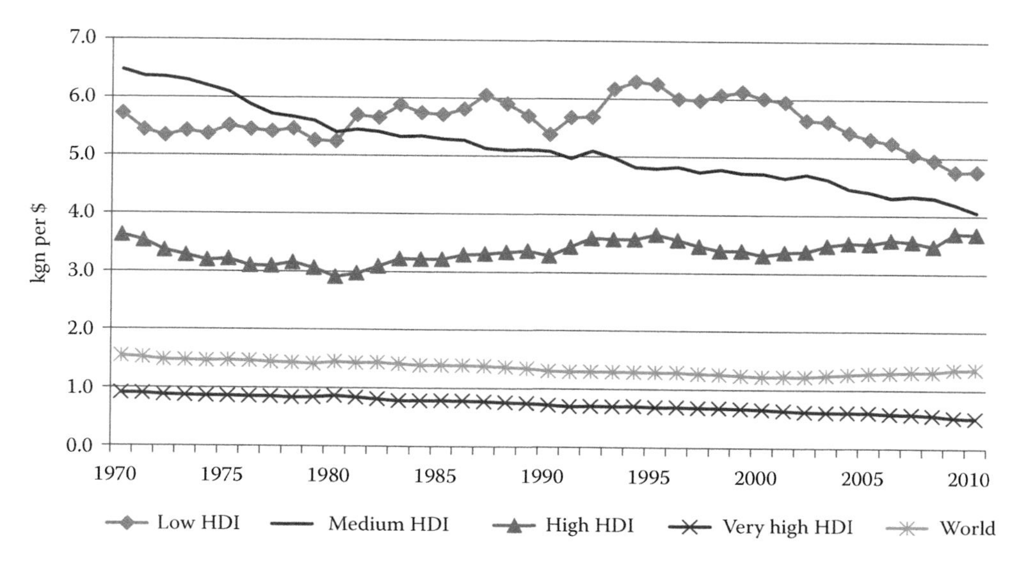

FIGURE 1.3
Per capita material footprint (MF) by HDI level, 1970–2010. (From Schandl et al., Global Material Flows and Resource Productivity: Summary for Policymakers, An Assessment Study of the International Resource Panel, UN Environment. Paris, United Nations Environment Programme, p. 26, 2016. With permission, online source: http://www.isa.org.usyd.edu.au/about/16-00271_LW_GlobalMaterialFlowsUNE_SUMMARY_FINAL_160701.pdf.)

with environmental degradation. If the environmental and social costs of development are considered, then the *true GDP* of the nations will be lower.

The Index of Sustainable Economic Welfare (ISEW) was developed in 1989 as an economic indicator intended to replace the GDP. Rather than simply adding together all expenditures like that done in the GDP, in ISEW the consumer expenditure includes

BOX 1.1 MATERIAL FOOTPRINT[3]

Material Footprint (MF) is the attribution of global material extraction to domestic final demand of a country. It is calculated as the raw material equivalent of imports (RMEIM) plus domestic extraction (DE) minus raw material equivalents of exports (RMEEX). For the attribution of the primary material needs of final demand, a global, multi-regional input-output (MRIO) framework is employed.

Material Footprint of consumption reports the amount of primary materials required to serve the final demand of a country and can be interpreted as an indicator for the material standard of living/level of capitalization of an economy. Per-capita MF describes the average material use for final demand. Direct material consumption (DMC) and MF need to be looked at in combination as they cover the two aspects of the economy, production, and consumption.

The DMC reports the actual amount of material in an economy and MF reports the virtual amount required across the whole supply chain to service final demand. A country can, for instance, have a very high DMC, because it has a large primary production sector for export or a very low DMC because it has outsourced most of the material intensive industrial processes to other countries. The material footprint corrects for both phenomena.

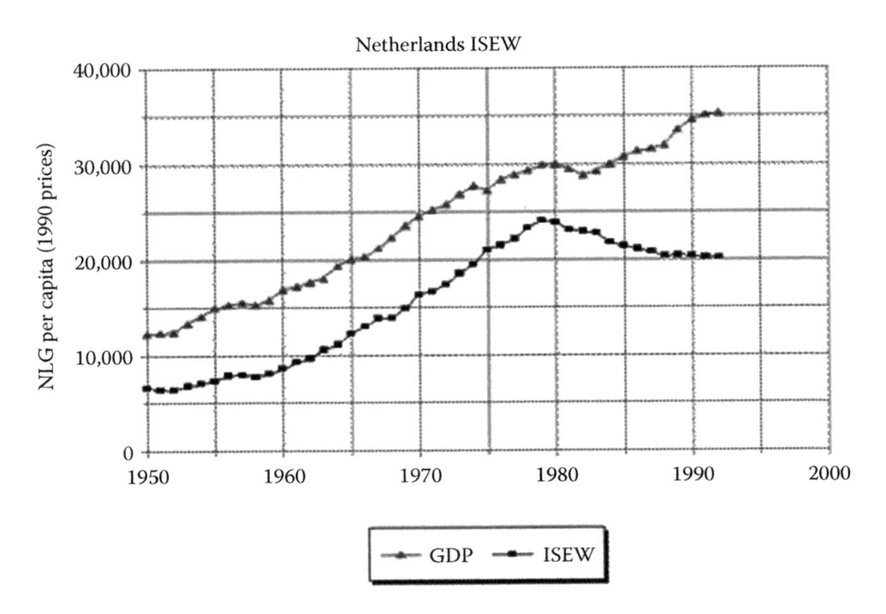

FIGURE 1.4
Relationship between GDP and ISEW for the Netherlands. NLG, Netherlands Guilders. (From Alberta and other International GPI/ISEW Graphs, Anielski Management—Building communities of genuine wealth. With permission, online source: http://www.anielski.com/alberta-international-gpiisew-graphs/.)

income distribution and cost associated with pollution amongst other negative externalities such as depreciation of natural capital. Figure 1.4 shows the relation between GDP and ISEW for the Netherlands.[4]

The Genuine Progress Indicator (GPI) was created in 1995, as an alternate to the GDP. The GPI is essentially an extension of ISEW that addresses welfare and the ecological sustainability of the economy to assess genuine and real progress of the society.

Unlike the GDP, the GPI considers both social and environment factors, as well as economic factors. It is widely considered as a better indicator of progress than the GDP, by scientists and by governmental and non-governmental organizations, globally. The difference between GDP and GPI is analogous to the difference between a company's gross profit and net profit, respectively. The net profit is the gross profit minus the costs incurred, in the same way that the GPI is the GDP minus the environmental and social costs incurred.

The GPI accounts for 26 indicators that are either negative or positive, such as: income distribution, value of household work and volunteer work, costs of crime, pollution, resource depletion, long-term environmental damage such as climate change, changes in leisure time, defense expenditures, life span of consumer durables and public infrastructure, and dependence on foreign assets, etc.[5] Some of the environmental indicators that are considered in calculating the GPI are: Cost of Water Pollution, Cost of Air Pollution, Cost of Noise Pollution, Loss of Wetlands, Loss of Farmland Quality or Degradation, Loss of Primary Forest and Damage from logging roads, Damage due to the increase in CO_2 emissions, Cost of Ozone Depletion, and the Depletion of Non-renewables. GPI has been calculated for Austria, Canada, Chile, France, Finland, Italy, the Netherlands, Scotland, and the rest of the UK.

Figure 1.5 shows a similar trend for Canada between 1970 and 1995.

More recently, the concept of Gross Ecosystem Product (GEP)[6] has evolved such that it is recognized that there is an importance for the ecosystem products and services for human survival, well-being, and development. GEP accounts for the indirect and direct

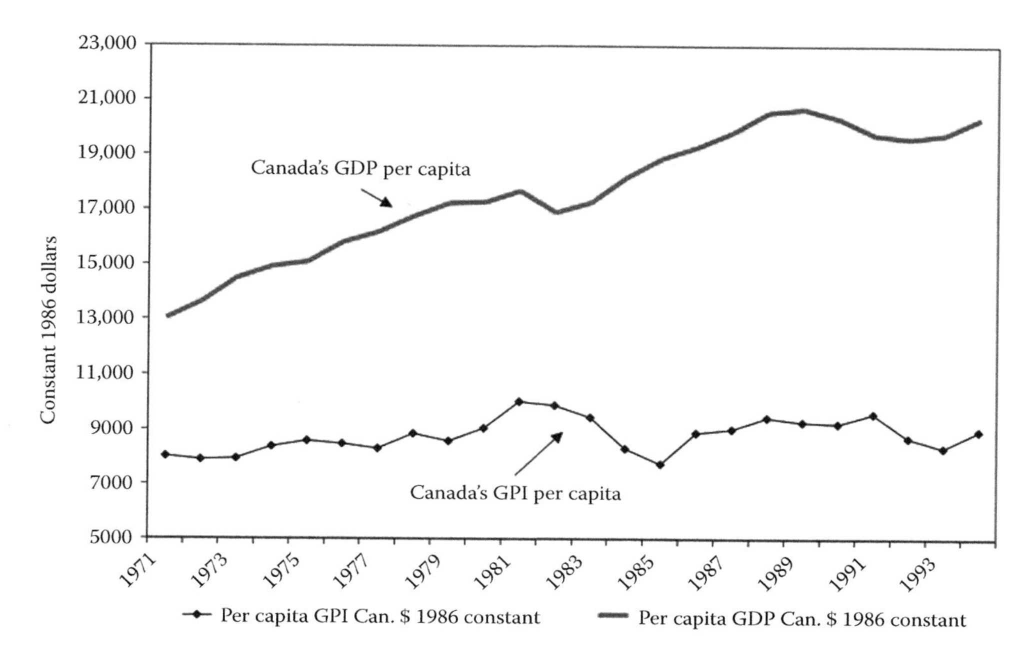

FIGURE 1.5
Relationship between GDP and GPI, from 1970 to 1995 in Canada. (From Alberta and other International GPI/ISEW Graphs, Anielski Management—Building communities of genuine wealth. With permission, online source: http://www.anielski.com/alberta-international-gpiisew-graphs/.)

values of ecosystem provisioning, ecological regulation services and ecological culture services. GEP is defined as: "the total value of final ecosystem goods and services supplied to human well-being in a region annually, and can be measured in terms of biophysical value and monetary value."[7] As part of China's new Eco-Civilization initiative, a pilot project was launched to test the country's first Gross Ecosystem Product (GEP).[8]

DISCUSSION QUESTIONS

- *Write a detailed note on the structure of the GPI and compare with the GEP. Illustrate application and comparison based on two countries.*
- *Prepare a scatter plot based on GPI and GDP based on data of 10 countries representative of different economies and geographical regions.*

In this context, it may be worthwhile to examine the concept of Ecological Footprint (EF). EF measures to what extent an activity uses productive land and sea resources versus how much land and sea resources are available. Land and sea resources are important as we use these resources for food, water, timber, mining, energy, and infrastructure. EF is measured in global hectares per person. The larger the footprint, the less ecologically or sustainably sustainable is the activity.

While computing the EF availability of resources, biocapacity or bioproductivity is also measured, which represents the productivity of its ecological assets.[9] Since both EF and

BOX 1.2 GLOBAL HECTARE[10]

Global hectare is a useful measure of biocapacity as it can convert things like human dietary requirements into a physical area, which can show how many people a certain region on earth can sustain, assuming current technologies and agricultural methods. It can be used as a way of determining the relative carrying capacity of the earth.

A given hectare of land may be measured in equivalent global hectares. For example, a hectare of lush area with high rainfall would be scaled higher in global hectares than would a hectare of desert.

Global hectares can also be used to show that consuming different foods may increase the earth's ability to support larger populations. To illustrate, producing meat generally requires more land and energy than what producing vegetables requires; sustaining a meat-based diet would require a less populated planet.

biocapacity are expressed in global hectares, these can be compared to each other. If a nation's EF exceeds the region's biocapacity, then that region is experiencing an ecological deficit, where the demand for the goods and services that the ecosystems can provide are far greater than what the ecosystems can renew. See Box 1.2 on explanation of the term global hectare.

The United Arab Emirates has the highest EF in the world measuring to 10.68 gha/person. Other countries outside of the Middle East include Denmark (8.26 gha/pers), Belgium (8 gha/pers), United States (8 gha/pers), Estonia (7.88 gha/pers), Canada (7.01 gha/pers), Australia (6.84 gha/pers), and Iceland (6.5 gha/pers). The worldwide total human ecological footprint is 2.6 global hectares per capita (gha/per) compared with a total worldwide biocapacity of only 1.8 gha/cap.[11,12] This overshoot means that we are already using 1.4 times resources that are sustainably available.

Another important and interesting indicator is the Environmental Performance Index.[13] EPI, developed at Yale University, is a framework that includes nine issues and more than 20 indicators. A number of key environmental issues have been, however, excluded from the calculations, as global data for them remain incomplete: freshwater quality, species loss, indoor air quality of residential, commercial buildings, and toxic chemical exposures, and municipal solid waste management, nuclear safety, wetlands loss, agricultural soil quality and degradation, recycling rates, adaptation, vulnerability, and resiliency to climate change.

DISCUSSION QUESTIONS

- *How is EPI aggregated from the sub-indicators? What are the limitations of this approach?*
- *Prepare a scatter plot between EPI and GDP for ten countries that represent different geographical regions and economies for a year. Discuss the diagram.*
- *Prepare a time series of EPI and GDP for Finland, United States of America, China, and India. Discuss the figure.*

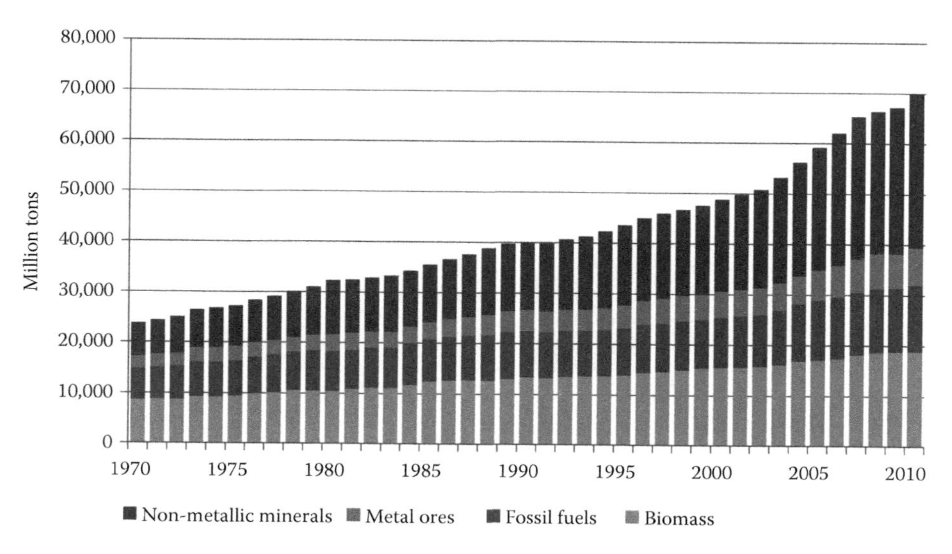

FIGURE 1.6
Global exports of materials by four material categories from 1970 to 2010 in million tons. (From Schandl et al., Global Material Flows and Resource Productivity: Summary for Policymakers, An Assessment Study of the International Resource Panel, UN Environment. Paris, United Nations Environment Programme, p. 26, 2016. With permission, online source: http://www.isa.org.usyd.edu.au/about/16-00271_LW_GlobalMaterialFlowsUNE_SUMMARY_FINAL_160701.pdf.)

1.3 Resource Security

With growing consumption, our limited reserve of natural resources has significantly declined and the non-renewable sources are extracted at a faster pace than the reserves. The extraction of primary materials has increased from 22 billion tons in 1970–1985 to 85 billion tons in 2015. This figure is expected to grow to 183 billion tons by 2050.[14] Figure 1.6 shows trends in direct extraction of four key resources such as non-metallic minerals, metal ores, fossil fuels and biomass on a global basis between 1970 and 2010. It may be observed that the non-metallic minerals are getting depleted at a faster rate. Figure 1.7 reports the statistics on Domestic Extraction (DE)[15] of resources across seven regions of the world. The Asia-Pacific region that is highly populated emerges as a region with high and rising DE.

Some of the challenges associated with the rising resource demand with regard to the environment are: greenhouse gas (GHG) emissions, pollution, unsustainable land use, biodiversity loss. The economic considerations include insecurity of supply, costs, and highly volatile prices. There are also equity-related issues such as inequality of resource consumption, conflicts over resource extraction, and social and human rights concerns. Here, more than availability, equitable distribution of resources becomes a matter of concern.

1.4 The Issues

We discuss in this section some of the key issues that are of concern. There are several reports available on these issues on both thematic as well as regional basis, especially by

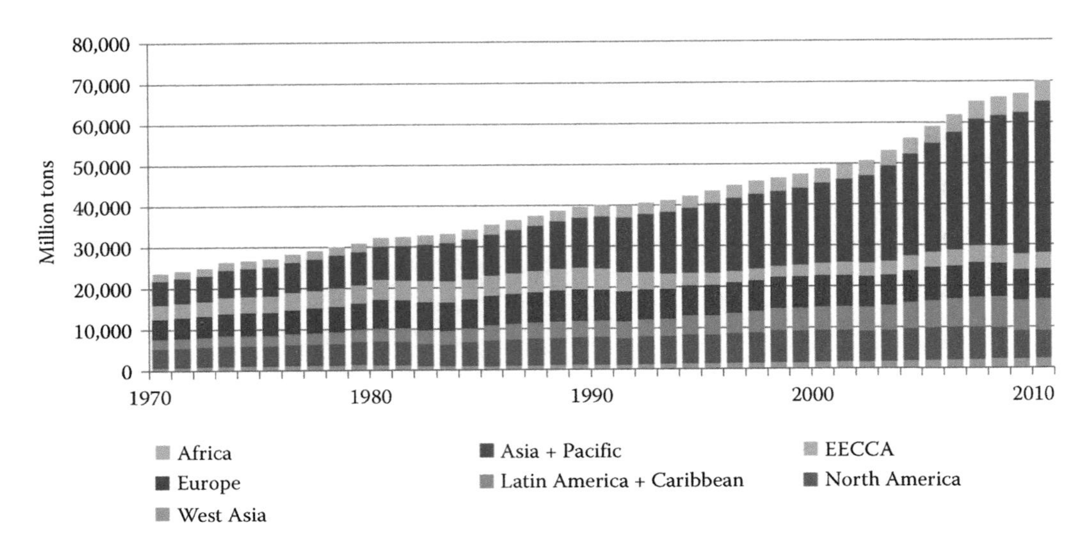

FIGURE 1.7
Domestic extraction (DE) by seven subregions from 1970 to 2010 in million tons. EECCA: Eastern Europe, Caucasus and Central Asia. (From Schandl et al., Global Material Flows and Resource Productivity: Summary for Policymakers, An Assessment Study of the International Resource Panel, UN Environment. Paris, United Nations Environment Programme, p. 26, 2016. With permission, online source: http://www.isa.org.usyd.edu.au/about/16-00271_LW_GlobalMaterialFlowsUNE_SUMMARY_FINAL_160701.pdf.)

the United Nations Environment Programme (UNEP). These reports are referred to as the Global Environment Outlooks (GEO). Readers are encouraged to access these reports for more information and analyses on a comprehensive basis. See Box 1.3. Another important report to look into is the Environmental Outlook for 2050 prepared by OECD.[16]

BOX 1.3 UNEP'S GLOBAL ENVIRONMENT OUTLOOKS[17,18]

Global Environment Outlook (GEO) is a series of reports on the environment issued periodically by the United Nations Environmental Program (UNEP). The GEO project was initiated in response to the environmental reporting requirements of UN Agenda 21 followed by the decision of May 1995 by the UNEP Governing Council.

The GEO reports are produced using a participatory and consultative approach. Input is solicited from a wide range of sources throughout the world, especially the Collaborating Center (CC) network, United Nations Organizations and independent experts.

During the preparation of the report, UNEP organizes consultations inviting policy makers and other stakeholders to review and comment on the drafts followed by extensive peer review. This iterative process is designed to ensure that the contents are scientifically accurate and relevant to users. The reports are developed for various regions supplemented by the Fact Sheets.

Six GEO reports have been published to date: GEO-1 in 1997; GEO-2000 in 1999; GEO-3 in 2002; GEO-4 in 2007; GEO-5 in 2012; and GEO-6 in 2016.

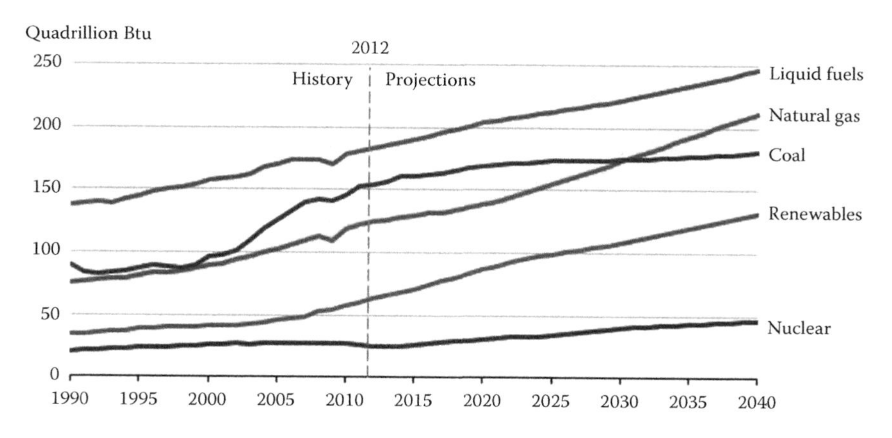

FIGURE 1.8
Sources of global energy consumption from 1990 to 2040. (Adapted from EIA projects 48% increase in world energy consumption by 2040, Today in Energy, Independent Statistics & Analysis—EIA: U.S. Energy Information Administration, 2016, online source: https://www.eia.gov/todayinenergy/detail.php?id=26212.)

1.4.1 Energy

According to the UN, globally, one in five people still lack access to modern electricity and 3 billion people still rely on traditional sources of energy, such as wood and charcoal, for cooking and heating.[19] So there is an extraordinary target to be met for energy access, especially to the poor.

Energy consumption has been a major driver for material extraction. The world's energy consumption is expected to grow by almost 48%.[20] See Figure 1.8, which shows trends in energy consumption for different sources of energy.

Clearly, we need to continue to focus on sources of renewable energy. In addition, we need to make significant efforts on improving energy efficiency as well as demand side management. We will also need to design products that have low material and energy intensity in the making of low energy requirements in the use phase. The entire life cycle will need to be looked at to ensure that energy inputs are reduced, and the larger share is taken by the renewable energy sources.

Cities of tomorrow are going to be energy guzzlers. Figures 1.9 and 1.10 show increasing energy use in cities as a function of the urbanization rate and a breakdown of energy usage across various cities. It may be observed that the energy consumption in cities of low income countries is still dominated by industrial use, while for cities in high income countries, energy demand from buildings dominates. Transport takes a major share of energy use in the cities belonging to middle income countries. Clearly a shift to public transport and promotion of low energy or green buildings is needed. Creating smart integrated urban energy systems is another option. But the importance of reducing the energy demand itself through less consumptive lifestyles must be underscored.

DISCUSSION QUESTION

- *Several cities have started increasing the use of renewable energy resources, especially based on solar power. Prepare case studies of cities where more than 50% of renewable energy is used and present the costs and benefits.*

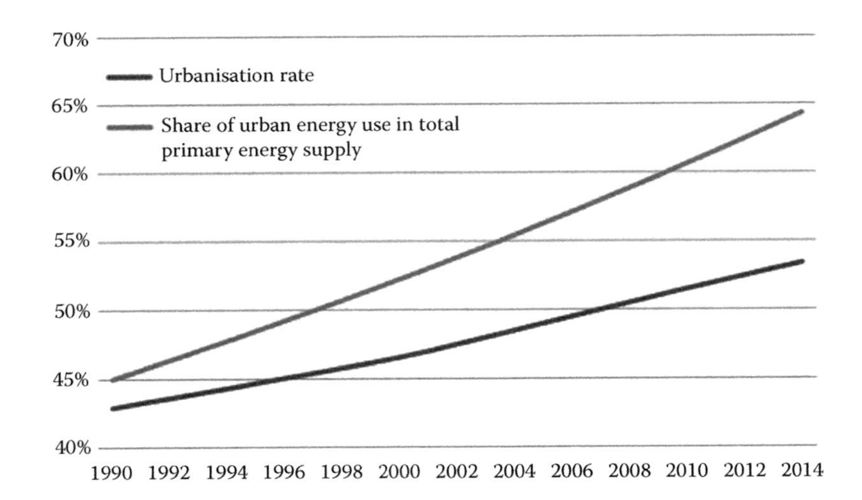

FIGURE 1.9
Urbanization rate and share of energy use in total primary energy supply. (From Rigter, J., Saygin,D., and Kieffer, G., IRENA, Renewable Energy in Cities, International Renewable Energy Agency (IRENA), p.11, 2016, Abu Dhabi. With permission, www. Irena.org, online source: http://www.irena.org/DocumentDownloads/Publications/IRENA_Renewable_Energy_in_Cities_2016.pdf.)

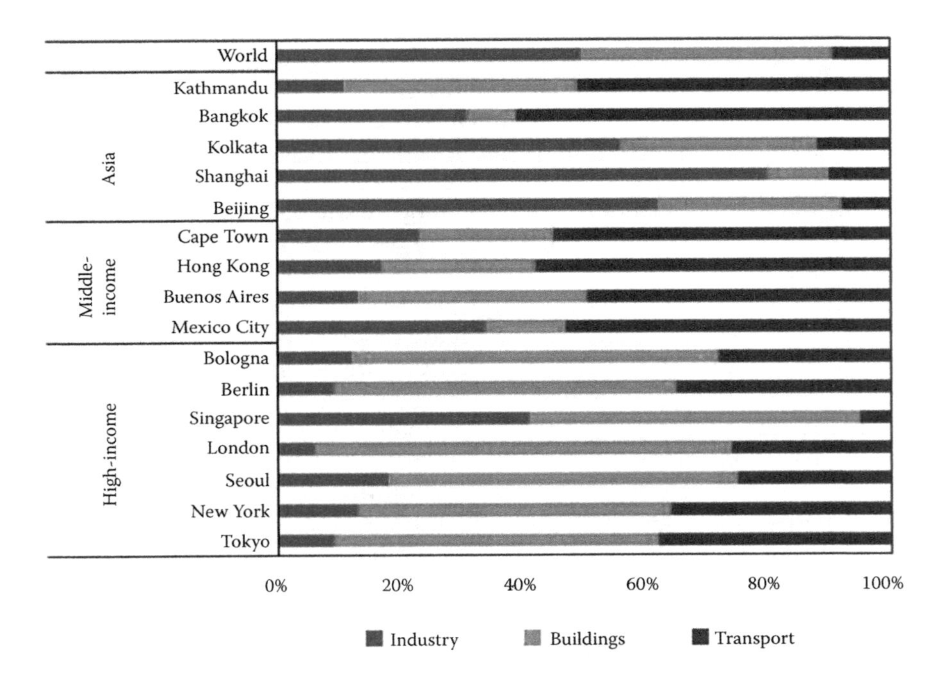

FIGURE 1.10
Breakdown of energy use by sector in selected cities. (From Rigter, J., Saygin,D., and Kieffer, G., IRENA, Renewable Energy in Cities, International Renewable Energy Agency (IRENA), p.11, 2016, Abu Dhabi. With permission, www.Irena.org, online source: http://www.irena.org/DocumentDownloads/Publications/IRENA_Renewable_Energy_in_Cities_2016.pdf.)

1.4.2 Overexploitation of Minerals and Fossil Fuels

Telephones are made from as many as 42 different minerals, including aluminum, beryllium, coal, copper, gold, iron, limestone, silica, silver, talc, and wollastonite. A television requires 35 different minerals, and a computer more than 30.[21] Production of mined metal commodities is expected to increase by 250% by 2030. Most of the demand will originate from small and medium-sized mining operators, particularly in developing countries.

In mining and quarrying, to gain access to the precious minerals, soil overburden and interburden have to be removed. The minerals are then separated from waste rock and tailings. In the extraction and separation processes, mining generates the largest quantities of unused extracted resources and are dumped as waste.

Figure 1.11 shows trend of world mining production between 1984 and 2014 by groups of minerals. It may be observed that extraction of mineral or fossil fuels dominates. The most common mineral fuel is coal, which is used to produce energy. Oil, natural gas, and uranium are used to produce electricity, or as fuel for transportation.

DISCUSSION QUESTION

- *Describe key manufacturing sectors that consume the world's major mineral resources. List the specific minerals that are important in manufacturing but their availability is under threat.*

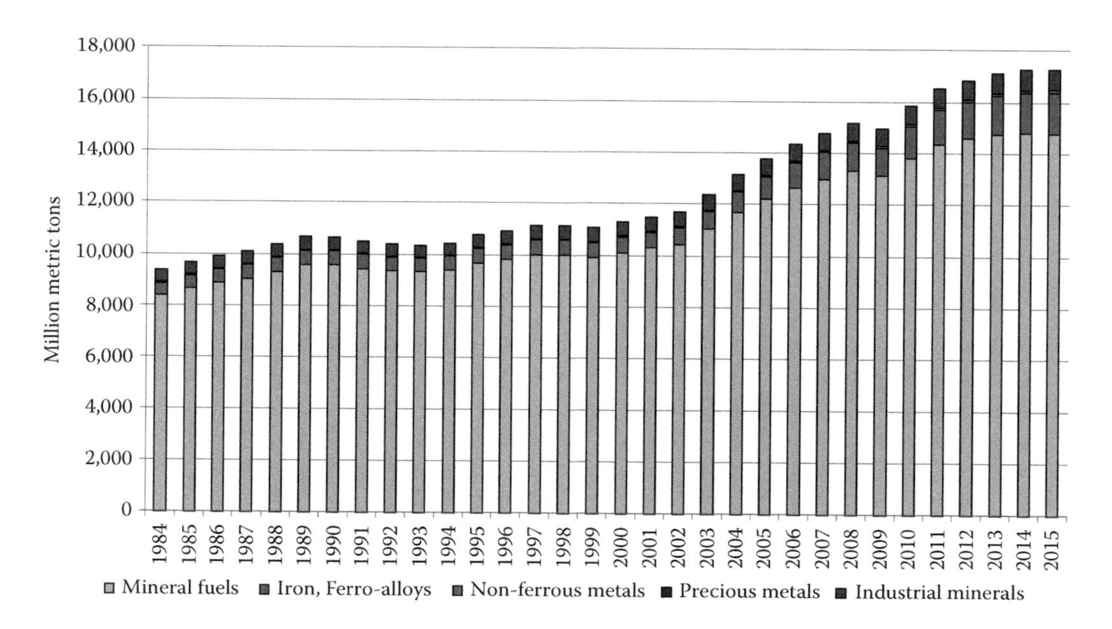

FIGURE 1.11
World mining production 1984 to 2015 by groups of minerals. (From Reicl, C., World Mining Data, Federal Ministry of Science, Research and Economy of the Republic of Austria & International Organizing Committee for the World Mining Congresses, Vienna, 2017. With permission, online source: http://www.wmc.org.pl/sites/default/files/WMD2017.pdf.)

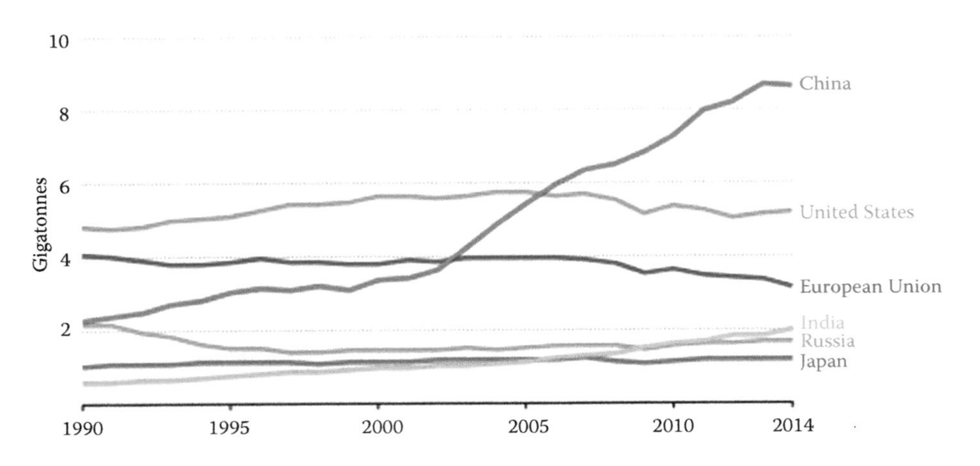

FIGURE 1.12
Energy related GHG emissions for various regions. (From © OECD/IEA 2015 World Energy Outlook Special Report: Energy and Climate Change, International Energy Agency (IEA) Publishing, p. 28, 2015. With permission. Licence: www.iea.org/t&c, online source: https://www.iea.org/publications/freepublications/publication/WEO2015SpecialReportonEnergyandClimateChange.pdf.)

Fossil fuels have constantly been the highest mined resource. Around 87% of the energy consumed globally comes from fossil fuels.[22] Fossil fuels remain the dominant form of energy powering the global expansion: providing around 60% of the additional energy and accounting for almost 80% of total energy supplies in 2035.[23]

In 2014, the consumption of coal for energy had reached 13,000 million tons of oil equivalent.

Africa is known to be the world's most mineral rich continent, but in terms of production, Asia is leading the way. This is mainly due to the high reserves of coal found in many Asian countries, such as China, India, and Siberia.[24]

Figure 1.12 shows the energy related emissions for various regions. It may be observed that GHG emissions from China dominate. Figure 1.13 shows the environmental impacts of fossil fuel consumption.

1.4.3 Housing

It is estimated that at present, buildings contribute as much as one third of total global GHG emissions, primarily through the use of fossil fuels during their operational phase and consume up to 40% of the global energy.[25]

The building sector is a highly resource intensive industry due to its continuous consumption of water, energy, land, and materials. These resources are consumed not just during the construction phase of a building but more so during its operational phase, that is, from cradle to grave.

A study by the United Nations Environment Program (UNEP) says that over an entire life-cycle, the building industry uses 30% of the fresh water available globally and generates 30% of the world's effluents.[26] While in EU countries, public water supply represents about 21% of total water use is used in buildings (Figure 1.14).[27]

When it comes to energy use it is estimated that, at present, buildings contribute as much as one third of total global GHG emissions, primarily through the use of fossil fuels during their operational phase and consume up to 40% of the global energy.[28]

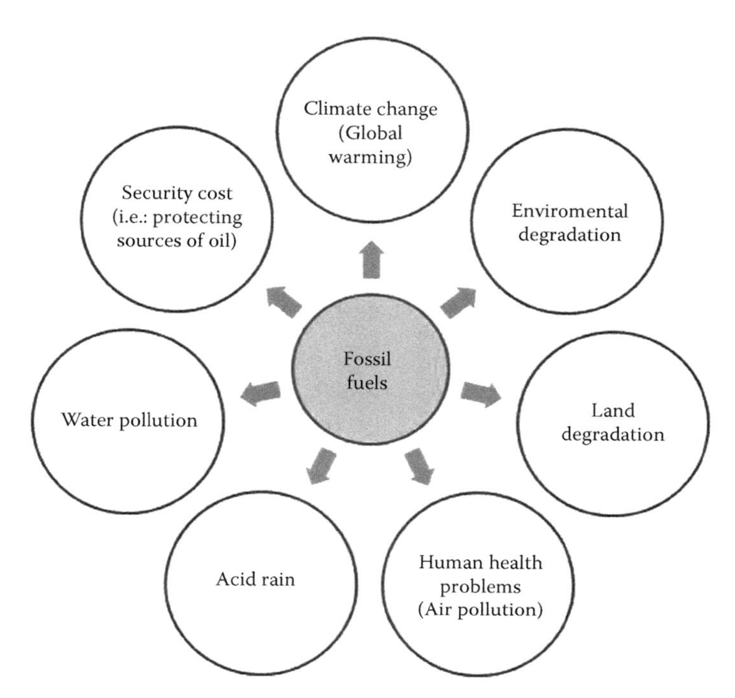

FIGURE 1.13
Impacts of fossil fuel consumption.

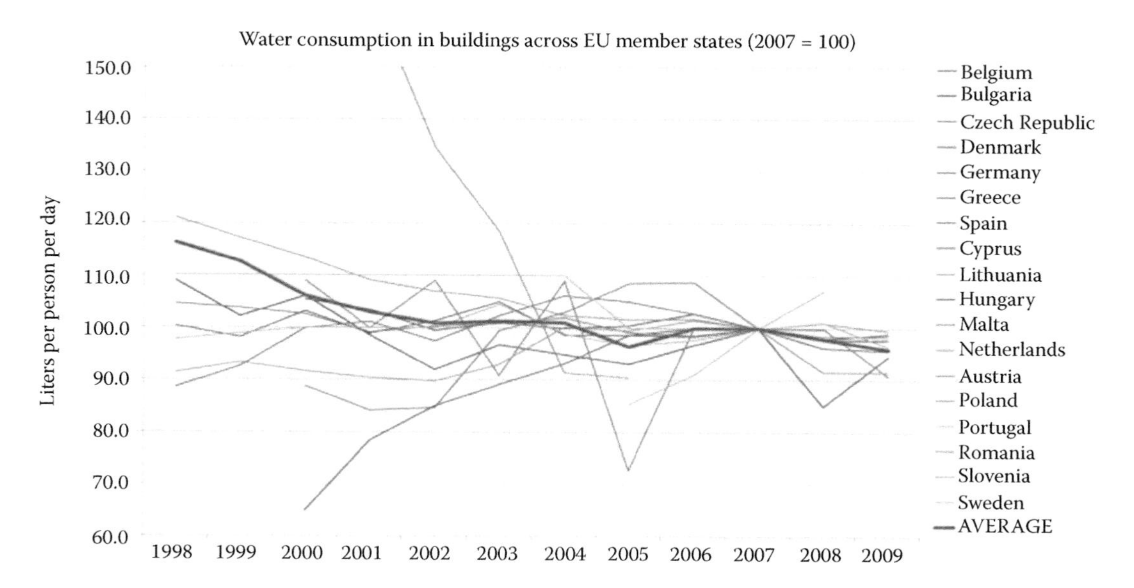

FIGURE 1.14
Water consumption in buildings across the different EU member states. (From Resource efficiency in the building sector, ECORYS, Copenhagen Resource Institute, p. 43, 2014. With permission, online source: http://ec.europa.eu/environment/eussd/pdf/Resource%20efficiency%20in%20the%20building%20sector.pdf.)

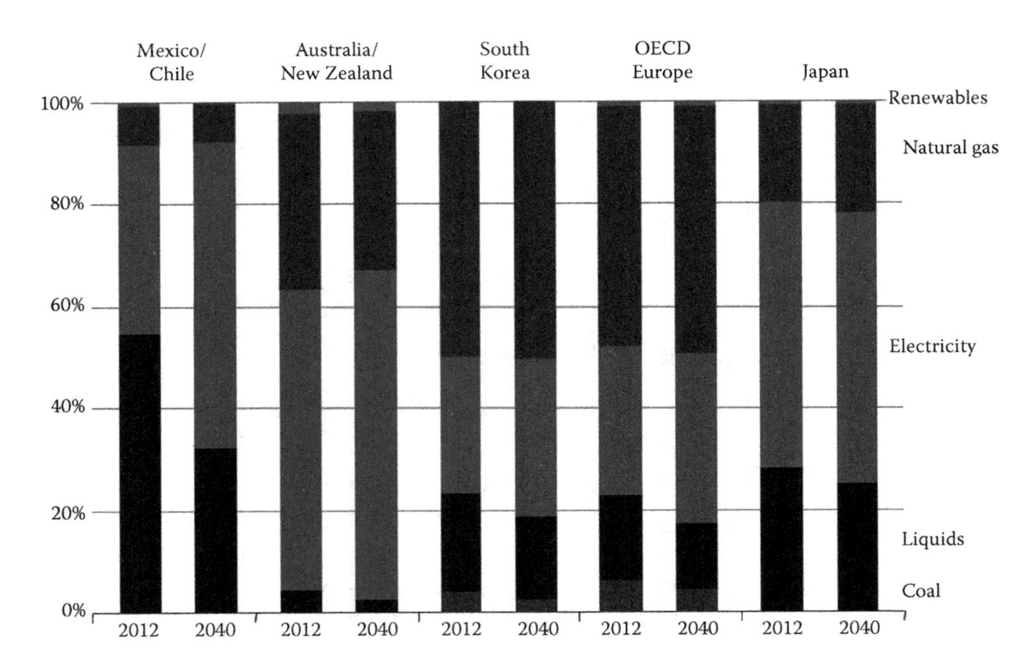

FIGURE 1.15
Residential sector energy consumption in selected OECD regions by energy source from 2012 to 2040. (Adapted from Chapter 6: Buildings Sector Energy Consumption, International Energy Outlook 2016, U.S Energy Information Administration, p. 102, online source: https://www.eia.gov/outlooks/ieo/pdf/0484(2016).pdf.)

According to the International Energy Outlook 2016 (IEO2016) for OECD nations, this energy consumption will increase by an average of 1.5%/year from 2012 to 2040, while for non-OECD nations, it grows by 2.1%/year from 2012 to 2040. Sources of energy consumption for some of the OECD and non-OECD countries are shown in Figures 1.15 and 1.16.

Figure 1.17 shows the trend of energy consumption in residential and commercial buildings across various countries in the world.[29]

The construction industry accounts for 24% of global raw materials removed from the earth.[30] In addition, the extraction, processing, transport and installation of materials associated with construction consume copious quantities of energy and water. It is estimated that building construction activities consume 3 billion tons of raw materials used each year. For example, the UK's annual construction output requires 170 million tons of primary materials and products, 125 million tons of quarry products and 70 million tons of secondary recycled and reclaimed products. To manufacture and deliver these products, 6 million tons of energy are consumed and 23 million tons of CO_2 are emitted.

The main materials used in construction are steel and concrete, both of which have a high embodied energy.[31] Not only these but every building material to some extent emits CO_2 during their production. Hence, innovative building materials, their production processes, re-use of materials and promoting use of eco-friendly techniques and material is vital to reduce environmental impacts of the building sector.

With the ever increasing population growth and demand for housing, the efficient extraction and management of resources to sustain this sector are a challenge. Thus, to improve environmental performance of the building sector, many initiatives have been devised by various countries to promote green buildings, eco-friendly materials, innovative design

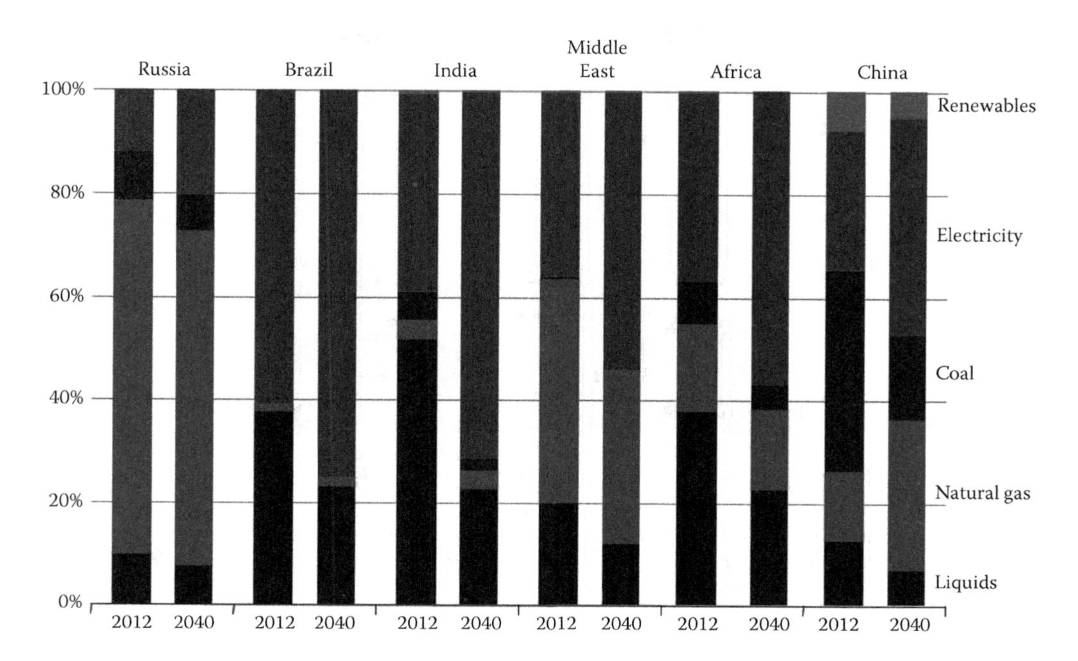

FIGURE 1.16
Residential sector energy consumption in selected non-OECD regions by energy source from 2012 to 2040. (Adapted from Chapter 6: Buildings Sector Energy Consumption, International Energy Outlook 2016, U.S Energy Information Administration, p. 104, online source: https://www.eia.gov/outlooks/ieo/pdf/0484(2016).pdf.)

and technologies by labelling them according to certain certification criteria. A few of them are Leadership in Energy and Environmental Design (LEED) in the USA and Canada, the Building Research Establishment Environmental Assessment Method (BREEAM) scheme of UK, Deutsche Gesellschaft fur Nachhaltiges Bauen (DGNB) in Germany, Haute Qualite Environmentale (HQE) in France, Green Star in Australia, The Energy and Resources

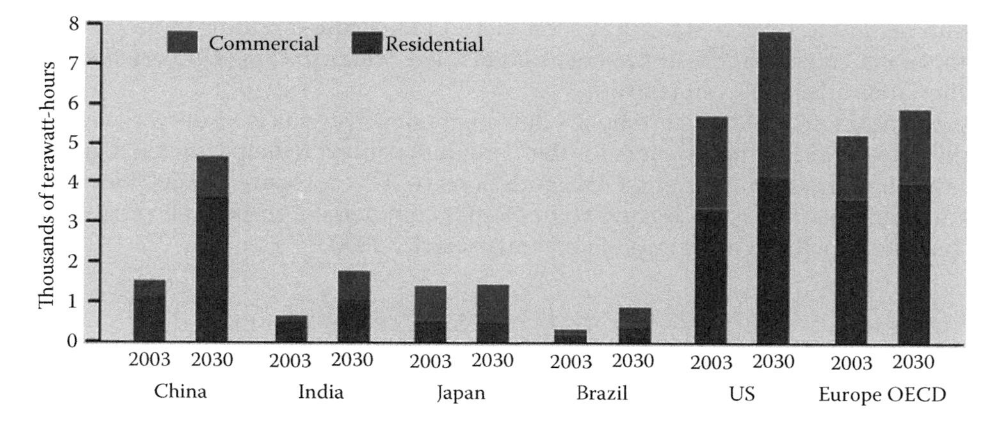

FIGURE 1.17
Projection of energy for buildings by region since 2003 and projected at 2030. (From Energy and Buildings, Centre for Science and Environment (CSE), India, 2014, p. 1. With permission, online source: http://www.cseindia.org/userfiles/Energy-and-%20buildings.pdf.)

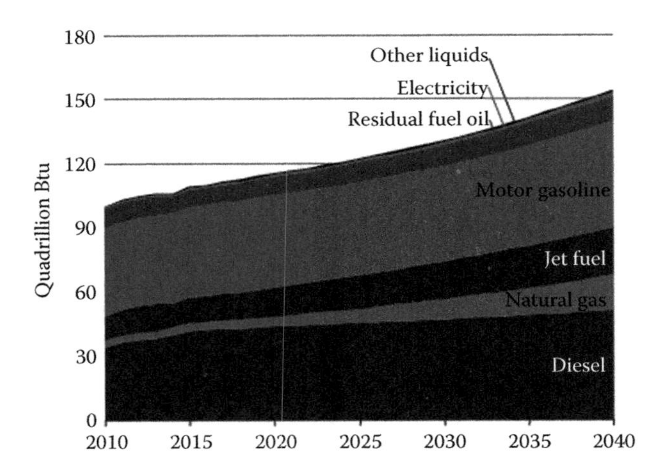

FIGURE 1.18
World energy consumption due to transport sector by energy source. (Adapted from U.S. Energy Information Administration (EIA), *International Energy Outlook—With Projections to 2040—International Energy Outlook*, Chapter 8: Transportation sector energy consumption, EIA, p. 127, 2016, online source: https://www.eia.gov/outlooks/ieo/pdf/transportation.pdf.)

Institute Green Rating for Integrated Habitat Assessment (TERI-GRIHA), and the Indian Green Building Council (IGBC) in India, etc.

DISCUSSION QUESTION

- *Compare the material intensities and GHG emissions of a conventional building with a green building. Use examples or case studies in making such a comparison. Do green buildings cost more? Why?*

For example, in the United States it is seen that LEED-certified buildings have 34% lower CO_2 emissions, consume 25% less energy and 11% less water, and have diverted more than 80 million tons of waste from landfills.[32]

Also, the promotion of Net Zero policy development in Canada is underway and CaGBC is working to define the parameters for the Canadian context to help Canada to achieve its 2030 emission targets of reducing GHG emissions by 17% compared to 2005 levels. Thus, it is estimated that if the policies and technologies complement these green rating criteria, it can halt the growth of building-related emissions by 2050.[33]

1.4.4 Transport

The transport sector is demanding different forms of energy: gasoline, jet fuel, natural gas, and diesel. Figure 1.18 provides the statistics on world energy consumption due to the transport sector by energy source.

Figure 1.19 presents EEA-32[34] countries contribution of the transport sector compared to total emissions of $PM_{2.5}$ and NOx in 2009. It may be observed that the emissions from the transport sector dominate and specifically road emissions.

DISCUSSION QUESTION

- *Is use of electric vehicles the right choice to reduce environmental impacts of automobiles? Why?*

1.4.4.1 Aviation

The global aviation industry produces around 2% of all human-induced carbon dioxide emissions. Aviation is responsible for 12% of CO_2 emissions from all transports sources, compared to 74% from road transport. In 2015, the flights produced 781 million tons of CO_2. Around 80% of aviation CO_2 emissions are emitted from flights over 1,500 kilometers, for which there is no practical alternative mode of transport.[35] We will in Chapter 5 present case studies from the aviation industry where efforts are taken to curb emissions.

1.4.4.2 Vehicles

Automobile ownership worldwide has been increasing at a higher rate than the global population and reached more than 1 billion units in 2010. This trend is especially noticeable in Asia and in Central and South America. Manufacturing of vehicles (four wheelers

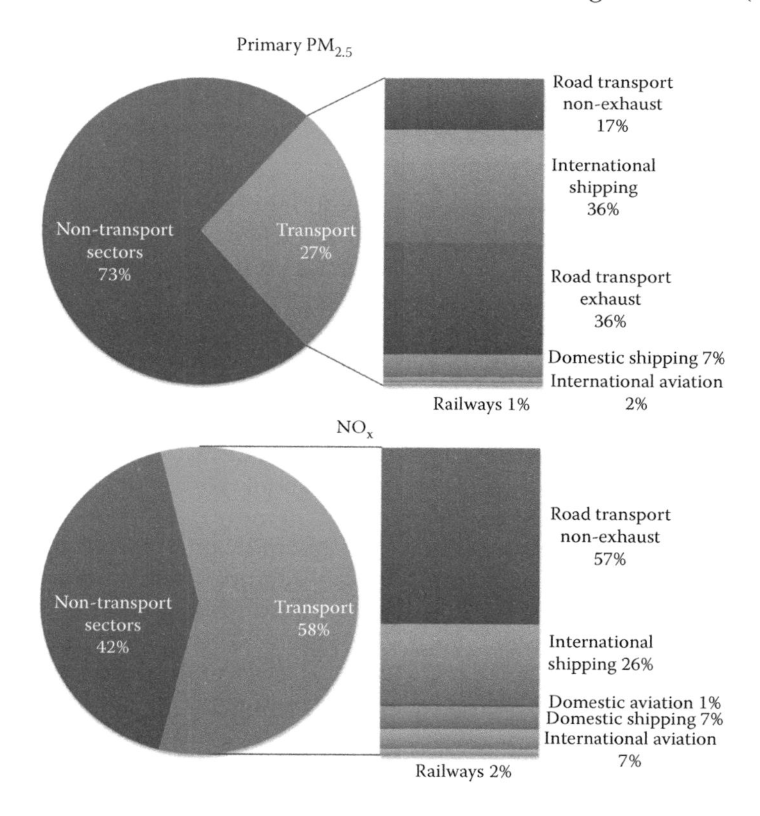

FIGURE 1.19
EEA-32 countries contribution of the transport sector to total emissions of $PM_{2.5}$ and NOx in 2009. (From CER & UIC, Rail Transport and Environment: Facts & Figures, 2015, p. 29. With permission, online source: http://www.cer.be/sites/default/files/publication/Facts%20and%20figures%202014.pdf.)

BOX 1.4 END-OF-LIFE VEHICLES[36]

End-of-life vehicles (ELV) have become a global concern as automobiles are increasingly used worldwide. Automobile ownership worldwide has been increasing at a higher rate than the global population growth and has crossed more than 1 billion units. This trend is especially noticeable in Asia and in Central and South America. ELVs are projected to reach 2.5 billion by 2050. Assuming that each vehicle produces about one ton of steel scrap, this gives an indication of the size of this market.[37]

The EU estimates that around 8–9 million tons of waste is generated from ELVs, out of which 25% belongs to hazardous waste (oil, tires, batteries, etc.) which ends up in landfill. This output of waste is expected to rise to 14 million tons by 2015.[38]

To address the challenge of managing ELVs, EU came up with EU-Directive 2000/53/EC in 2000 to reuse/recycle ELVs. Since then the rate of recycling of ELVs has shown an upward trend. Here, the hazardous waste is delivered to hazardous waste management companies, while the remaining 75% of useful metal waste is recycled at various industries, and the remaining parts are sold for reuse.

DISCUSSION QUESTION

- *One of the strategies to reduce fuel consumption of the vehicles is to reduce the weight by increasing the use of plastic components. While this strategy addresses the issue of air emissions, recycling of plastic from the ELVs becomes very important. Discuss the pros and cons of introducing more plastic components in the vehicles from the* life cycle *perspective.*

as well as two wheelers) requires substantial material inputs, especially of metals (steel) and composites (including plastic). But only 16% of the overall CO_2 emissions over the lifecycle of an average car occur in the production phase and over 80% occur in the use phase. This is the hidden environmental cost of driving.

Efforts are now being made to reduce vehicular emissions using technology and cleaner fuels. Apart from these interventions, options such as increasing public transport infrastructure, levying taxes, and promoting collaborative approaches such as car-pooling are considered.

Apart from the material extraction and use and emissions due to fuels, management of End-of-Life Vehicles (ELV) has become a global concern. ELVs are projected to reach 2.5 billion by 2050. On an average, each vehicle produces about one ton of steel scrap and plastic waste (Box 1.4). Box 1.4 explains the concept of ELV in more detail.

Thus, the recycling of ELVs has benefited from freeing valuable land and parking space as well as retrieving the precious materials and saving future extraction of non-renewable minerals for producing new cars. But it is seen that the legislative framework on ELV recycling is still limited to only the EU, the EFTA, Japan, Korea, China, and Taiwan. Thus, there is an urgency to develop such frameworks elsewhere. In many countries ELVs are processed by the informal sector that is difficult to regulate.

1.4.4.3 Rail Transport

The transport sector emitted 7.5 billion tCO_2 in 2013. In 2013, 3.5% of transport GHG emissions was due to the rail sector, while railways transported 8% of the world's passengers and goods.[39] A doubling of rail freight transport, with the freight shifted from the road sector, could result in a reduction of GHG emissions of around 45–55 million tons of CO_2 a year.[39]

The transport sector emitted 7.5 billion tCO_2 in 2013. The share of CO_2 emissions from transport has continuously increased since 2010 from 22.7% to 23.4% in 2013. In 2013, 3.5% of transport CO_2 emissions were due to the rail sector, while railways transported 8% of the world's passengers and goods.

Global railway passenger activity grew by 133% between 1975 and 2013 while freight activity has increased by 78% since 1975. An almost doubling of rail freight transport, with the freight shifted from the road sector, could result in a reduction of GHG emissions of around 45–55 million tons of CO_2 a year.[40]

The rail sector was 57% fueled by oil products and 36.4% by electricity in 2013. This share of electrified railways increased the use of electricity from 17.2% to 36.4% and also renewable electricity sources from 3.4% to 8.7% in 1990 to 2013, while coal consumption had dramatically declined. Specific CO_2 emissions in the rail sector have been following a similar improvement rate: they dropped by 60% in passenger services and by 38% in freight services between 1975 and 2013. Between 2000 and 2013, high-speed passenger activity has almost doubled. On average, a high-speed line uses 3.2 hectare per kilometer (ha/km), while an average motorway uses 9.3 ha/km.

Today, European railways have already achieved the EU Climate Package target of using 20% renewable energy by 2020. The main reasons have been electrification and green procurement. Thus, railways are more energy and CO_2 efficient per traffic unit than competitor transport modes.

1.4.4.4 Sea Routes

Maritime shipping is the world's most carbon-efficient form of transporting goods – far more efficient than road or air transport. Studies undertaken at the International Maritime Organization (IMO) conclude that total GHG emissions from international shipping decreased by 10% between 2007 and 2012, while cargo carried by the world fleet increased during that period. Sea routes have, however, a high risk of oil pollution.

1.4.5 Water Stress

Globally, agriculture consumes the most amount of water (70%), followed by industry (20%) and domestic use (10%). In the last 50 years, freshwater withdrawals have tripled, and the demand for freshwater is increasing by 64 billion cubic meters a year.[41] Clearly water demand from BRIICS (Brazil, Russia, India, Indonesia, China and South Africa) dominates.

Global water withdrawals are projected to increase by 55% through 2050.[42] Around 15% of the world's total water withdrawals are used for energy production. Meat production requires 8–10 times more water than cereal production.[43] In the irrigated agriculture sector, which represents 70% of freshwater withdrawals worldwide, the potential efficiency savings from increased water productivity could be as high as $115 billion annually by 2030 (based on 2011 prices).[44] Water is an important sector supporting livelihoods and about 78% of jobs constituting the global workforce are dependent on water.

Box 1.5 provides definitions of the terms water scarcity, water stress, and water risk. Water stress at global, national, and local levels has been an issue of great concern. In 2030, 47% of world population will be living in areas of high water stress. Unfortunately, these areas correspond to high population, most of which consist of poor and vulnerable people with limited access to safe drinking water and adequate sanitation facilities.

Figure 1.20 shows projections of water stress by 2050 as prepared by the World Resources Institute (WRI) indicating that nearly 50% of the countries in the world are under high and extremely high risks.

BOX 1.5 DEFINING WATER SCARCITY, WATER STRESS, AND WATER RISK[45]

WATER SCARCITY

Water scarcity refers to the volumetric abundance, or lack thereof, of water supply. This is typically calculated as a ratio of human water consumption to available water supply in a given area. Water scarcity is a physical, objective reality that can be measured consistently across regions and over time.

WATER STRESS

Water stress refers to the ability, or lack thereof, to meet human and ecological demand for water. Compared to scarcity, water stress is a more inclusive and broader concept. It considers several physical aspects related to water resources, including water scarcity, but also water quality, environmental flows, and the accessibility of water.

WATER RISK

Water risk refers to the probability of an entity experiencing a deleterious water-related event. Water risk is felt differently by every sector of society and the organizations within them and thus is defined and interpreted differently (even when they experience the same degree of water scarcity or water stress). That notwithstanding, many water-related conditions, such as water scarcity, pollution, poor governance, inadequate infrastructure, climate change, and others, create risk for many different sectors and organizations simultaneously.

Attempts have also been made to develop water stress indices at national levels. The Environmental Management Center LLP developed a map of Water Stress Index (WSI) across all the districts of India. Water Stress was defined as "unavailability of freshwater" or "water scarcity" for different consumptive uses. Today, Water Stress is not limited to water availability but there are many other aspects such as deteriorating water quality and uncertainties due to climate change that need to be considered. In this perspective, the study defined Water Stress in a more comprehensive manner such as

$$\text{Water Stress} = f(\text{Quantity, Quality, Uncertainty})$$

The Water Stress index proposed above was further expressed in ten indicators. These included availability for meeting the demand, groundwater draft, stage of groundwater development, future availability of groundwater for agriculture, pollution due to agriculture runoff and sewage disposal, groundwater quality, climate change, and disaster proneness due to droughts, floods, and cyclones.

Based on the above, a mathematical model was derived to calculate the Water Stress Index for India at the district level. Data analysis included – defining the Stress Function for each indicator, application of an Analytical Hierarchy Process to assign weights and use of GIS for spatial analyses. Accordingly, Water Stress Maps for water quantity, quality, uncertainty and for the Composite Water Stress Index were developed.

The idea of the map of water stress was to come up with interventions to address the issue. For identifying interventions at the block level (under the districts), aspects such as the contiguous nature of the water problems, impact on the beneficiaries, access to safe

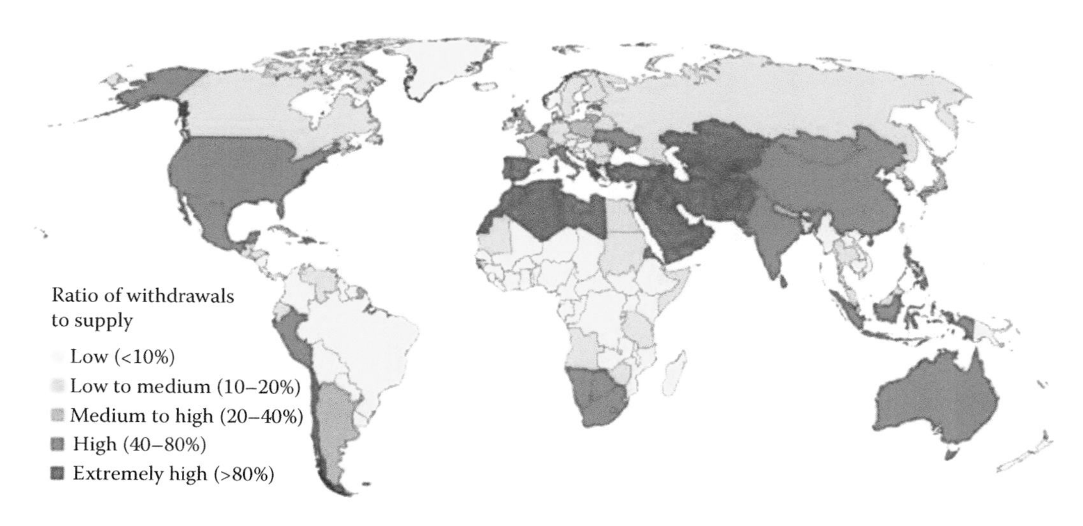

FIGURE 1.20
Projection of Water Stress by Country in 2040. (Adapted from WRI Aqueduct Projected Water Stress Country Rankings, 2015, World Resources Institute, online Source: http://www.wri.org/resources/data-sets/aqueduct-projected-water-stress-country-rankings. Link: aqueduct.wri.org.)

drinking water, and adequacy of sanitation and irrigation facilities, were also assessed. Such a study helped a large corporation in India to identify the water stressed areas for their Corporate Social Responsibility. Figure 1.21 shows application of WSI in India. Mapping of WSI helps in prioritizing and deciding the needed interventions.

1.4.5.1 Virtual Water[46]

It is important that we understand the concept of virtual water to understand the water stress and especially the water transfers that take place due to trade of goods across the world. Virtual water is the water embedded in the products, which needs to be accounted for. A country may choose to reduce the burden on the natural resources within its borders by importing water intensive products. For water-scarce countries it can sometimes be attractive to import virtual water (through import of water-intensive products), thus relieving the pressure on the domestic water resources. Mexico, for example, imports maize and, in doing so, it saves 12 billion cubic meters per year of its national water resources. This is the volume of water that Mexico would need domestically if it had to produce the imported maize within the country.

Globally, the major gross virtual water exporters are the USA, China, India, Brazil, Argentina, Canada, Australia, Indonesia, France, and Germany, and the major gross virtual water importers are the USA, Japan, Germany, China, Italy, Mexico, France, the UK, and the Netherlands. In Europe as a whole, 40% of the water footprint lies outside of its borders.

DISCUSSION QUESTION

- *Compile information on virtual water embedded in the products. Calculate the true per capita water consumption of a British citizen in London on this basis and compare with a citizen of Mumbai.*

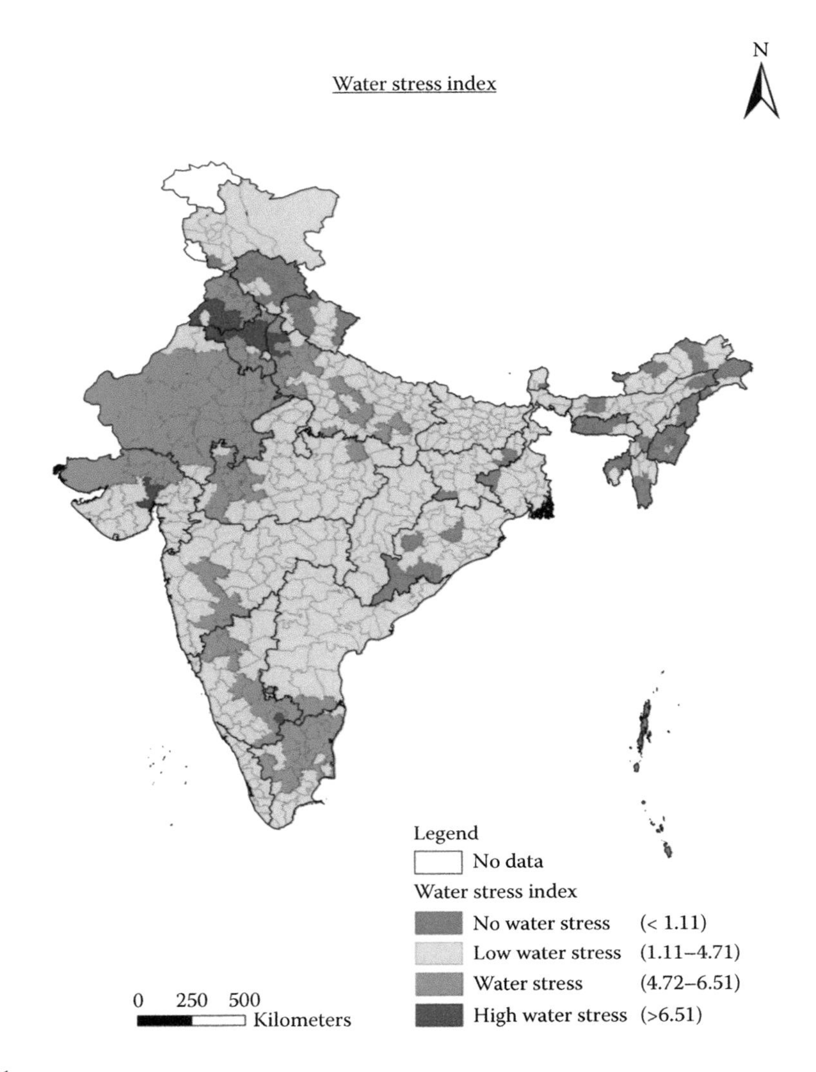

FIGURE 1.21
The water stress map for India. (Prepared by Prasad Modak, Sonal Kaushik, Krupa Desai, Kiran Apsunde and Tausif Farooqui at Environmental Management Centre LLP, India. For more details contact prasad.modak@emcentre.com.)

1.4.6 Sanitation & Health

At least 2.4 billion people still lack improved sanitation facilities in 2015.[47] It is estimated that 1.7 billion people need treatment and care for neglected tropical diseases[48] resulting from poor water, sanitation, and unhygienic conditions. These diseases include Guinea Worm Disease, Buruli Ulcer, Trachoma, Schistosomiasis, and diarrheal diseases.

Since 1990, 2.1 billion people have gained access to an "improved" form of sanitation, such as flush toilets or latrine with a slab. But it still falls as a deficit to the 2015 Millennium Development Goal target by 700 million people (Figure 1.22).

One out of eight people worldwide practice open defecation.[49] The open defecation numbers have decreased by around 21%, that is, from 771 million in 1990 to 610 million in 2015, mainly in the Sub-Saharan Africa region.

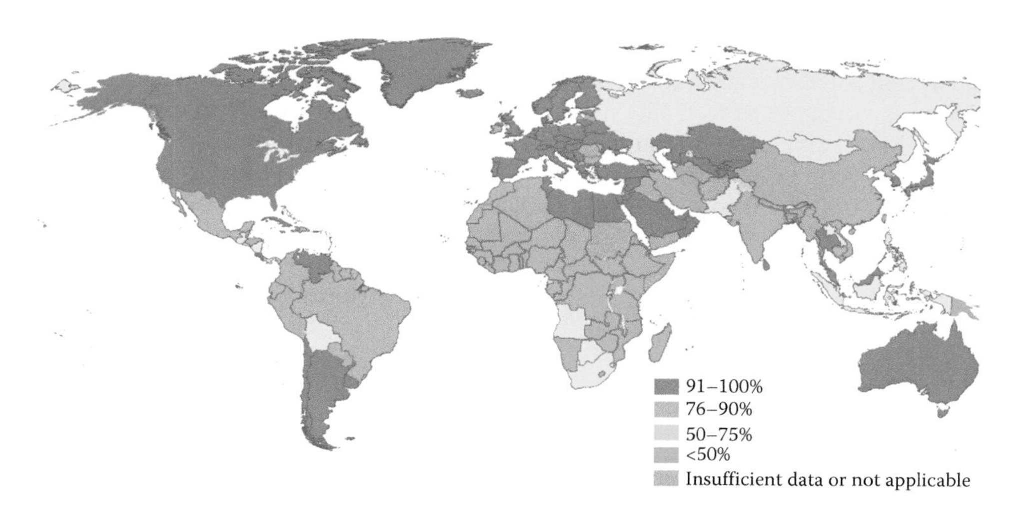

FIGURE 1.22
Spatial distribution of access to improved sanitation. (From WHO/UNICEF JMP Progress on Sanitation and Drinking Water: 2015 Update and MDG assessment, ISBN 978 92 4 150914 5, Figure 16, Page 12 . With permission, online source: http://files.unicef.org/publications/files/Progress_on_Sanitation_and_Drinking_Water_2015_Update_.pdf.)

Improved services in water, sanitation and hygiene have the potential to prevent at least 9.1% of the global disease burden and 6.3% of all deaths.[50] For instance, provision of improved sanitation can reduce diarrhea morbidity by 37.5%. Every $1 invested in water and sanitation is estimated to provide a $4 economic return.[51]

1.4.7 Air Pollution

Nearly 92% of the world's population lives in areas where air quality does not meet WHO standards.[52]

Air pollution is the 4th highest ranking risk factor for death globally. Air pollution was responsible for 5.5 million deaths in 2013.[53]

Household air pollution is caused by burning solid fuels for heating and cooking: coal, wood, and dung. In India, for example, recent research estimates that almost 30% of outdoor air pollution is from household sources. Clean household energy is therefore important for reducing the heavy burden of disease from outdoor air pollution. Women experience higher personal exposure levels than men, owing to their greater involvement in daily cooking and other domestic activities. Sixty percent of all premature deaths attributed to household air pollution occur in women and children.[8]

The single biggest killer of children aged under five years worldwide is pneumonia. This disease cuts short almost a million young lives each year. More than 50% of those pneumonia deaths are caused by exposure to Hazardous Air Pollutants (HAP). Emerging evidence links exposure to HAP with risk for other adverse health outcomes, such as low birth weight and stillbirths, cervical cancer, tuberculosis, asthma, ear and upper respiratory infections, and nasopharyngeal and laryngeal cancers.

Ground level ozone has been another matter of concern. Methane is a precursor in ground level ozone formation that is generated from disposal of untreated organic wastes in urban areas. Apart from causing direct health effects, ground level ozone reduces agricultural productivity.[54]

With relatively-weak policies to manage urban emissions, and increasing economic and transportation activity, the situation is likely to get worse in urban areas where most of the population is expected to live. The air pollutant linked most closely to excess death and disease is $PM_{2.5}$ (particulate matter less than 2.5 micrometers in diameter), heavily emitted by both diesel vehicles and the combustion of biomass, coal, and kerosene. The ambient concentrations of $PM_{2.5}$ from transport sources alone are expected to double by 2030, if no action is taken.[55] While exposure to air pollution is a risk factor common to both rural and urban populations, the routine monitoring of air quality is generally limited to large cities. This makes the task of understanding the nature and distribution of population exposures much more difficult.

DISCUSSION QUESTION

- *Air quality data is now increasingly shared online. Illustrate three examples of such data sharing on a global and a national basis.*

1.4.7.1 Short-Lived Climate Pollutants

It is important that we understand the Short-Lived Climate Pollutants (SLCPs) considering their local as well as global context.[56] Inefficient use and burning of biomass and fossil fuels in transport, housing, power production, waste disposal, and industry cause most SLCP emissions. SLCPs include black carbon (soot) and methane that are a major contribution to global warming. SLCPs persist in the atmosphere for weeks or months while CO_2 emissions persist for years. Reductions of SLCP emissions could thus greatly help to reduce near-term climate change by as much as 0.5°C before 2050. Reducing black carbon emissions could also help mitigate regional disruption of traditional rainfall patterns.

BOX 1.6 ASIA BROWN CLOUD[57,58]

The combustion of fossil fuels and biomass, in Asia, has led to large amounts of aerosols, such as soot and dust, causing the Asian Brown Cloud (ABC). This large brown cloud occurs annually from November to May over eastern China and southern Asia. It has been linked to decreased rainfall in India during the monsoon season since 1930, shift of the summer monsoon in eastern China southward, decrease in agricultural production in the region and an increase in respiratory and cardiovascular problems in the inhabitants of this region. Due to the ABC, India and China are 6% dimmer today at the surface, than they were in preindustrial times. It was found that India and China can affect climate thousands of kilometers away, warming the United States by up to 0.4°C by 2024, while cooling other countries.

DISCUSSION QUESTIONS

- *Have there been any action plans developed to address the menace of Asia Brown Cloud (ABC) at the national level? Is there any regional cooperation?*
- *Are there any linkages of ABC to global warming?*

BOX 1.7 OZONE-DEPLETING SUBSTANCES AND GLOBAL WARMING[59]

As a result of the 1987 Montreal Protocol that was ratified by 196 countries, the overall quantity of Ozone-Depleting Substances (ODS) in the lower atmosphere (troposphere) has declined by approximately 8% from peak values that occurred during the 1990s. These reductions were due to a 95% decrease in the use of ozone-depleting substances from 1986 to 2008.

The CFCs were substituted for by hydrochlorofluorocarbons (HCFCs), and, hence, while global emissions of ODS such as chlorofluorocarbons (CFCs), halons, carbon tetrachloride and methyl bromide reduced global emissions, HCFC (specifically HCFC-22 and HCFC-142b) increased. This increase in emissions of HCFCs was primarily driven by increases in production and consumption of HCFCs among developing countries.

From 1989 to 2004, production of HCFCs increased 1184% among developing countries compared to 5% among industrialized countries. Consumption of HCFCs increased from 745% among developing countries compared to a decline of 6% among industrialized countries from 1989 to 2004. Although HCFCs have significantly lower ozone-depleting potential than CFCs, they are likely to remain in the stratosphere for a long time, and these chemicals have a very high global warming potential. Refer to Box 5.4 in Chapter 5 to know more about the ODP and GWP of chemicals used to phase out ODS in refrigerants used in automobile mobile air conditioning (MAC).

 **DISCUSSION QUESTION**

- *Discuss how to address the dichotomy between the phasing out of ODS and global warming. Are there any solutions?*

According to recent estimates by the United Nations Environment Program, reducing methane and black carbon emissions could avoid annual crop losses of over 30 million tons annually. See Box 1.6 on Asia Brown Cloud that has been a serious matter of concern. There are complexity and conflicting situations when we address global issues, such as management of global warming and reducing depletion of ozone layer. To address the problem of ozone layer depletion, for instance, we substitute the Ozone-Depleting Substances (ODS). These chemical substitutes, however, lead to increased global warming. We illustrate this dichotomy in Box 1.7.

1.4.8 Degradation of Land and Agriculture

The world land area dedicated to agriculture is 38.5%. Agriculture and forestry account for 2% of total world energy used. The total world water withdrawal used for agriculture is 95%.[60]

Around 75 billion tons of fertile soil disappear yearly.[61] Around 12 million hectares are lost each year specifically to desertification and drought, where 20 million tons of grain could have been grown.[62] Over the next 25 years, land degradation could reduce global food productivity by as much as 12%, leading to a 30% increase in world food prices.

Nearly 52% of the land used for agriculture is moderately or severely affected by land degradation affecting 1.5 billion people across the globe. Excluding the current deserts,

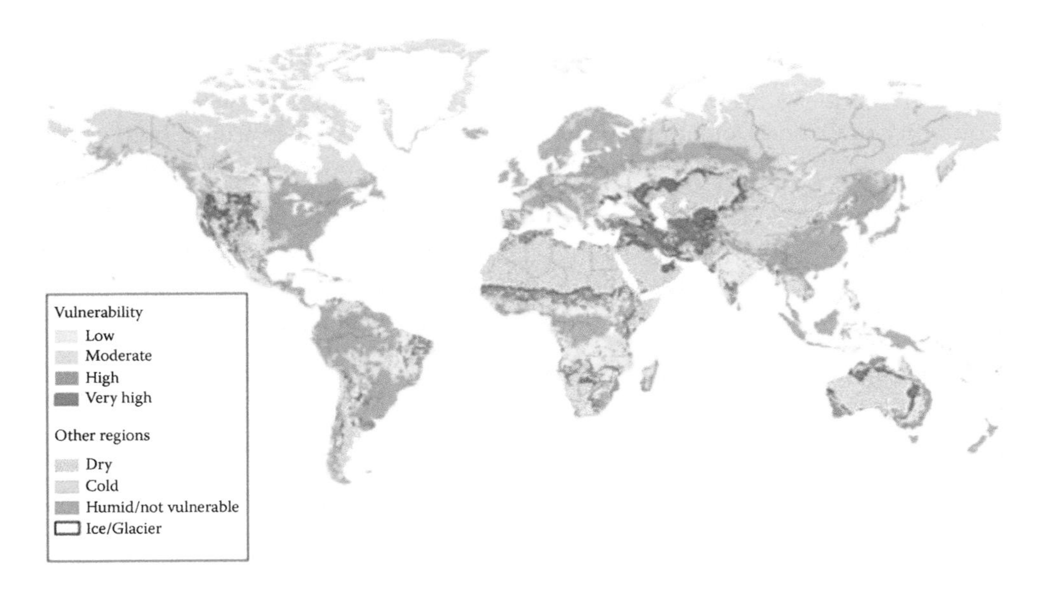

FIGURE 1.23
Areas vulnerable to desertification. (From Climate and Land Degradation, World Meteorological Organization, No. 989, p. 7, 2005. With permission, online source: http://www.wamis.org/agm/pubs/brochures/wmo989e.pdf.)

which occupy about 46% of the landmass, about 25% of the land is prone to water erosion and about 22% to wind erosion.

Of the 2 billion hectares of drylands in Asia, more than half are affected by desertification. Countries in Asia and the Pacific region rank among the highest globally, with the most affected being China (457 million hectares), followed by India (177 million hectares), Indonesia (86 million hectares) and Bangladesh (72 million hectares).[63] Figure 1.23 shows the regions vulnerable to desertification across the world.

About 40% of West Asia's land area is subject to desertification. It is reported that 83% of the marginal lands of West Asia are considered to be in danger of desertification. The underlying causes of desertification include low and variable amounts of precipitation, increases in temperature, evapotranspiration rates, changes in land-use patterns and practices, and recent trends in climate change.[64]

Climatic stresses account for 62.5% of all the stresses on land degradation in Africa. These climatic stresses include high soil temperature, seasonal excess water, short duration low temperatures, seasonal moisture stress and extended moisture stress, and affect 18.5 million km^2 of the land in Africa.

1.4.9 Biodiversity

Biodiversity provides the raw materials and combinations of genes, that produce the plant varieties and animal breeds upon which agriculture depends. Healthy ecosystems provide humans with a variety of services and goods including the provision of food, clean water, fuel, timber and forests that produce benefits such as natural medicines. Ecosystems also help to regulate disease, provide protection against natural disasters, improve air quality, and support agriculture through pollination and soil formation.

The Millennium Ecosystem Assessment (MA), led by the United Nations, mapped out the nature and extent of human impact on ecosystem services. It was found that the growing demand for natural resources over the last 50 years has resulted in the degradation or unsustainable use of 60% of the ecosystem services that were assessed. A great deal of this impact was attributed to industrial and anthropological activities.[65] Industrial activities and infrastructure development contribute to deforestation, overexploitation of bio-resources, introduction of invasive alien species and pollution leading to reduced flow of ecosystem services.

Loss of habitat due to destruction, fragmentation or degradation is the primary driver of wildlife population decline and species extinction worldwide. Extractive activities (such as mining, quarrying, oil and gas through drilling and fracking) and construction of large-scale infrastructure (dams, ports, power plants) and linear infrastructure (roads and railway lines) in ecologically sensitive areas can cause habitat loss. Finally, pollution and accidental introduction of invasive species can cause degradation of habitat. For example, nitrate runoff creates anoxic areas called dead zones around the mouths of some rivers, which in turn depletes fish stocks.

Commercial agriculture favors genetic uniformity. Monoculture is practiced with a single, high-yielding crop using expensive inputs such as water, fertilizer and pesticides to maximize production. In the process, not only traditional crop varieties, but long-established farming ecosystems are adversely affected. The uniformity leads to risks and makes a crop vulnerable to pest attack and diseases. Thus, although planting a single, genetically uniform crop might increase short-term yields, there could be undesirable side effects on a long-term basis. When natural habitats are lost, species lose the physical space and resources they need to sustain themselves. When habitats are stressed, they can no longer support the number of species that live there and species begin to disappear until the habitat reaches a new state.

Trade in forest products has increased significantly over the past 50 years. The processed wood products such as sawn timber, pulpwood, board, and wood-based panels are heavily traded across the world. According to the United Nation's Food and Agriculture Organization, wood-based panel trade has increased over 800% in the past three decades.[66] Each year around 2.5 million acres of land is converted to fast-wood forests (Figure 1.24).

DISCUSSION QUESTION

- *Can loss to biodiversity be monetized? Is it possible? Have efforts been made to assess the* natural capital?

1.4.10 Natural Disasters

Responding and managing the aftermath of natural disasters has become an issue. The social impacts and the economic damage due to disasters is extremely high. It is estimated that the 15-year average economic losses per year due to natural disasters is around US$175 billion. Table 1.1 shows the statistics for 2015. Apart from loss of life, property damage, spread of diseases, etc., management of disaster waste has become a serious issue.

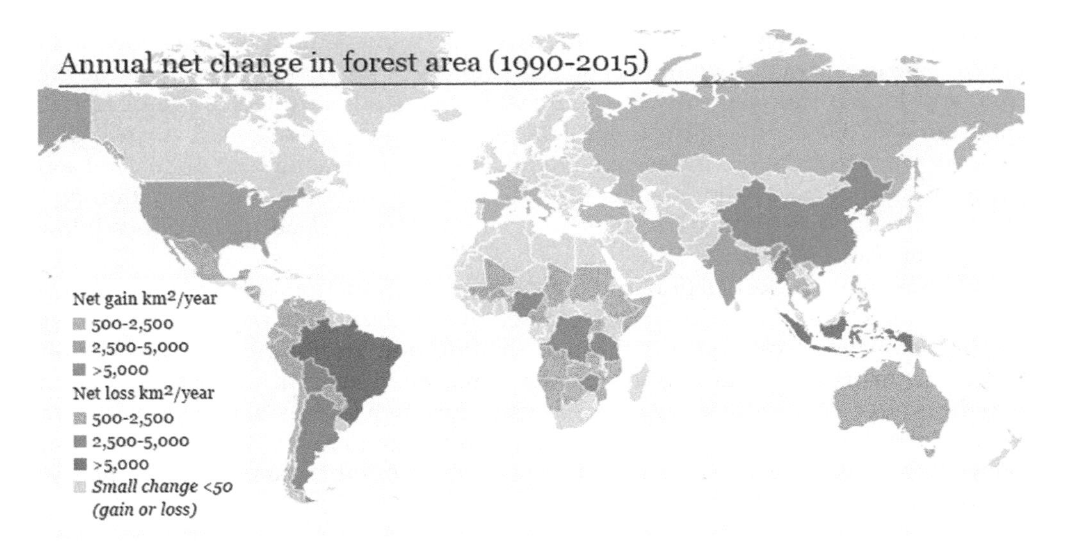

FIGURE 1.24
Change in the forest area by region 1990 to 2015. (From Food and Agriculture Organization of the United Nations, Global Forest Resources Assessment, 2015, p.18. Reproduced with permission, online source: http://www.fao.org/3/a-i4793e.pdf.)

Guidelines have been developed to assist in disaster waste management, building the element of recovery of materials.[67]

DISCUSSION QUESTIONS

- *Do you think that in the last decade the frequency and the severity of natural disasters have increased? Is there a role for climate change?*
- *Describe the work carried out on disaster waste management in the 2010 earthquake in Haiti and in Tohoku earthquake and tsunami in 2001 in Japan.*

1.4.11 The Nexus

As underscored earlier, it is important to recognize the nexus between the issues. Issues cannot be addressed in isolation. The nexus gets into complexities with uncertainties because of factors such as climate change.

Figure 1.25 shows a nexus between water and agriculture covering water availability, water quality, inputs such as seeds, energy, pesticides and fertilizers, cropping patterns, and extension services including post-harvest facilities and market linkages. A rounded approach to address all these *connectors*, operating in a nexus, is required if the objective of climate resilient agriculture is to be achieved.

The nexus gets better understood when we start showing the linkages in more details based on *cause-effect* relationship with evidence or science and empirical or field data.

TABLE 1.1

Global Economic Losses estimated in 2015

Date(s)	Event	Location	Deaths	Economic Loss (USD)	Insured Loss (USD)
April 14 & 16	Earthquake	Japan	154	38 billion	5.5 billion
Summer	Flooding	China	475	28 billion	750 million
Sept. 28–Oct. 10	Hurricane Matthew	US, Caribbean	605	15 billion	5.0 billion
August	Flooding	United States	13	10 to 15 billion	3.0 billion
Yearlong	Drought	China	N/A	6.0 billion	200 million
May/June	Flooding	Western/Central Europe	20	5.5 billion	3.4 billion
Yearlong	Drought	India	N/A	5.0 billion	750 million
August 24	Earthquake	Italy	299	5.0 billion	100 million
July	Flooding	China	289	4.7 billion	200 million
May	Wildfire	Canada	0	4.5 billion	2.8 billion
			All other events	83 billion	33 billion
			Totals	**210 billion**[a]	**54 billion**[a,b]

[a] Subject to change as loss estimates are further developed.
[b] Indudes losses sustained by private insurers and government sponsored programs.
Source: From 2015 Annual Global Climate and Catastrophe Report—Forecasting, 2015, AON, p. 2. With permission, online source: http://thoughtleadership.aonbenfield.com/Documents/20160113-ab-if-annual-climate-catastrophe-report.pdf

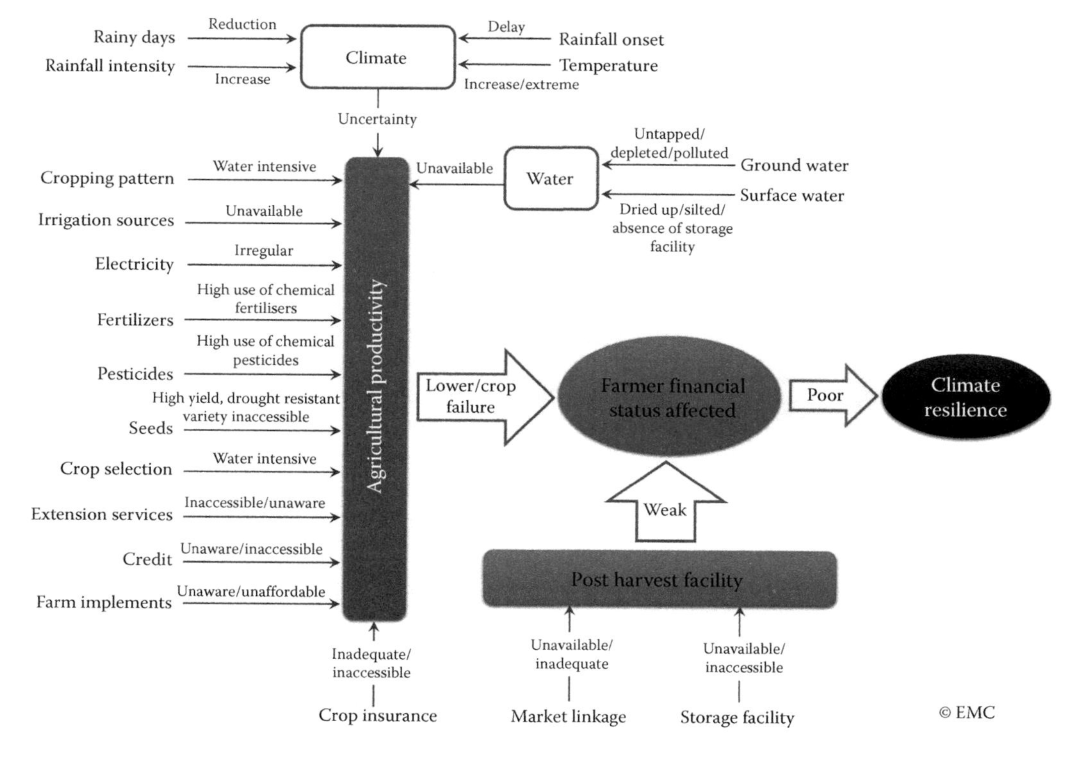

FIGURE 1.25
Nexus between agriculture and water and associated systems to be recognized for achieving climate resilience.

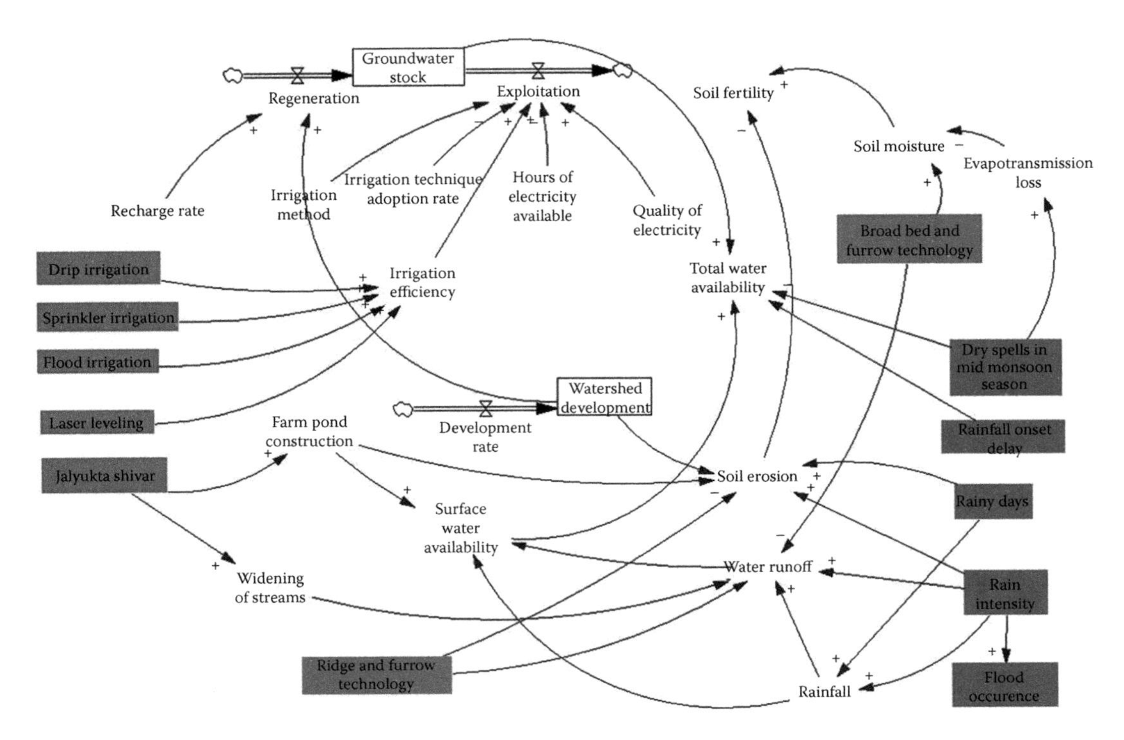

FIGURE 1.26
Illustration of linkages of the nexus focusing on watershed development.

Figures 1.26 through 1.28 show such diagrams prepared for a project on Climate Resilience in Agriculture in the State of Maharashtra in India. The diagrams depict various *stock flows* and feedback loops.

These cause-effect depictions could be simulated using platforms such as VenSIM to help recognize the nexus between issues and come up with a suite of interventions. The interventions include watershed planning, policy measures, technology-based solutions and measures to establish market linkages and financing.

We will be discussing the Sustainable Development Goals (SDGs) in the next section. The SDGs address the various issues described in this chapter. The SDGs are interlinked to one another and meeting the goals requires strategies that are holistic recognizing the nexus. Some of the questions policy makers will need to answer to successfully achieve the SDGs will be: How to coordinate and harmonize investments in different areas; How to leverage on the positive synergies among interventions and limit the adverse or undesired effects, etc.

To help policy makers answer such daunting questions, the Millennium Institute created a model called the Integrate Simulation Tool or iSDG. The iSDG uses the nexus approach to deal with the sustainable development goals rather than looking at them individually, or in silos.

The iSDG model is based on the Threshold 21 (T21) model which is a *dynamic simulation tool* designed to support comprehensive, integrated long-term national development planning. T21 supports comparative analysis of different policy options, and helps users

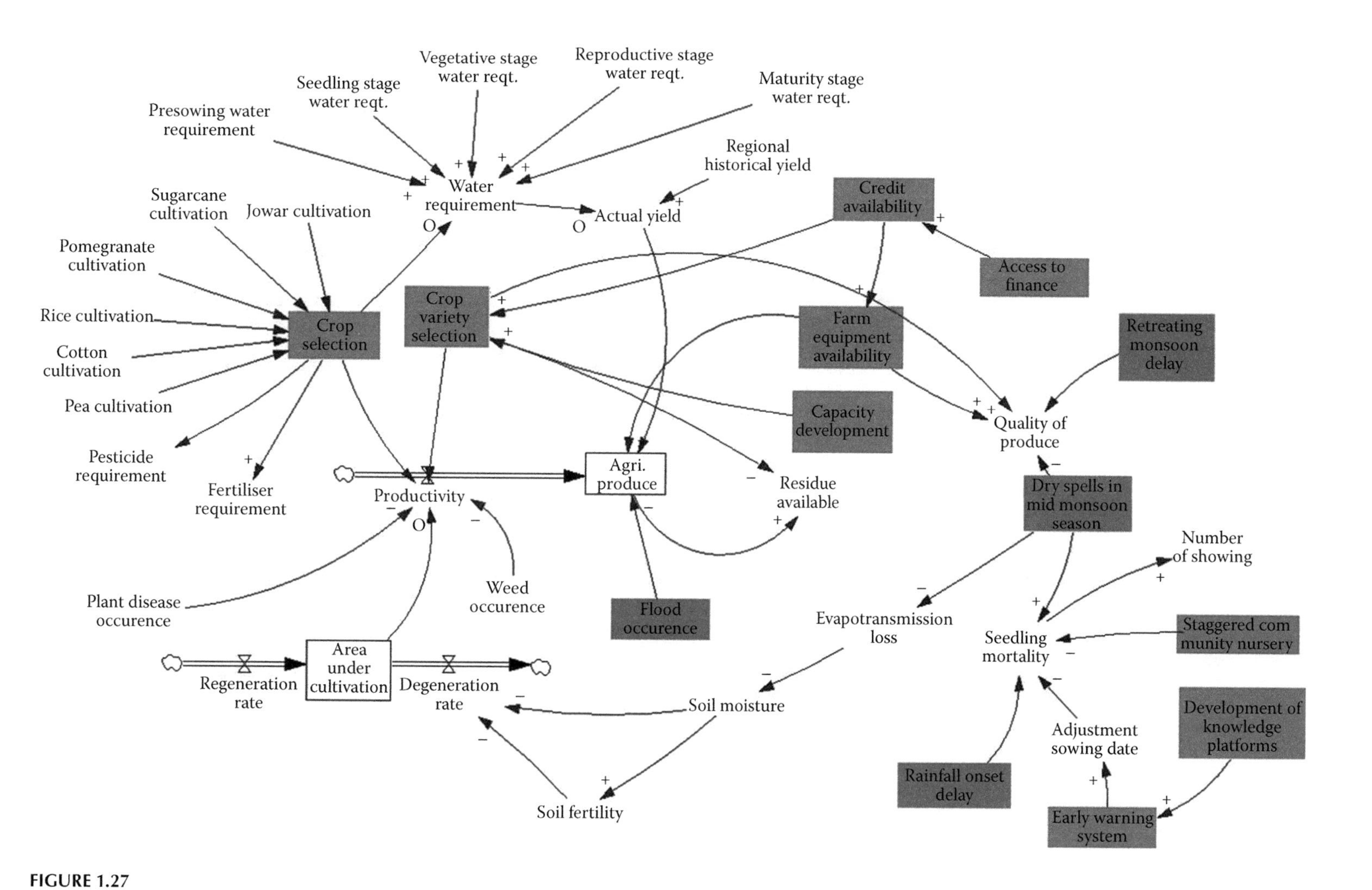

FIGURE 1.27
Illustration of linkages of the nexus focusing on agriculture produce.

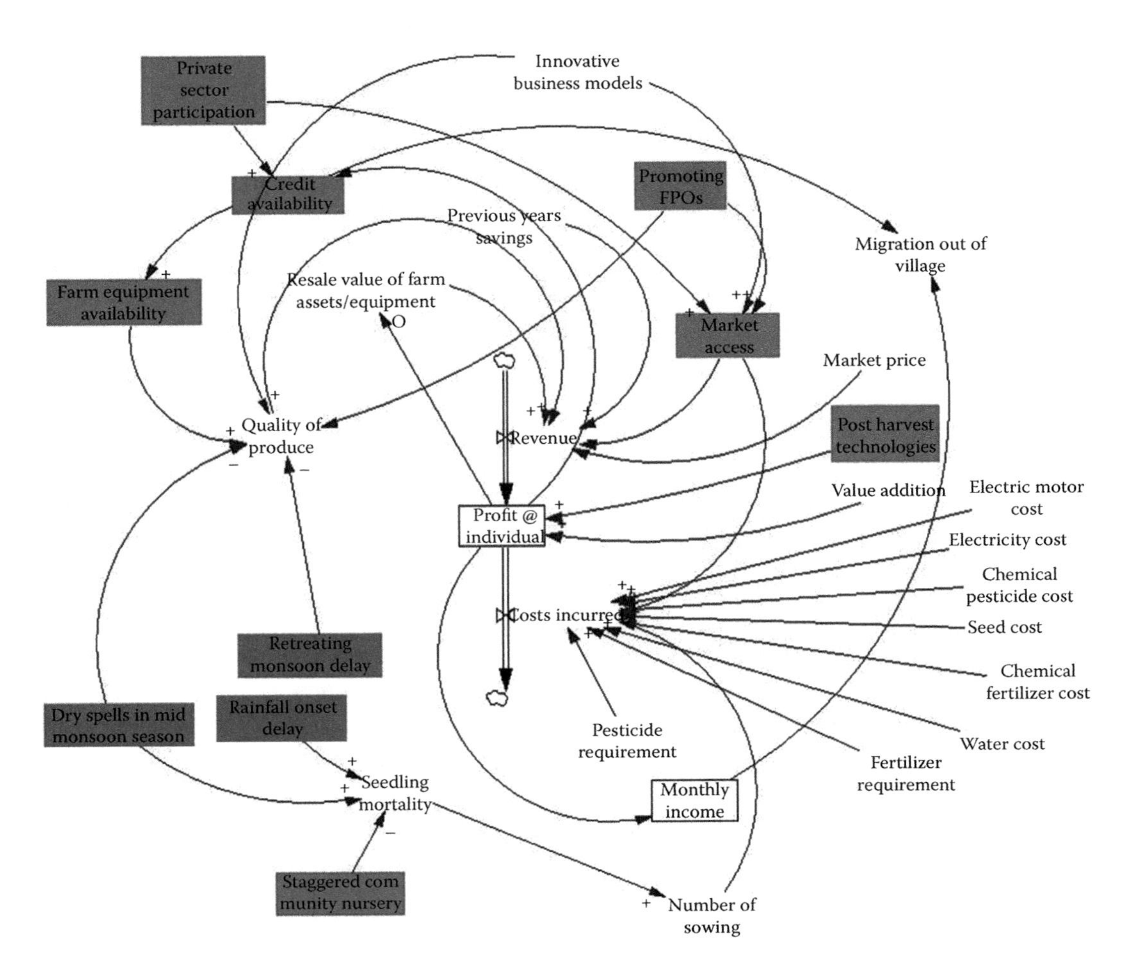

FIGURE 1.28
Illustration of linkages of the nexus focusing on livelihoods and financing.

to identify the set of policies that tend to lead towards a desired goal. Simulations made using T21 provide insight on how different indicators of sustainable development are interlinked and interact with one another to achieve desired outcomes while meeting the development challenges.[68]

An integrated analysis is required to successfully address complex development issues that balance social, economic, and environmental development. By bringing together the three dimensions of sustainable development into one framework, the iSDG model enables broad, cross-sector and long-term analyses of the impacts of alternative policies. The T21 model may not be accurate because of limitations on data or capture the complexity of the real world, but its use can greatly help to build scenarios and importantly lead to discussions and understanding of the nexus. Figure 1.29 shows a screenshot of the model where nexus between water, agriculture and food is considered.

See Box 1.8 for more details on iSDG.

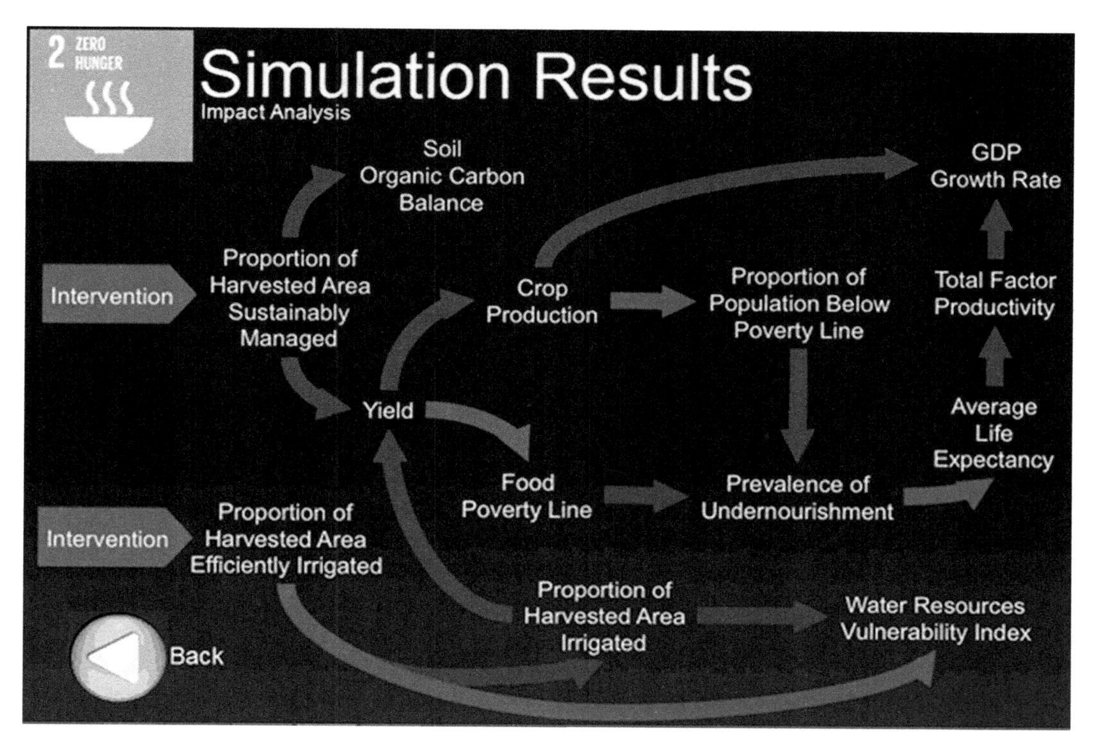

FIGURE 1.29
A screenshot of iSDG showing nexus between water, agriculture and food. (From Simulation Results, Millennium Institute, 2016. With permission, online source: http://www.millennium-institute.org/images/isdg_Slide4.png.)

BOX 1.8 THE iSDG MODEL[69]

The iSDG model enables policy makers and planning officials at all levels of governance to understand the interconnectedness of policies designed to achieve the SDGs and test their likely impacts before adopting them.

The T21 model is customized to the specific issues faced by a country on their path to sustainable development. This customized model, called the Starting Framework (SF), which covers a broad range of issues faced by most developed and developing countries such as: poverty, environmental degradation, education, healthcare, economic growth, and demographic shifts. It is a large model, consisting of more than a thousand equations, about 60 stock variables, and several thousand feedback loops.

The model simulates the fundamental trends for SDGs until 2030 under a business-as-usual scenario, and supports the analysis of relevant alternative scenarios. It also traces the trends beyond the SDGs' time span to 2050. The iSDG model is especially useful both in the early stages of policy design, to support scenario exploration, and in its advanced stage, when specific interventions designed for various sectors can be jointly simulated to assess their combined effect.

The iSDG model was constructed starting from the well-vetted, time tested and validated Threshold21 (T21)® model, covers all the 17 Goals, and supports a better understanding of the interconnections of the goals and targets, in order to develop synergetic strategies to achieve them.

DISCUSSION QUESTION

- *Other examples of nexus are Poverty and Biodiversity, Fossil Fuels, Air Emissions and Health, Waste Dumping, Site Contamination and Risks to Human Health and Ecosystems. Develop cause-effect diagrams on the lines of Figures 1.26 through 1.28 and identify the interventions that may be considered as a system.*

1.5 Key Drivers and Challenges

The complexity of the issues, apart from the nexus, is the scale. The impact of issues occurs on various scales, for example, local, regional, national, and global, covering economic, environmental, and social dimensions. Climate change adds further to the uncertainties on the outcomes. Our consumptive and fossil fuel-based development has escalated local pollution issues at the regional, national, and global scales affecting a wide cross section of stakeholders.

There are two important concerns that need to be addressed as the root cause of the problem. These are

- Polarization of Population due to Migration to Urban Areas

  and
- Skew in the Global Material Flows

We will discuss in this section these two drivers.

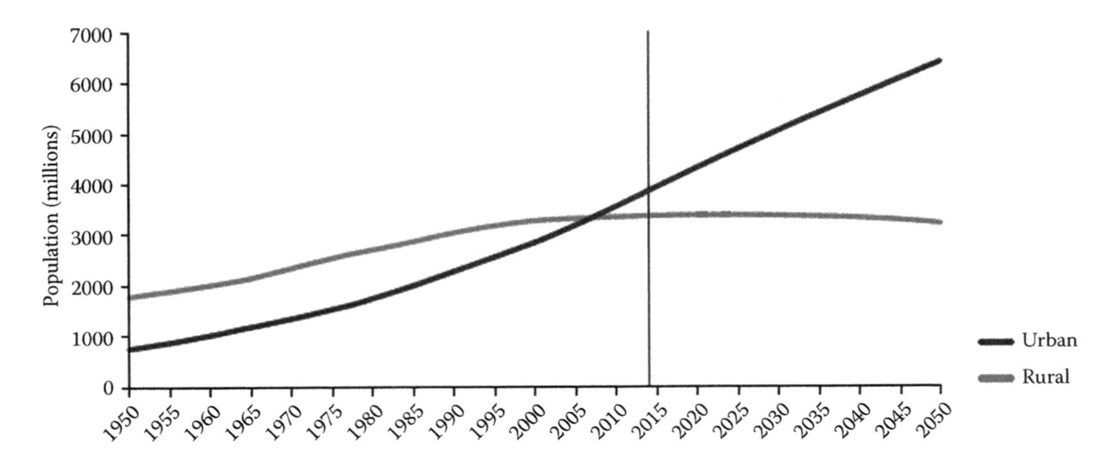

FIGURE 1.30
Urban and rural population of the world from 1950 to 2050. (From World Urbanization Prospects, by Department of Economic and Social Affairs, © 2014 Revision, United Nations, p. 7, New York, Reprinted with the permission of the United Nations, online source: https://esa.un.org/unpd/wup/publications/files/wup2014-highlights.Pdf.)

1.5.1 Urban–Rural Polarization

In the future, most people will be living in cities. Figure 1.30 shows the increase in urban and rural populations over the years.

Cities have high footprints of resources such as water, energy and materials. Cities produce a huge amount of waste and emissions. This has led to risks to the security of resources and poor quality of life to millions of people living in the cities.

It is estimated that fast paced urbanization, especially in emerging economies, will account for 70% of global growth in energy use up to 2030.[70] Cities are considered to be the major contributors to climate change as they are estimated to generate more than 60% of all carbon dioxide and significant amounts of other GHG emissions, mainly through energy generation, vehicle use and industrial production. Use of biomass in the slums and peri-urban areas is another important contributing factor.

It is interesting to note that at the same time, cities are highly vulnerable to climate change. Hundreds of millions of people in urban areas across the world will be affected by rising sea levels, increased precipitation, inland floods, more frequent and stronger cyclones and storms, and extreme weather.

Many major coastal cities with populations of more than 10 million people are already under threat. The most affected populations are the urban poor, that is, slum dwellers in developing countries who tend to live along river banks, on hillsides and slopes prone to landslides and near contaminated areas, in unsafe structures vulnerable to earthquakes. Despite these risks, many cities have not yet addressed climate change.[71] Programs have been undertaken now for ensuring climate resiliency in the cities by forming networks.[72] Initiatives like the Green Cities of the Asian Development Bank,[73] and the Guidebook on ECO2 Cities published by the World Bank[74] are other examples. Many national governments have taken up schemes and programmes to upgrade the city infrastructure notably in China and India with considerations for urban sustainability. Apart from the launch of the Smart Cities program, the Government of India has been operating a scheme called Provision of Urban Amenities in Rural Areas (PURA) to reduce in-migration of population from rural areas to cities.[75]

DISCUSSION QUESTIONS

- *Do Eco-Cities and Green Cities address climate resilience?*
- *Are there programs like PURA undertaken in other countries?*
- *To what extent has PURA in India been successful?*

1.5.2 Skew in the Global Material Flows

Population increase and highly consumptive lifestyles have resulted in high material extraction, transportation, and consumption. It is estimated that the annual per capita material footprints for the Asia-Pacific, Latin America, and the Caribbean, and West Asia are between nine and ten tons/person. These footprints for Europe and North America are around 20–25 tons per person. In contrast, Africa has an average material footprint of below 3 tons per capita. See Figure 1.31.

For Europe and North America, meeting the high material requirements is not possible only through Domestic Material Extraction (DME). Materials need to be, therefore,

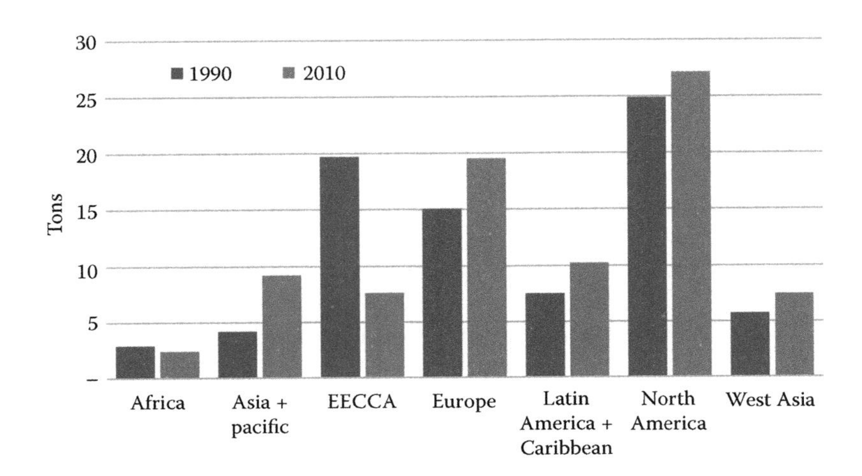

FIGURE 1.31
Per-Capita Material Footprint by Seven World Regions, from 1990 to 2010, in tons. (From Schandl et al., Global Material Flows and Resource Productivity: Summary for Policymakers, An Assessment Study of the International Resource Panel, UN Environment. Paris, United Nations Environment Programme, p. 26, 2016. With permission, online source: http://www.isa.org.usyd.edu.au/about/16-00271_LW_GlobalMaterialFlowsUNE_SUMMARY_FINAL_160701.pdf.)

imported from nations where these resources are available and cheap. Trade in materials is thus rising, driven mainly by consumption.

Trade has grown faster than domestic extraction and direct trade in materials has expanded fourfold since 1970. The new indicators of raw material equivalents of imports and exports show that trade mobilizes much greater amounts of materials than direct traded flows indicate. In 2010, 30 billion tons of materials extracted globally were required to produce 10 billion tons of directly traded goods. A raw material trade balance based on the attribution of globally extracted materials to traded goods shows that only Europe and North America have remained net importers of materials. By contrast, the Asia-Pacific region has changed into a net exporter of materials through large exports of manufactured goods which are mostly consumed in Europe and North America. See Figure 1.32.

Exports of developing economies have increased from $1,284 billion from 1995 to $8,072 billion in 2014. While, in comparison, during the same time period, for developed countries, the increase has been from $3,536 billion to $9,686 billion. Figure 1.33 shows China as the world's top exporter in 2014, followed by the United States and Germany.

The unevenness in the material consumption has thus led to a skew in material flows leading to economic, environmental, and social impacts across the world. Importing countries have strong incentives to invest in Resource Efficiency (RE) strategies and increase Secondary Materials Recycling (SMR). If these countries progress, then the global material flows will change. Unfortunately, these strategies are not in place or are rather in the infancy in the exporting countries, especially of the developing world.

Both net importing and exporting countries are affected by global resource price changes but in very different ways. Countries relying on material imports profit from low world market prices and their economic performance is harmed by high prices. Material exporters make major gains when natural resource prices are high but experience a hit to their balance of trade when prices fall and also the production contracts. Production has

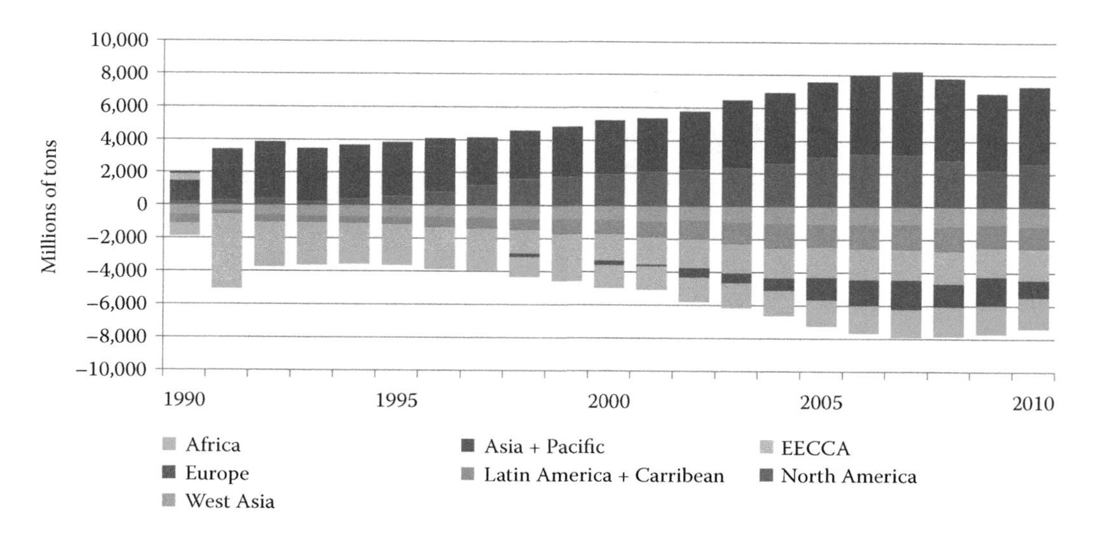

FIGURE 1.32
Raw Material Trade Balance (RTB) by Seven Sub-Regions, from 1990 to 2010. (From Schandl et al., Global Material Flows and Resource Productivity: Summary for Policymakers, An Assessment Study of the International Resource Panel, UN Environment. Paris, United Nations Environment Programme, p. 26, 2016. With permission, online source: http://www.isa.org.usyd.edu.au/about/16-00271_LW_GlobalMaterialFlowsUNE_SUMMARY_FINAL_160701.pdf.)

thus shifted from very material-efficient countries to countries that have lower material efficiency, which has resulted in an overall decline in material efficiency and, importantly, increased generation of wastes and emissions. This is a matter of concern.

Policy settings in countries that grow their infrastructure and consumption substantially need to be tailored towards achieving rapid and short-term gains in resource efficiency to offset some of the growth and to build cities and infrastructure in ways in which natural resource requirements, waste, and emissions can be minimized. Achieving human development and improved well-being at lower levels of material consumption will require a complex policy mix of RE and SMR and incentives supported by investments for high level research and development fostering innovation.

Increasingly, developing countries are mainstreaming Sustainable Consumption and Production (SCP) and green economy policies into their national development plans acknowledging the need to decouple their human development efforts from ever-increasing natural resource use, emissions, and waste. But there is a lot of work to be done.

1.6 Global Response

Realizing the complexity, the nexus, and the escalation of the issues on different scales, the national governments responded to the challenge by framing policies and regulations. Institutions for implementation were also created with needed finance and capacity building measures. It was soon realized that response at the national level was not sufficient and efforts were needed at the level of the United Nations to ensure a congruence on a global platform.

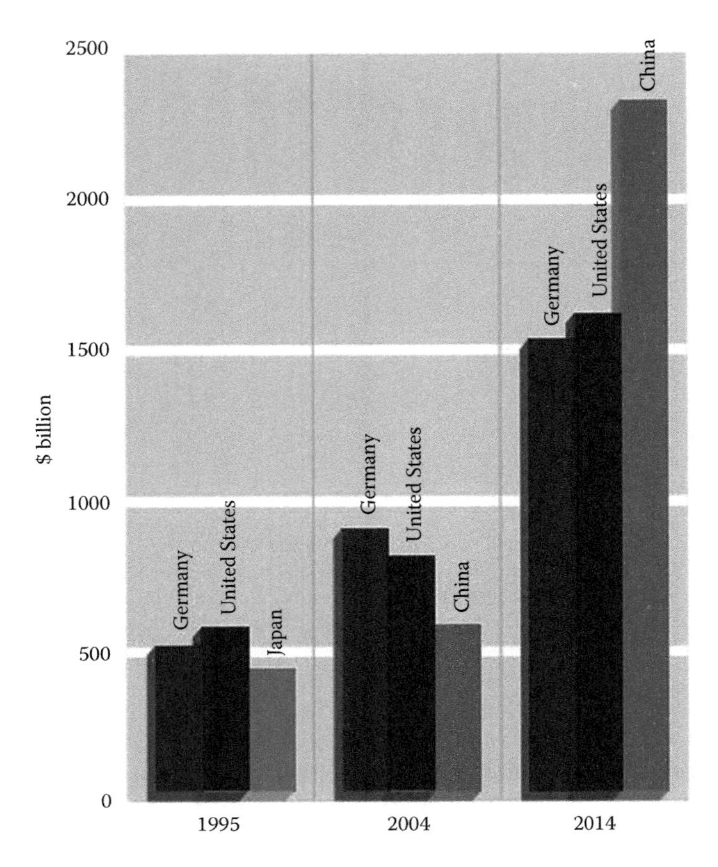

FIGURE 1.33
World's top exporters between 1995 and 2014. (From © World Trade Organization (WTO) 2017, International Trade Statistics 2015, World Trade Organization, p.25. With permission, online source: https://www.wto.org/english/res_e/statis_e/its2015_e/its2015_e.pdf.)

1.6.1 United Nations

The United Nations Conference on the Human Environment was held in Stockholm, Sweden from June 5–16 in 1972. The meeting agreed upon a Declaration containing 26 principles concerning the environment and development; an Action Plan with 109 recommendations, and a Resolution. Box 1.9 shows the Principles of the Stockholm Declaration.

One of the seminal issues that emerged from the conference is the recognition for poverty alleviation for protecting the environment. The Prime Minister of India, Indira Gandhi, in her seminal speech brought forward the nexus between environmental management and poverty alleviation that could threaten the sustainability of the planet.

To facilitate the implementation of these principles, the UN General Assembly formed the United Nations Environment Program. UNEP was charged with the responsibility to coordinate UN's environmental activities and assist developing countries in implementing environmentally sound policies and practices.

Taking a cue from the Conference, in 1973, the EU created the Environmental and Consumer Protection Directorate, and prepared the first Environmental Action Program.

BOX 1.9 PRINCIPLES AGREED IN THE STOCKHOLM DECLARATION OF 1972

1. Human rights must be asserted, apartheid and colonialism condemned
2. Natural resources must be safeguarded
3. The Earth's capacity to produce renewable resources must be maintained
4. Wildlife must be safeguarded
5. Non-renewable resources must be shared and not exhausted
6. Pollution must not exceed the environment's capacity to clean itself
7. Damaging oceanic pollution must be prevented
8. Development is needed to improve the environment
9. Developing countries therefore need assistance
10. Developing countries need reasonable prices for exports to carry out environmental management
11. Environment policy must not hamper development
12. Developing countries need money to develop environmental safeguards
13. Integrated development planning is needed
14. Rational planning should resolve conflicts between environment and development
15. Human settlements must be planned to eliminate environmental problems
16. Governments should plan their own appropriate population policies
17. National institutions must plan development of states' natural resources
18. Science and technology must be used to improve the environment
19. Environmental education is essential
20. Environmental research must be promoted, particularly in developing countries
21. States may exploit their resources as they wish but must not endanger others
22. Compensation is due to states thus endangered
23. Each nation must establish its own standards
24. There must be cooperation on international issues
25. International organizations should help to improve the environment
26. Weapons of mass destruction must be eliminated

In 1983, the UN General Assembly realized that there was a heavy deterioration of the human environment and natural resources. To rally countries to work and pursue sustainable development together, the UN decided to establish the Brundtland Commission, formally known as the World Commission on Environment and Development (WCED). The mission of the Brundtland Commission was to unite countries to pursue sustainable development together. The Brundtland Commission was officially dissolved in December 1987 after releasing Our Common Future, also known as the Brundtland Report, in October

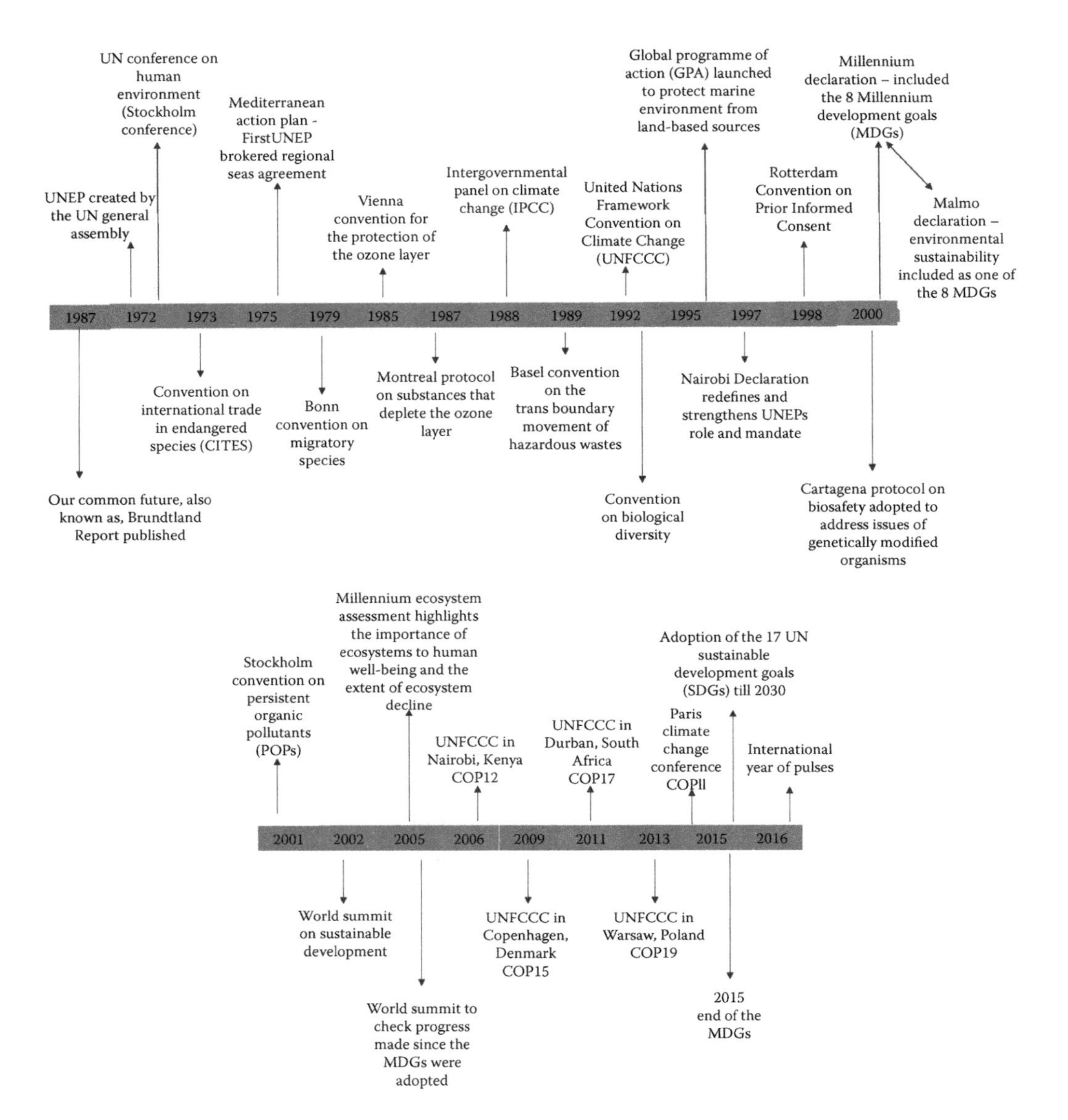

FIGURE 1.34
Key milestones of the sustainability journey of the United Nations.

1987. The term sustainable development was coined, representing the economic, environment, and social interests. The journey towards better environmental management in the interest of sustainability began. Figure 1.34 shows the key milestones of this journey over last four decades. Milestones such as Earth Summit of 1992, World Summit on Sustainable Development of 2002, and various Multilateral Environmental Agreements (e.g., on ozone, hazardous wastes, biodiversity, and global warming) are important as they guided the national policies and action plans, influenced the business, technologies, and investments and led to behavioral change, albeit not being able to address concerns such as unsustainable consumption.

1.6.2 Sustainable Development Goals

In 2005, world leaders gathered in New York, at the United Nations headquarters, for the Millennium Summit. At this summit, the Millennium Declaration was adopted by 189 counties, where they committed to the Millennium Development Goals (MDGs) which were formulated with the aim of reducing extreme poverty by 2015.

The Millennium Ecosystem Assessment (MA), led by the United Nations, was launched a year later in 2001 to assess human impact on the environment. It was established to help provide the knowledge base for improved decision-making and to build capacity for analyzing and supplying information. The findings were published in 2005, and it was found that the growing demand of natural resources over the last 50 years had led to the unsustainable use of 60% of the ecosystem services that were assessed. Moreover, it was found that a great deal of this degradation was due to industrial activities.

Both MAs and MDGs have been instrumental in recognizing the importance of living in a sustainable world, in which environment sustainability, social inclusion, and economic development are of equal value.

In 2012, the Rio + 20 Conference, which is the World Conference on Sustainable Development, was held in Rio de Janeiro. The member states developed a new set of goals called Sustainable Development Goals (SDGs) that were meant to build on the MDGs beyond 2015 till 2030. To arrive at the 17 goals, a participatory approach was used. The SDGs were launched in 2015 after accounting for the inputs from citizens, civil society organizations, scientists, academics, and the private sector from around the world. While the MDGs were mapped with the new SDGs, the 17 SDGs have gone into more detail and have taken a more holistic approach. See Table 1.2.

1.6.3 Addressing Resource Efficiency on Global Platform

The SDGs comprehensively address the importance of conserving natural resources to ensure both economic growth and human development. Goal 6 is concerned with water use, Goal 7 with energy, Goal 12 with materials and waste, and Goal 13 addresses carbon emissions and climate change. Very importantly Goal 8, which has a focus on economic growth, specifically addresses resource efficiency in its Target 8.4. This target asks countries to continuously improve their resource efficiency of production and reduce consumption over time.

The International Resource Panel (IRP) set up by UNEP tracks the consumption of resources. The data and indicators presented in the IRP study provide pressure indicators for a number of targets of the SDGs, including Targets 8.4, 12.2, and 12.5. The pressure

TABLE 1.2

The Sustainable Development Goals

SDG 1: No Poverty	SDG 6: Clean Water and Sanitation	SDG 5: Gender Equality
SDG 2: Zero Hunger	SDG 7: Affordable and Clean Energy	SDG 10: Reduced Inequalities
SDG 3: Good health and Well-being	SDG 9: Industry, Innovation and Infrastructure	SDG 16: Peace, Justice and Strong Institutions
SDG 4: Quality Education	SDG 11: Sustainable cities and communities	SDG 17: Partnerships for the Goals
SDG 8: Decent work and economic growth	SDG 12: Responsible Consumption and Production	
	SDG 13: Climate Action	
	SDG 14: Life below water	
	SDG 15: Life on Land	

BOX 1.10 RESOURCE USE AND SUSTAINABLE DEVELOPMENT GOALS[76]

Target 8.4: Improve progressively, through 2030, global resource efficiency in consumption and production and endeavor to decouple economic growth from environmental degradation, in accordance with the 10-year framework of programs on sustainable consumption and production, with developed countries taking the lead.

Target 9.4: By 2030, upgrade infrastructure and retrofit industries to make them sustainable, with increased resource-use efficiency and greater adoption of clean and environmentally sound technologies and industrial processes, with all countries taking action in accordance with their respective capabilities.

Target 12.2: By 2030, achieve the sustainable management and efficient use of natural resources

indicators relate to production and consumption. The production is expressed as Domestic Material Consumption (DMC) per unit of GDP, and Material Footprint (MF) per unit of GDP represents consumption. Target 12.2 asks for a measure of metabolic performance at the national economy level and can be expressed as DMC per capita and MF per capita, reflecting both production and consumption perspectives. Target 12.5 is about waste reduction and can be expressed as Domestic Material Consumption (DMC) per unit of land area to demonstrate the ecological pressure of waste disposal. We have discussed these concepts and the statistics earlier in this chapter.

More sustainable resource use is central to achievement of 12 of 17 SDGs and amongst these Goals 8.4, 9.4 and 12.2 are more specific. Box 1.10 provides the details.

The Group of Seven (G7) and Group of Twenty (G20) are international summits where leaders from the world's advanced and emerging economies meet to discuss critical global issues. The G20 is dedicated primarily to international economic cooperation, whereas the G7 focuses on a wider array of economic and political challenges. We will discuss in this section the role played by G7/G20 towards sustainable development.

1.6.3.1 The G7/G8

The Group of 7 (G7) is a group consisting of Canada, France, Germany, Italy, Japan, the United Kingdom, and the United States. These countries are the seven major advanced economies and represent more than 64% of the net global wealth ($263 trillion). The European Union is also represented within the G7. The G7 Summit meets annually to discuss global, economic, political, and social issues of mutual interest. From 1998 through 2014, Russia was also included in the annual forum, making it the G8 or Group of Eight.

Since 2000, the Environmental Ministers of G7 countries have been meeting once a year on themes such as climate change, energy, and biodiversity. After a lapse of nearly 7 years from 2009, the Environmental Ministers met once again at Toyama in Japan. The focus of this meeting was on resource efficiency and the 3Rs. Box 1.11 summarizes the agreed upon actions.

The leaders of the Group of 7 (G7) met in Germany in June 2015 and decided on ambitious action to improve resource efficiency as a main element of a broader strategy to promote sustainable materials management and the circular economy. Most notably Japan, the European Union and China have instituted high-level policy agendas for reducing

BOX 1.11 THE TOYAMA FRAMEWORK[77]

The G7 Ministers and high representatives, and the European Commissioner responsible for the environment, met to discuss resource efficiency and the 3Rs in Toyama, May 15–16, 2016. The following actions were agreed upon:

- Recognizing that the global population is estimated to exceed 9 billion by 2050 and rising demand for resources has caused an increase in resource consumption and waste generation, and that these trends contribute to deterioration of natural environment, including air, soil, and water pollution due to hazardous materials and climate change that affect our future generations;
- Understanding that appropriate policies on resource efficiency and the 3Rs with the consideration of a resource nexus, can contribute not only to environmental conservation but also to sustainable use of resources, avoidance of business risks, innovation, job creation, and green growth;
- Emphasizing that the G7 Alliance on Resource Efficiency is a dynamic voluntary platform that benefits from actively engaging relevant stakeholders and supporting networks;
- Reconfirming that material life cycles (extraction, design, manufacturing, use, and recycling or disposal) and transactions of materials including secondary ones are often global; and, therefore, it is increasingly important to ensure cooperation with relevant countries and stakeholders including businesses;
- Noting the significance of the 3Rs (reduce, reuse, and recycle) plus other concepts on efficient and cyclical use of resources including sustainable use of renewables, and noting also the significance of sustainable materials management, material-cycles societies and circular economies;
- Commit to take the following actions, building upon the Kobe 3R Action Plan, G7 Alliance on Resource Efficiency and other existing initiatives, reflecting the new challenges we face today while also respecting the role of each country to determine policies and other actions in accordance with its own specific circumstances.

material extraction and use and increasing the circular behavior of their economies through remanufacturing, recycling, and reuse. Japan's Sound Material Cycle policy objective, the European Strategy for Sustainable Natural Resource Management, and the Chinese circular economy promotion law are key examples of the growing interest in resource efficiency and sustainable materials management.

Some of the leading members of G7, especially Germany, have been working on the theme of increasing Resource Efficiency (RE) and Secondary Material Flows since 2012. Box 1.12 shows some of the accomplishments made by Germany. Figure 1.35 describes the thematic areas that are well aligned with the SDGs. Under its theme of international cooperation, Germany has been assisting the Government of India to set up a Resource Panel on the lines of the United Nations Environment Program. The Indian Resource Panel, one of the first of this kind set up in the world, has been working since 2015 and has developed a strategic plan for resources from the Construction & Demolition (C&D) and the Automobile sector (ELVs).

BOX 1.12 RESOURCE EFFICIENCY AND SECONDARY MATERIAL RECYCLING IN GERMANY

Germany is an export-oriented economy with a strong industrial base. Germany, however, depends in a major way on imports of raw materials. Although the country is rich in minerals, 66.8% of the metals are imported. Materials account for 44% of costs in the German manufacturing sector.

Germany is a leading country regarding green or recycling technologies. Saving and efficient use of resources for the whole value chain is of a high priority for German environmental and economic policies.

In 2012, the German Resource Efficiency Program (ProgRess) was adopted by the Federal Ministry of Environment, Nature Conservation, Building, and Nuclear Safety. Implementation of ProgRess led to major improvements in Resource Efficiency (RE) and increased Secondary Material Recycling (SMR) reducing the import of resources. As a next step, ProgRess II was adopted in 2016 by the Parliament, securing top political support. ProgRess II emphasizes:

- Expansion of efficiency advice for small and medium-sized enterprises
- The support of environmental management systems
- A stronger transfer of technology and knowledge to developing and newly industrialized countries

The program has a stronger focus on sustainable building and sustainable urban development. Resource-efficient information and communication technology is focusing on mobile phones and data centers as a new field of action.

DISCUSSION QUESTIONS

- *The Government of India set up an Indian Resource Panel (InRP) in 2014. The InRP was assisted by the Government of Germany. Study the progress made by InRP and the challenges that it will address.*
- *Read the report on India Circular Economy prepared by the Ellen MacArthur Foundation in this context.*[78]

1.6.3.2 G20

Founded in 1999, the Group of Twenty (G20) brings together the world's largest and leading economies, representing two-thirds of the global population. It represents 85% of the global GDP and accounts for 75% of global trade.[79] The member countries are: Argentina, Brazil, Canada, China, France, Germany, India, Indonesia, Italy, Japan, Republic of Korea, Mexico, Russia, Saudi Arabia, South Africa, Turkey, the United Kingdom, the United States, and the European Union. See Figure 1.36.

The G20 engages with several groups of stakeholders on an annual basis that includes Business (B20), Civil Society (C20), Labour (L20), Think Tanks (T20), and Youth (Y20). All these engagements help in decision-making of the G20 leaders. G20 provides an opportunity for collaboration on the SDGs focusing on resource efficiency. The Business 20 or B20 has recommended that the G20 should advance resource and energy efficiency by establishing a "Resource Efficiency Dialogue."[80]

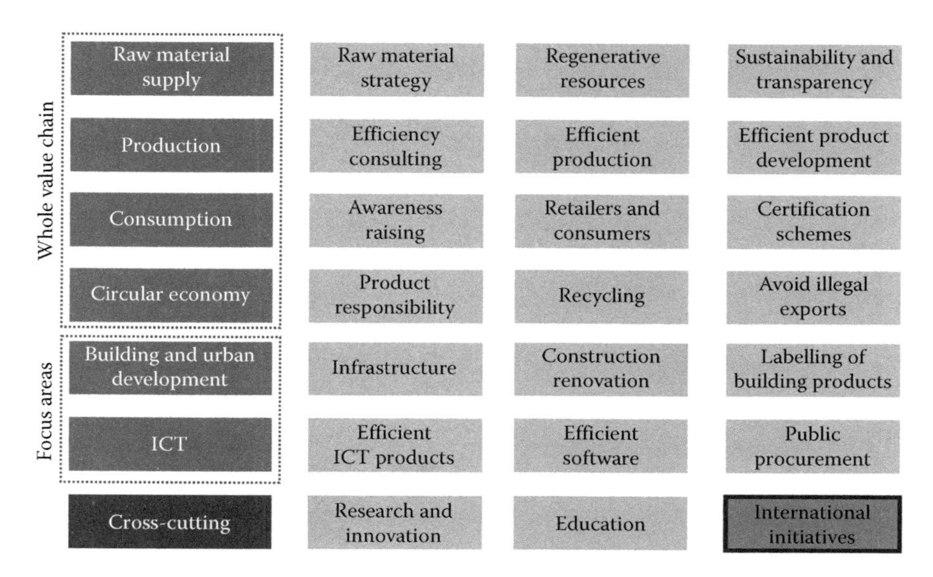

FIGURE 1.35
Thematic areas in ProgRess II. (From Federal Ministry for the Environment and Construction (BMUB), Germany, 2017. With permission.)

1.6.4 Circular Economy

The conventional models of development and economy follow a *take, make and dispose* approach. This is essentially an extractive industrial model of development often called a linear economy as it does not emphasize the importance of recycling waste streams that are essentially secondary materials or used resources. In the last decade,

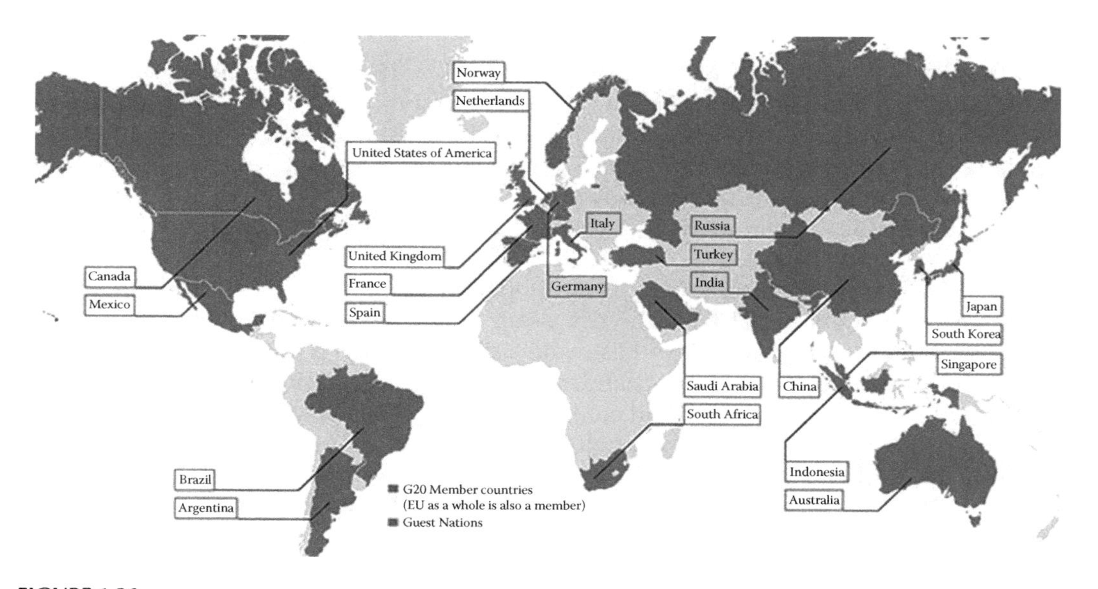

FIGURE 1.36
Members of the G20 alliance. (G20—a meeting at the highest level, Participants of the G20 Summit in Hamburg in 2017, online source: https://www.g20.org/Webs/G20/EN/G20/Participants/participants_node.html.)

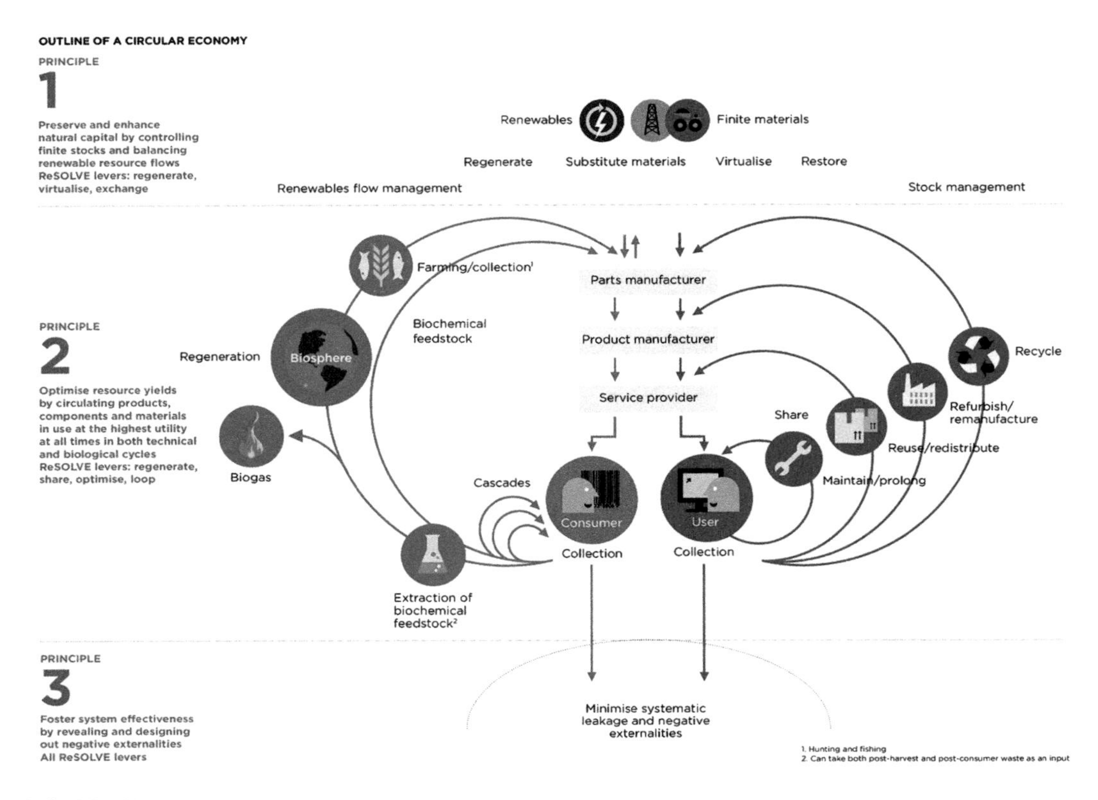

FIGURE 1.37
The concept of circular economy. (From Circular Economy System Diagram, Ellen MacArthur Foundation, 2015. With permission, online source: https://www.ellenmacarthurfoundation.org/circular-economy/interactive-diagram.)

the concept of Circular Economy (CE) has emerged. CE stresses recycling of the secondary material flows and leads to restoration and regeneration of resources by design. CE promotes innovation and redefines products and services keeping the life cycle perspective. CE encourages use of materials with low intensity and reliance on environmentally benign sources of energy while minimizing the negative impacts. CE therefore leads to sustainable development and accordingly builds economic, natural, and social capital.

Figure 1.37 shows the concept of CE as developed by the Ellen MacArthur Foundation.

To practice the concept of CE, it is necessary that the national governments, businesses, financing institutions as well as communities work in partnership. Instruments such as policy measures; laws, regulations and standards; market forces, investments (especially in the infrastructure), and community participation are necessary to achieve the desired outcomes. A multi-stakeholder and multi-level engagement is necessary. We delve in Chapter 2 into the steps taken by the national governments. Chapter 3 is devoted to the role played by the financing institutions. We then describe the responses from business in Chapter 4 and discuss the lead taken by business across various sectors in Chapter 5. We end with Chapter 6, where we bring in the role played by communities in partnership with national governments, financing institutions, and business.

Endnotes

1. Based on: World Population Prospects—The 2015 revision: Key findings & advance tables, UN ESA (United Nations Economic and Social Affairs), ESA/P/WP 241, New York, p. 9, 2015, online source: https://esa.un.org/unpd/wpp/publications/files/key_findings_wpp_2015.pdf
2. The 2010 Human Development Report introduced an Inequality-adjusted Human Development Index (IHDI). While the simple HDI remains useful, it stated that "the IHDI is the actual level of human development (accounting for inequality)," and "the HDI can be viewed as an index of 'potential' human development (or the maximum IHDI that could be achieved if there were no inequality)." See Human Development Index. https://en.wikipedia.org/wiki/Human_Development_Index, last modified November 1, 2017.
3. Taken from: Indicator 12.2.1—Material footprint, material footprint per capita, and material footprint per GDP, Goal 12—Ensure sustainable consumption and production patterns, online source: https://gsa.github.io/sdg-indicators/12-2-1/
4. Taken from: Alberta and other International GPI/ISEW Graphs, Anielski Management—Building communities of genuine wealth, online source, http://www.anielski.com/alberta-international-gpiisew-graphs/
5. Based on: The Genuine Progress Indicator, Redefining Progress, online source: http://www.sustainwellbeing.net/gpi.html
6. Based on: Gross ecosystem product: Concept, accounting framework and case study, *Acta Ecologica Sinica*, Vol. 33, no. 21, pp. 6747–6761, 2013, online source: https://www.researchgate.net/publication/270668905_Gross_ecosystem_product_Concept_accounting_framework_and_case_study
7. Based on: Gross Ecosystem Product (GEP), IUCN, 2017, online source: https://www.iucn.org/asia/countries/china/gross-ecosystem-product-gep%EF%BC%89
8. For further reading: IUCN China takes lead in measuring the true value of nature, 2013, online source: https://www.iucn.org/content/iucn-china-takes-lead-measuring-true-value-nature
9. The biocapacity of a particular surface represents its ability to renew what people demand. Biocapacity is therefore the ecosystems' capacity to produce biological materials used by people and to absorb waste material generated by humans, under current management schemes and extraction technologies. Biocapacity can change from year to year due to climate, management, and also what portions are considered useful inputs to the human economy. In the National Footprint Accounts, the biocapacity of an area is calculated by multiplying the actual physical area by the yield factor and the appropriate equivalence factor. Biocapacity is usually expressed in global hectares. See http://www.footprintnetwork.org/resources/glossary/
10. Taken from: Global hectare: Applications, online source: https://en.wikipedia.org/wiki/Global_hectare, last modified May 8, 2017.
11. For further reading: Ewing, B., Goldfinger, S., Oursler, A., Reed, A., Moore, D. and Wackernagel, M., Ecological Footprint Atlas 2009, Oakland: Global Footprint Network, online source: http://www.footprintnetwork.org/content/images/uploads/Ecological_Footprint_Atlas_2009.pdf
12. For further reading: Niccolucci, V., Tiezzi, E., Pulselli, F.M. and Capineri, C., Biocapacity vs Ecological Footprint of world regions: A geopolitical interpretation, *Ecological Indicators*, Vol. 16, pp. 23–30, 2012, online source: http://www.sciencedirect.com/science/article/pii/S1470160X11002743
13. Based on: Hsu, A. et al., *2016 Environment Performance Index: Global Metrics for the Environment*, Yale University, New Haven, CT, pp. 13, 27, 2016, online source: http://epi.yale.edu/sites/default/files/2016EPI_Full_Report_opt.pdf
14. See: Mosbergen, D., *Our Consumption of Earth's Natural Resources Has More Than Tripled in 40 Years*, Huffpost, 2016, online source: http://www.huffingtonpost.in/entry/natural-resource-use-tripled_us_57a05c3ae4b0693164c273a8

15. Based on: Domestic extraction, abbreviated as DE, is the input from the natural environment to be used in the economy. DE is the annual amount of raw material (except for water and air) extracted from the natural environment, online Source: http://ec.europa.eu/eurostat/statistics-explained/index.php/Glossary:Domestic_extraction_(DE)
16. See: OECD Environmental Outlook to 2050—The Consequences of Inaction, OECD, 2012, online source: http://www.oecd.org/env/indicators-modelling-outlooks/oecd-environmental-outlook-1999155x.htm
17. Taken from: Global Environment Outlook, online source: https://en.wikipedia.org/wiki/Global_Environment_Outlook, last modified November 10, 2016.
18. See: GEO: Global Environment Outlook, UNEP, online source: http://web.unep.org/geo/
19. Based on: Wibby, B., 5 facts related to affordable and clean energy, Michigan State University Extension, 2016, online source: http://msue.anr.msu.edu/news/5_facts_related_to_affordable_and_clean_energy
20. Taken from: EIA projects 48% increase in world energy consumption by 2040, Today in Energy, Independent Statistics & Analysis—EIA: U.S. Energy Information Administration, 2016, online source: https://www.eia.gov/todayinenergy/detail.php?id=26212
21. See: Depletion of natural resources statistics, The World Counts, 2016, online source: http://www.theworldcounts.com/counters/environmental_effect_of_mining/depletion_of_natural_resources_statistics
22. See: Gonzalez, G. and Lucky, M., Fossil Fuels Dominate Primary Energy Consumption, Worldwatch Institute—Vision for a Sustainable World, 2013, source online: http://www.worldwatch.org/fossil-fuels-dominate-primary-energy-consumption-1
23. For further reading: BP Statistical Review of World Energy, pp. 12 and 42, 2016, online source: https://www.bp.com/content/dam/bp/pdf/energy-economics/statistical-review-2016/bp-statistical-review-of-world-energy-2016-full-report.pdf
24. Taken from: WHO/UNICEF JMP Progress on Sanitation and Drinking Water: 2015 Update and MDG assessment, online source: http://files.unicef.org/publications/files/Progress_on_Sanitation_and_Drinking_Water_2015_Update_.pdf.
25. Taken from: Building and Climate Change—Summary for Decision –Makers, UNEP SBCI, p. 9, 2009, online source: http://www.unep.org/sbci/pdfs/SBCI-BCCSummary.pdf
26. Taken from: Bardhan, S., Assessment of water resource consumption in building construction in India, *WIT Transactions on Ecology and the Environment*, Vol. 144, online source: https://www.witpress.com/Secure/elibrary/papers/ECO11/ECO11008FU1.pdf
27. See: BIO Intelligence Services, Water performance of Buildings, Final report prepared for European Commission, DG Environment, 2012, http://ec.europa.eu/environment/consultations/pdf/background_water_efficiency.pdf
28. See: Building and Climate Change—Summary for Decision –Makers, UNEP SBCI, p. 9, 2009, online source: http://www.unep.org/sbci/pdfs/SBCI-BCCSummary.pdf
29. Taken from: Energy and Buildings, Centre for Science and Environment (CSE), India, p. 1, 2014, online source: http://www.cseindia.org/userfiles/Energy-and-%20buildings.pdf
30. Based on: Reducing the environmental impact of building materials, Science for Environment Policy, European Commission, 2011, online source: http://ec.europa.eu/environment/integration/research/newsalert/pdf/232na1_en.pdf
31. See: Environmental Impact of Construction Materials, The University of Nottingham, online source: https://equella.nottingham.ac.uk/uon/file/1c4d7433-74db-9779-b605-7681374bc79a/1/Eng_sustainability.zip/Engineering%20Sustailability/62_environmental_impact_of_construction_materials.html
32. See: Benefits of Green Building, U.S. Green Building Council, 2016, online source: http://in.usgbc.org/articles/green-building-facts
33. For further reading: IPCC report: Building sector emissions growing, green buildings can be solution, LEED, USA, 2014, online source: http://www.usgbc.org/articles/ipcc-report-building-sector-emissions-growing-green-buildings-can-be-solution

34. CER& UIC, Rail Transport and Environment: Facts & Figures, p. 25, 2015, online source: http://www.cer.be/sites/default/files/publication/Facts%20and%20figures%202014.pdf
35. Facts & Figures, ATAG: Air Transport Action Group, online source: http://www.atag.org/facts-and-figures.html
36. For further reading: Sakai, S. et al., An international comparative study of end-of-life vehicle (ELV) recycling systems, *J Mater Cycles Waste Manag*, Vol. 16, no. 1, pp. 1–20, 2014, online source: http://link.springer.com/article/10.1007/s10163-013-0173-2
37. Based on: End of life vehicles, Harsco: Metals & Minerals, 2017, online source: http://harsco-m.com/221/End-Of-Life-Vehicles.aspx
38. Based on: End Life Cycle Vehicles (ELV), Hellenic Recycling Agency, 2017, online source: https://www.eoan.gr/en/content/330/end-life-cycle-vehicles-elv
39. Based on: UIC-IEA Railway Handbook: Energy Consumption and CO2 Emissions of World Railway Sector, UIC, 2017, online source: http://www.uic.org/energy-and-co2-emissions#UIC-IEA-Railway-Handbook-Energy-Consumption-and-CO2-emissions-of-World-nbsp
40. Based on: CER& UIC, Rail Transport and Environment: Facts & Figures, p. 25, 2015, online source: http://www.cer.be/sites/default/files/publication/Facts%20and%20figures%202014.pdf
41. See: Water consumed this year (million of liters), Worldometers, 2017, online source: http://www.worldometers.info/water/
42. Based on: WWAP (United Nations World Water Assessment Programme*), The United Nations World Water Development Report 2014: Water and Energy*, Pairs, UNESCO, Vol. 1, 2014, online source: http://unesdoc.unesco.org/images/0022/002257/225741e.pdf
43. Based on: UN-Water Annual Report, UN Water, 2012, online source: http://www.unwater.org/downloads/UN-Water_Annual_Report_2012.pdf
44. For further reading: WWAP (United Nations World Water Assessment Programme), *The United Nations World Water Development Report 2016: Water and Jobs, Paris*, UNESCO, online source: http://unesdoc.unesco.org/images/0024/002439/243938e.pdf
45. Taken from: Schulte, P., Defining Water Scarcity, Water Stress, and Water Risk: It's Not Just Semantics, Pacific Institute, 2014, online source: http://pacinst.org/water-definitions/
46. Taken from: Virtual water trade, Water footprint network, 2014, online source: http://water-footprint.org/en/water-footprint/national-water-footprint/virtual-water-trade/
47. Based on: Progress on Sanitation and Drinking Water: 2015 Update and MDG Assessment, p. 12, 2015, UNICEF/WHO, online source: http://files.unicef.org/publications/files/Progress_on_Sanitation_and_Drinking_Water_2015_Update_.pdf
48. Global Health Observatory (GHO) data, WHO, 2015, online source: http://www.who.int/gho/en/
49. *Progress on Sanitation and Drinking Water: 2015 Update and MDG Assessment*, UNICEF/WHO, 2015, online source: http://files.unicef.org/publications/files/Progress_on_Sanitation_and_Drinking_Water_2015_Update_.pdf, Accessed on: October 1, 2016.
50. Based on: Prüss-Üstün, A., Bos, R., Gore, F. and Bartram, J, Safer water, better health: costs, benefits and sustainability of interventions to protect and promote health, 2008, online source: https://www.cdc.gov/healthywater/global/wash_statistics.html
51. Based on: Hutton, G., Global costs and benefits of drinking-water supply and sanitation interventions to reach the MDG target and universal coverage, *World Health Organization*, p. 4, 2012, online source: http://www.who.int/water_sanitation_health/publications/2012/globalcosts.pdf
52. WHO releases country estimates on air pollution exposure and health impact, WHO: World Health Organization, 2016, online source: http://www.who.int/mediacentre/news/releases/2016/air-pollution-estimates/en/
53. University of British Columbia, Poor air quality kills 5.5 million worldwide annually, ScienceDaily, 2016, online source: www.sciencedaily.com/releases/2016/02/160212140912.htm
54. Taken from: Air Pollution, Climate and Health, WHO and CCAC, online source: http://who.int/sustainable-development/AirPollution_Climate_Health_Factsheet.pdf

55. Taken from: Report of the Steering Committee on Air Pollution and Health Related Issus, p. 8, 2015, Government of India, online source: http://www.mohfw.nic.in/showfile.php?lid=3650
56. Taken from: Public health, environmental and social determinants of health (PHE), WHO, 2016, online source: http://www.who.int/phe/health_topics/outdoorair/COP19_airpollution_sideevents/en/
57. Based on: Asian Brown Cloud, Atmospheric Science, Encyclopedia Britannica, online source: https://www.britannica.com/science/Asian-brown-cloud
58. Based on: Perkins, S., 'Asian Brown Cloud' Threatens U.S., Science, 2012, online source: http://www.sciencemag.org/news/2012/05/asian-brown-cloud-threatens-us
59. Taken from: GCO Global Chemicals Outlook—Towards Sound Management of Chemicals, UNEP, p. 63, 2013, online source: http://www.unep.org/hazardoussubstances/Portals/9/Mainstreaming/GCO/The%20Global%20Chemical%20Outlook_Full%20report_15Feb2013.pdf
60. Taken from: Agri-Environment statistics, FAO, online source: http://www.fao.org/economic/ess/environment/en/
61. Taken from: Roser, M., *Land Use in Agriculture*, Our world in data, 2016, online source: https://ourworldindata.org/land-use-in-agriculture/
62. Taken from: Healthy soil is set to become the next HOT commodity, 2012, online source: http://www.unccd.int/Lists/SiteDocumentLibrary/Publications/DrylandsSoilUNCCDBrochureFinal.pdf
63. Taken from: Geo-6, Regional Assessment for Asia and the Pacific, UNEP, 2016, online source: http://apps.unep.org/publications/index.php?option=com_pub&task=download&file=012227_en
64. Taken from: GEO-6, Regional Assessment of West Asia, UNEP, 2016, online source: http://apps.unep.org/publications/index.php?option=com_pub&task=download&file=012097_en
65. Visit website of Millennium Ecosystem Assessment at http://www.millenniumassessment.org/en/About.html
66. See overview presented by World Wild Life https://www.worldwildlife.org/industries/timber
67. Read the Disaster Waste Management Guidelines prepared by United Nations Office for the Coordination of Humanitarian Affairs – Emergency Preparedness Section, online source: https://www.msb.se/Upload/English/news/Disaster_Waste_Management.pdf
68. Taken from: Threshold 21 Model, Millennium Institute, 2015, online source: http://www.millennium-institute.org/integrated_planning/tools/T21/
69. Based on: iSDG Model, Millennium Institute, online source: http://www.millennium-institute.org/integrated_planning/tools/SDG/index.htmll
70. Renewable Energy in Cities, IRENA: International Renewable Energy Agency, Abu Dhabi, p. 11, 2016, online source: http://www.irena.org/DocumentDownloads/Publications/IRENA_Renewable_Energy_in_Cities_2016.pdf
71. Taken from: Climate Change, UN Habitat—For a Better Urban Future, 2012, online source: http://unhabitat.org/urban-themes/climate-change/
72. Visit http://www.100resilientcities.org to read about the 100 Resilient Cities Program
73. See Green Cities program in South-East Asia by the Asian Development Bank at https://www.adb.org/publications/green-cities
74. Access the ECO2Cities Guide of the World Bank from http://siteresources.worldbank.org/INTURBANDEVELOPMENT/Resources/336387-1270074782769/Eco2_Cities_Guide-web.pdf
75. Learn about the Provision of Urban Amenities in Rural Areas (PURA) scheme of the Ministry of Rural Development of Government of India at http://rural.nic.in/sites/pura.asp
76. Learn more about the 10-Year Framework of UNEP at http://web.unep.org/10yfp
77. Taken from: Toyama Framework on Material Cycles, Ministry of Environment—Japan, online source: http://www.mofa.go.jp/files/000159928.pdf
78. Read India Circular Economy Report online at https://www.ellenmacarthurfoundation.org/assets/downloads/publications/Circular-economy-in-India_5-Dec_2016.pdf
79. G20 members, online source: http://g20.org.tr/about-g20/g20-members/
80. Annex to G20 Leaders Declaration G20 Resource Efficiency Dialogue, Online source at https://www.g20.org/Content/DE/_Anlagen/G7_G20/2017-g20-resource-efficiency-dialogue-en.pdf?__blob=publicationFile&v=4

2

Response from the National Governments

2.1 Introduction

As discussed in Chapter 1, the world has been facing complex environmental and social challenges since the early 1900s. These challenges have arisen due to unabated industrial pollution, intense urbanization and exploitation of natural ecosystems across the globe. In Chapter 1, we highlighted that the issues we face cannot be managed in a silo, and we need to respond to them in a more holistic, cross-cutting and integrated manner recognizing the nexus between these challenges.

In Chapter 2, we will introduce the response by governments and international organizations at the global, regional, national, and sub-national scales towards addressing these challenges. We do this by highlighting some of the major milestones and initiatives that have been taken.

2.2 Key Stakeholders in Environmental Management

Efforts to combat environmental challenges have been gaining momentum in various parts of the world. Initially the responses came about in isolation and on case to case basis and were limited to a few countries. However in 1972, a collective international response evolved in the form of the United Nations Conference on the Human Environment in Stockholm. The Stockholm Convention brought together 114 governments and led to the formation of the United Nations Environment Program (UNEP). UNEP was envisaged as the agency to coordinate the environmental activities of the United Nations and assist developing countries in implementing environmentally sound policies and practices.[1]

Prior to the Stockholm Convention, countries such as the United States, Canada, and some other Organization for Economic Co-operation and Development (OECD) countries had policies in place for sound environmental management and also the implementation experience. Most of the policies in these countries followed the Command and Control (CAC) approach.

The CAC approach entails regulatory environmental policies and supporting laws commanding and controlling pollution related activities by stipulating environmental standards, restricting resource use or consumption and banning certain developmental activities in ecologically sensitive zones. All involved stakeholders have to seek relevant permits or consents from the regulatory agencies and comply with the prescribed standards and procedures. In some cases, the CAC approach directed phasing out of harmful substances and technologies, hence, influencing existing manufacturing operations.

Institutions were set up for policy formulation, planning, enforcement and monitoring. Economic incentives and disincentives were framed to encourage the polluters towards meeting the compliance standards and even go beyond.

Many of the policies, laws and regulations of developed nations could not be applied on an *as is* basis to other developed or developing nations. Other nations, especially from the developing world, started adapting the experience of the developed nations to local conditions reflecting their national priorities and availability of financial and technological resources.

In due course, nations realized that the intrinsic nature of environmental challenges is so complex that the government alone cannot resolve the issues. Partnerships with international organizations, businesses and communities are critical for success. So, in addition to enforcement, market mechanisms and economic instruments were introduced to encourage involvement of businesses and investors. Environmental policies, laws and regulations were expanded to address the social dimension, especially the concerns of vulnerable communities. It was observed that businesses better practice environmental management if they experience pressure from the markets, investors, and community (see Chapters 4, 5 and 6).

Amongst the international organizations, the Development Financing Institutions (DFIs) played a prominent role in influencing national policies in this direction (see Chapter 3). Communities supported by environmental activists, Community Based Organizations (CBOs) and Non-Governmental Organizations (NGOs) assumed the role of *informal regulators* and in many cases influenced the national environmental policies. We will discuss the important role played by the communities in the environmental governance in Chapter 6.

David Wheeler, an economist with World Bank, postulated the need for a collective response of Government, Businesses and Community (G-B-C). In the World Bank publication, *Greening Industries: New Role For Communities, Markets and Governments* (1999)[91], Wheeler and other economists proposed a *regulatory triangle* (see Figure 2.1) wherein regulators (government), leverage the communities and businesses (markets) to improve the environmental management in economic and social perspectives.

Over the last few decades, as stated earlier, DFIs such as the World Bank (WB), German Development Bank (KfW), European Investment Bank (EIB), the Asian Development Bank (ADB), and others became a fourth dimension to this regulatory triangle due to their work and influence. These DFIs influenced the national governnance and helped to build the capacity of environmental management at public and private institutions. The Private Sector Financing Institutions (PSFI) also played an important role by specifying environmental and social convenants while lending monies to businesses. Figure 2.2 presents this wider canvas of the *change makers* or key stakeholders. We will discuss each of these key stakeholders in chapters to follow, especially their role and the *zone of influence*.

2.3 Guiding Principles

Stakeholders described in Figure 2.2 follow different approaches and principles while dealing with environmental challenges. It is important to use a principle-based approach for better results. The principles, often called Guiding Principles, provide a framework to formulate various environmental management strategies. In this section, we will introduce

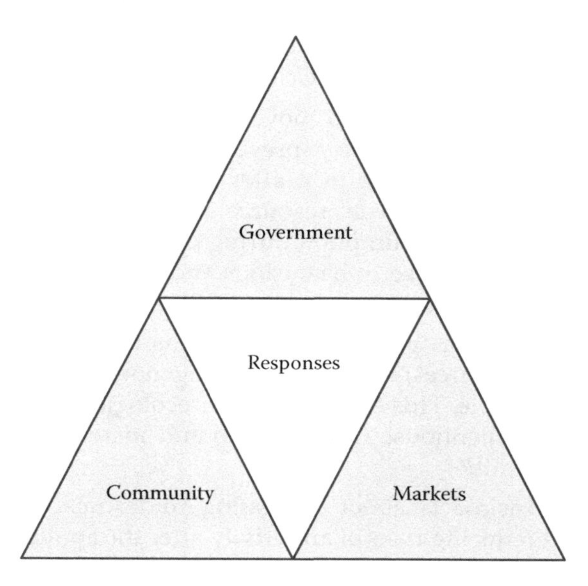

FIGURE 2.1
Regulatory triangle of G-B-C postulated by the World Bank. (Adapted from Development Research Group, World Bank, *Greening Industry: New Markets for Communities, Markets and Governments, A World Bank Research Report*, Published by Oxford University Press, 1999. This figure is an adaptation of the World Bank's original figure.)

some of the key Guiding Principles followed by stakeholders, while devising environmental and social policies and frameworks.

1. *Precautionary*: The Precautionary or *Do no harm* principle states that necessary precautions must be taken to prevent harm to humans and the eco-systems. Precautionary measures are to be taken even if the cause and effect of the impact

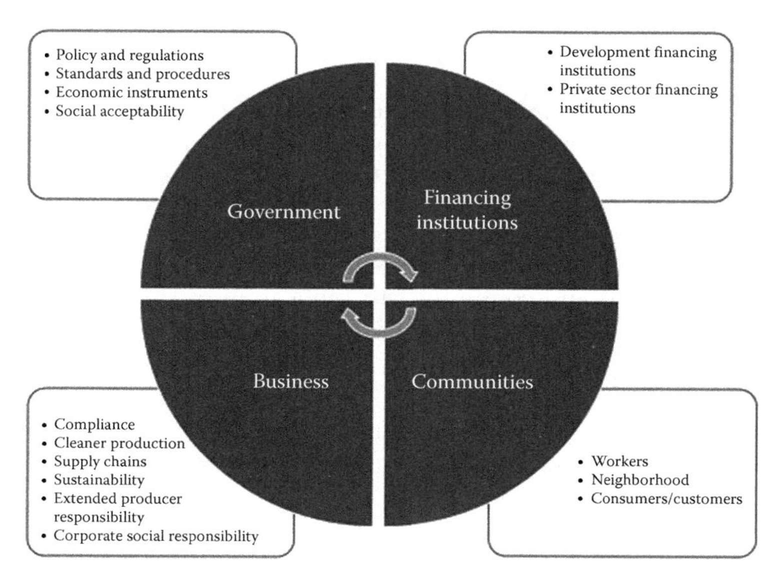

FIGURE 2.2
New paradigm of G-B-C-FI.

of the activity may not be fully understood or established on scientific grounds. This principle often leads to *Go* or *No Go* decisions.

2. *Preventive*: If a decision is taken to move ahead with a project/plan then the Preventive Principle emphasizes that preventing environmental harm is easier and more effective than reacting to it after the damage is done. For example, actions should be taken to prevent resource extraction to the maximum extent possible, reduce resource consumption during operations and reduce generation of emissions and residues. Use of hazardous and non-biodegradable substances should be avoided to the maximum extent possible.
3. *Preference to Proximity*: This principle recommends that resource use (materials, energy and human resources) from local or indigenous sources should be prioritized as much as possible. This ensures a lower ecological footprint, for example, lower emissions of Greenhouse Gases (GHG) and increased generation of local employment opportunities.
4. *Mitigative*: This principle is about mitigating undesirable environmental and social impacts and reducing risks of an activity after the application of the preventive principle. Mitigation is often achieved through a *4P* approach that consists of Policy measures, Plans, Programs and Projects that help reduce the severity, extent and nature of the impact/risk. The motive behind this principle is to achieve regulatory compliance.
5. *Participatory*: The Participatory principle encourages consultation and participation of stakeholders like communities, NGOs, and business employees into the decision-making of governments or organizations. This approach helps bring in transparency, disclosure of the application of preventive, and mitigative practices undertaken and proposed environmental and social management plans to address the residual impacts/risks. This principle also ensures ownership of the stakeholders and a project's sustainability.
6. *Compensatory*: This principle is about compensating the affected stakeholders and the natural resources and ecosystems that experienced adverse impacts due to developmental activities. This principle is to be applied only after all the above principles are applied. The compensation could be on a monetary basis, estimated on a scientific basis, appropriate, equitable, and delivered on time and in all fairness. This principle is relevant for mining projects, thermal power, dams or reservoirs and transportation projects where there is significant displacement and loss of land. In cases like uncontrolled oil spills or natural disasters that contaminate the natural resources and disrupt the livelihoods of local environment and communities, application of this principle becomes critical.
7. *Polluter pays*: This principle makes the polluter responsible to pay for the damage caused to humans and the environment. If several polluters are responsible then the costs of compensation, rehabilitation and restoration are to be apportioned based on scientific data, sound economic assessment, and after conducting stakeholder consultation.
8. *Restorative*: This principle is about restoring the natural environmental and ecological conditions of the site that has been disturbed due to natural disasters or anthropogenic activities. Restoration may involve conservation and enhancement related measures.

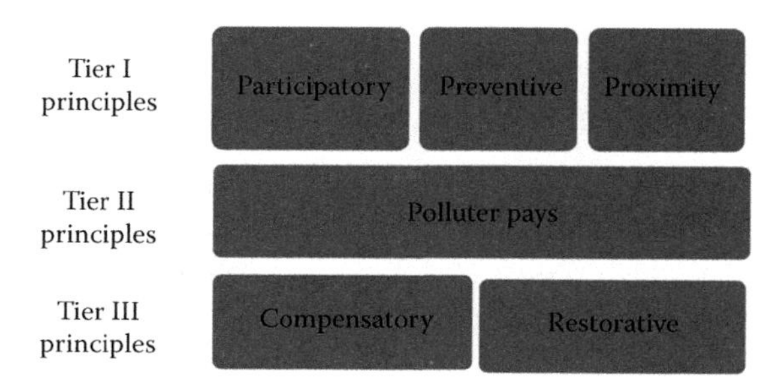

FIGURE 2.3
Hierarchy and interlinkages between the guiding principles.

Figure 2.3 shows a three-tier hierarchy and interconnections between the above Guiding Principles.

The *Precautionary, Preventive* and *Proximity* principles should be prioritized while developing a response to environmental issues. Innovation plays an important role in the application of preventive and proximity principles.

Mitigative principles should be applied on a comprehensive basis and not limited only to project based measures. Programmatic approaches and planning should also be looked into. It is also important to address institutional capacities and financial outlays to ensure that a satisfactory and timely implementation of mitigative measures takes place. Necessary strengthening, reforms and harmonization of the policies should also be on the agenda to ensure that the mitigations remain effective, replicable, and sustainable. The *Participatory* principle helps so that all key stakeholders are involved in achieving a rounded and sustainable solution at the relevant stages based on their relevance, expertise, and stakes in the activity.

Compensatory and Restorative principles should be implemented only as the last resort following the principle of *Polluter Pays.*

2.4 Need for Multi-Level and Multi-Stakeholder Response

The world responded to the environmental challenges at many levels—global, regional, national, sectoral, sub-national, and local. A multi-level approach was necessary because often natural resource depletion or degradation in a local area has national, regional or global implications. Communities typically respond at local and sub-national levels while the Financing Institutions (FIs) respond at sectoral and regional levels. The business operates on levels such as large, medium and small or sometimes in a form of a cluster or a supply chain and respond to the environmental challenges differently. The busines response is sometimes on an individual basis or in some cases collectively through associations and chambers of commerce.

Using emissions from vehicular and industrial sources as an example, Box 2.1 illustrates how a local problem has regional and global implications and, hence, a need for a multi-level response.

Recognizing the importance of multi-level interventions and engagements, most governments set up institutions at various levels for effective environmental management. Some

BOX 2.1 MULTI-SCALE IMPACT OF VEHICULAR AND INDUSTRIAL AIR EMISSIONS[2]

Let us understand the multi-level environmental repercussions of vehicular and industrial emissions. The vehicular and industrial emissions result in release of pollutants like Nitrogen Oxides (NO_x) and Volatile Organic Compounds (VOCs) that generate *ground-level or bad* ozone (O_3) on exposure to the sunlight. Pollutants like carbon monoxide (CO) and carbon dioxide (CO_2) are also released. These gaseous emissions have multi-level impacts like:

- Ground-level ozone, VOCs, and CO cause adverse health impacts (*local impact*)
- NO_x causes acid rain (*regional impact*) apart from local impact
- CO_2 as a greenhouse gas causes global warming (*global impact*)

See Figure 2.4 that explains the above.

Figure 2.5 illustrates the impacts due to vehicular emission. These impacts are progressive and could take some time to result in the change. Immediate effect on the local level will be deterioration of air quality followed by impact on health, especially in children and adults first on an acute basis during adverse meteorological conditions. Chronic health impacts may take a few years. Changes in the acidity levels of the waters in the region will take more time (say 5–10 years) and the global warming related contribution will be noticed over a decade. As the impacts escalate from local to global scale, attribution or contribution of the local emission on a relative basis becomes more difficult to assess.

There are numerous ways of responding to this multi-level challenge of air emissions. Governments can introduce emission standards that require business to manufacture efficient vehicles or follow cleaner production processes. The government

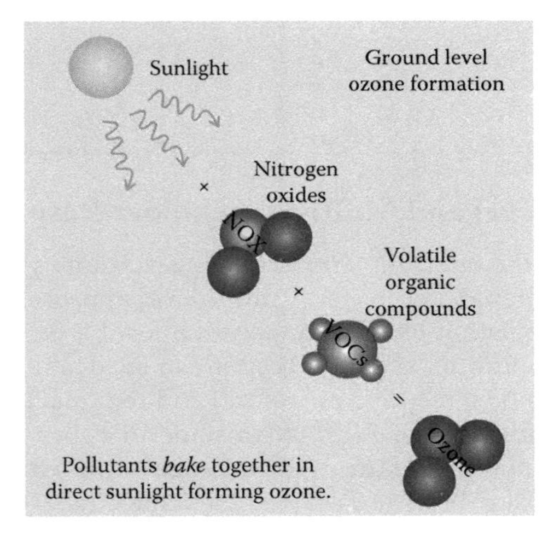

FIGURE 2.4
Ground level formation of ozone. (Based on the data from Recommended Binational Phosphorus Targets—Recommended Binational Phosphorus Targets to Combat Lake Erie Algal Blooms, EPA: United States Environmental Protection Agency, 2016, online source: https://www.epa.gov/glwqa/recommended-binational-phosphorus-targets.)

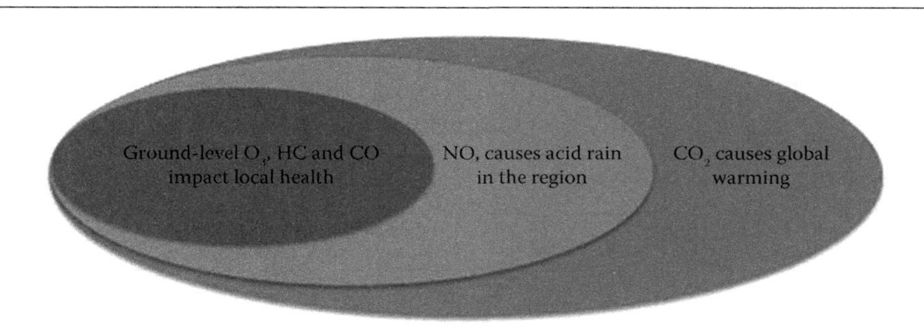

FIGURE 2.5
Multi-level impact of vehicular and industrial emissions.

may stipulate the type of fuel that must be used (e.g., Compressed Natural Gas (CNG)) or discourage use of coal in manufacturing through a *carbon tax*. The government could provide infrastructure like affordable public transportation (e.g., Metros) with improved connectivity that could reduce vehicular transportation and, hence, vehicular emissions. Industries may be asked to install air pollution control devices and maintain green belts to attenuate the transport of air emissions. Automobile manufacturers can come up with low emission vehicles, use cleaner fuels, and introduce electric vehicles.

We will now take the case of acidification of freshwater bodies in the United States and Canada to understand the multi-level impact of emissions from industrial emissions.

Acid rain is largely the result of discharges of sulfur dioxide (SO_2) and nitrogen oxide (NO) into the atmosphere that react with atmospheric moisture and oxygen. Although natural sources of sulfur and nitrogen oxides are present, industrial and residential anthropogenic activities contribute in a major way to sulfur and nitrogen that occur in ambient air.

The Great Lakes are the largest group of freshwater lakes in the world and consist of Lake Michigan, Erie, Huron, Ontario, and Superior. The lakes span across the Midwest region of North America and into Canada.

Emissions of SO_2 from coal-fired power plants in the Northeastern part of the US and industries based in both countries like automotive, steel, and glass caused acidification of fresh water bodies in the Great Lakes Region due to transportation by strong winds.[3,4] Acidification of the surface water bodies impacted aquatic life, fisheries, and human health due to consumption of contaminated fish, water meant for human consumption, industrial use of polluted water, and recreation activities by the lake (Figure 2.6).

To respond to this challenge, the Governments of Canada and USA entered into an Air Quality Management Agreement to mitigate acid rains under the Long-Range Transport of Air Pollutants Program in 1972 and 1980. In addition to the air pollution control program, the countries signed the Great Lakes Water Quality Agreement (GLWQA). GLWQA is a commitment between Canada and the USA to protect and restore the damage done to the Great Lakes. The Agreement lists bi-national priorities and action plans to design, implement, and monitor programs to control municipal and industrial pollution.

The Agreement has defined objectives. It entailed ten focus areas to deal with specific issues, which include clean-up of contaminated sites within the basin, improving

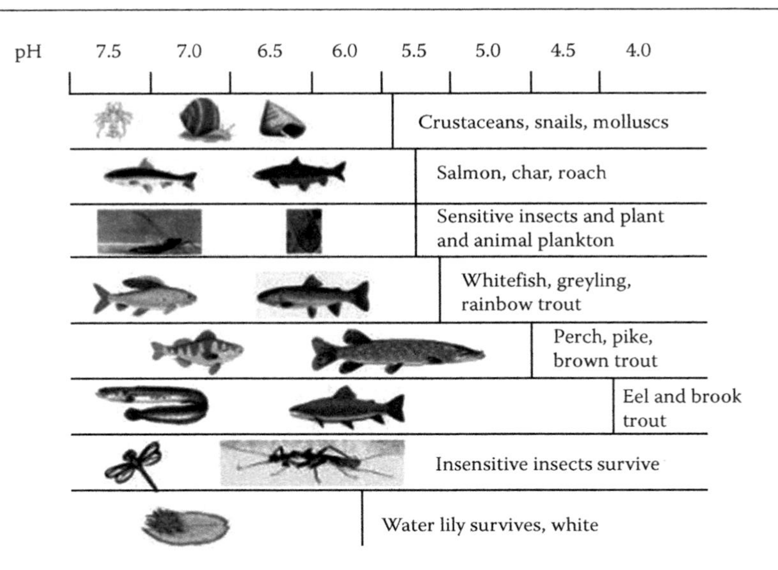

FIGURE 2.6
Tolerance ranges of pH of various aquatic species in water bodies. (Based on the data from Recommended Binational Phosphorus Targets—Recommended Binational Phosphorus Targets to Combat Lake Erie Algal Blooms, EPA: United States Environmental Protection Agency, 2016, online source: https://www.epa.gov/glwqa/recommended-binational-phosphorus-targets.)

water quality, reducing chemical release, preventing harmful discharge from ships, protecting native species, and preventing introduction of invasive species. Towards each of these objectives, the two countries have successfully taken action through regulations, research, and execution. In addition, other programs like the Great Lakes Toxic Substance Control Agreement, Great Water Program of the Clean Air Act and Great Lakes Air Deposition (GLAD) Program addressed the issue of atmospheric deposition of toxic pollutants. For instance, through the GLWQA program, contaminated sites were restored[5] and radioactive waste was cleaned up. Through the GLAD program, sources of pollutants and points of entry have been identified that can help develop strategies to abate the issue.[6] The implementation of actions is however reported to be slow.

DISCUSSION QUESTIONS

- *What are the options for clean fuels in curbing vehicular emissions?*
- *Discuss two low emission technologies that have come up in the Thermal Power and Cement sectors.*
- *Discuss the issue of algal bloom in the Great Lakes. What measures were taken under GLWQA to tackle this issue?*
- *Write a four page note on the GLAD program providing the current status of implementation and the outcomes.*

institutions were created with specific tasks to address protection of natural resources and control pollution (e.g., Ministries of Environment and Forests) and in some cases in addition to the above, line ministries (explained later) were also charged with environmental responsibilities. We will discuss in the next section examples of the institutional setups created by different countries in the world.

2.5 Legislative Framework and Institutions

Environmental governance at the sub-national, national, regional and global levels are critical for the achievement of environmental sustainability and sustainable development. To put the policies and plans in practice therefore, national governments responded by establishing institutions at various levels for enforcement and monitoring of laws.

Institutions were formed for management of resources (upstream) and residues or polluted streams (downstream). The Ministries of Environment generally focused on the management of residues and the resource management was handled by *Line Ministries* such as the Ministries of Water, Energy, Agriculture, Chemicals and Fertilizers, Mining, etc.

Most countries responded to environmental challenges via a downstream approach, that is, by framing policies and laws to prevent and control pollution of resources like water bodies, land or air, and projects of human health and the eco-systems. Although the aspect of prevention was included, the approach undertaken by the Ministries of Environment essentially focused on the residue management. These governments issued authorization or permits for developmental activities in the form of a *consent or a license* to pollute within acceptable limits.

Few countries followed an upstream approach to address resource consumption in addition to the management of residues. This approach guided judicious utilization of resources like rivers, lakes, forests, land, minerals, and fossil fuels. Here, the respective line ministries played a key role in coordination with the Ministries of Environment. A *resource consent* was issued to the project developer restricting the use of the natural resources to ensure local sustainability. Box 2.2 presents the residue and resource management approach with examples.

Environmental governance in a country is therefore managed by both Environment and Line Ministries and further by departments and institutions at sub-national and local levels. This makes the governance rather complex with overlapping jurisdiction and procedures, which can lead to ambiguities during enforcement.

We present case studies in Box 2.3 illustrating institutional structures operating at national levels covering both the Environment and the Line Ministries.

Box 2.4 provides examples of institutional structures operating at sub-national and local levels.

When it comes to implementation, it becomes necessary that the multi-level approach is formalized with responsibilities defined across the various institutions involved. We discuss this multi-level management approach in Box 2.5 with the case of Partnerships in Environmental Management for the Seas of East Asia (PEMSEA).

Having introduced examples of various institutional structures operating at multiple levels, we will now look at the constitutional provisions that stay at the apex and guide the national policies and legislation.

BOX 2.2 RESIDUE AND RESOURCE MANAGEMENT APPROACH

CASE OF POLLUTION CONTROL ACTS, INDIA

In India, the Ministry of Environment & Forests & Climate Change (MoEFCC) has framed laws to address prevention and control of water or air pollution with an umbrella act, the Environmental Protection Act (EPA) that was passed in 1986. Under the EPA, rules are stipulated to address hazardous substances, various waste streams and residues (such as Municipal Solid Waste, Biomedical Waste, E-Waste, Hazardous Waste, etc.).

RESOURCE MANAGEMENT ACT, NEW ZEALAND

The Resource Management Act (RMA) of New Zealand was promulgated in 1991 with the aim to promote sustainable management of natural and physical resources. Its purpose is to promote sustainable management of natural and physical resources. The Act lays restrictions on use of land, coastal marine areas, rivers and lake beds, use of water and discharges of contaminants into the environment. The RMA aimed to integrate management of air, land, water and coastal areas under one legislation.

For instance, with respect to water use, the RMA mandates a resource content to be obtained for taking, using, or diverting water from coastal, fresh, or geothermal sources. Exceptions were added to these rules, such as if the activity was permitted by national environmental standard or a regional rule, or fresh water was meant for reasonable domestic needs of humans or animals, or for firefighting purposes then the resource consent was not required. Resources consent provided individuals or organizations with permission to carry out activities other than those restricted under the RMA.[7]

The RMA was a revolutionary approach and was lauded when it was initially introduced. It, however, received criticism from stakeholders who stated that integration of land, water, and air uses is complex because of the differences in their management. The complexity of the resource consent procedure, cost and time delays and involvement of too many regulatory agencies were some of the other aspects criticized.[8]

DISCUSSION QUESTIONS

- *Discuss the process of obtaining a Resource Consent under the Resource Management Act (RMA) of New Zealand.*
- *Discuss the pros and cons of a resource conservation-based enforcement vs a residue controlled legislative approach.*
- *As an environmental policy maker, one should be using both the approaches. New Zealand, for example, uses both resource and residue related policy and legislative instruments. Write a four page review on the ground level experience of such an approach in New Zealand. Compare this with the experience in the European Union, taking the case of Germany.*

BOX 2.3 EXAMPLES OF INSTITUTIONAL STRUCTURES OPERATING AT NATIONAL LEVELS (ENVIRONMENT AND LINE MINISTRIES)

LEVEL: FEDERAL (NATIONAL), ENVIRONMENT

Swedish Environmental Protection Agency

Established in 1967, the Swedish Environmental Protection Agency (Swedish EPA) is the world's oldest environmental authority that was established with a mission to work for a better environment and for sustainable development. The organization developed national environmental policies for conservation of biodiversity elements like wetlands, forests, mountains, diverse animal and plant life, clean air and water sources like lakes, streams, and groundwater, ozone layer protection, and management of hazardous substances and waste amongst many others.[9]

One of the Swedish EPA's initiatives is to cater to the public's right to information about environment pollutants. Towards this, the agency maintains a Swedish Pollutant Release and Transfer Register, which is a repository of the quantities of chemical substances that large facilities with heavy environmental impact emit on an annual basis. This registry also helps the agency to monitor progress towards meeting the national environmental quality and pollution reduction objectives. This is done by setting emissions permits for the polluting companies whose operations are deemed to have a large environmental impact.[10]

LEVEL: FEDERAL (NATIONAL) LINE MINISTRY

Ministry of Agriculture, Land and Fisheries, Trinidad and Tobago (Port of Spain)

The Ministry's Vision is to "be an environmentally responsible organization delivering integrated services aligned to a changing food and agriculture system." To meet this objective the ministry provides support, agricultural information, and training to the local farming community to enhance efficiency and productivity. The Ministry provides necessary infrastructure like access roads to farmlands, suitable irrigation water, soil investigations, and materials testing lab, and sells high quality planting material to farmers and the general public. The Ministry provides guidelines and manuals on how to grow common crops like avocado, cabbage, lettuce, field corn, and many others, and on seed selection, and pest management and control.

The Ministry launched an interesting program called the Commercial Large Farms Program (CLFP). Under the CLFP program, the State partners with private investors under a public-private agreement to set-up large commercial agricultural farms in Trinidad. The Program aims to utilize new technologies for efficiency and improved productivity and achieve economies of scale, supply high quality raw material to the agro-processing and food manufacturing sector, improve reliability of domestic food supply, stabilize food prices and supply nutritious food at affordable prices, transfer the successful technologies to small-scale farmers, and induct entrepreneurs into the agricultural sector with a larger objective of improving the national food security, nutrition, and health. Under the program, 12 commercial farms were operating successfully in 2015.[11]

LEVEL: FEDERAL (NATIONAL), LINE MINISTRY

Ministry of Energy and Minerals, Tanzania

The Ministry of Energy and Minerals of Tanzania is vested with the role of facilitating development of energy and mineral sectors in Tanzania while promoting socio-economic development. The Ministry's mission is to set policies, strategies and laws for sustainability of energy and minerals resources to enhance growth and development of the economy.[12]

One example of the Ministry's work is the deployment of a semi-autonomous agency, the Tanzania Minerals Audit Agency. The agency aims to maximize national revenues from the mineral sector. To achieve this objective, amongst other functions, it monitors and audits quality, quantity, and value of minerals produced and exported by large-, medium-, and small-scale mining companies. The agency provides this information to the Tanzania Revenue Authority, advises the Government on administrative matters and also conducts research in the mining sector. It promotes research and development in the mineral sector.[13]

In 2015, the Ministry launched an online portal "Online Mining Cadastre Transactional Portal" as a platform where stakeholders in the sector could engage directly with the Ministry. The portal facilitates transactions like stakeholder registration, mining license applications, renewal and extensions, payments and interactions with mining sector stakeholders, thus streamlining procedures, monitoring, and accountability of businesses in the sector.[14]

LEVEL: FEDERAL (NATIONAL), LINE MINISTRY

Ministry of Water Resources, River Development & Ganga Rejuvenation, India

The Ministry of Water Resources (MoWR) is responsible for laying down policy guidelines and programs for the development and regulation of the country's water resources. Water was managed under various other ministries since 1951 and eventually MoWR was formed in 1985. It was renamed as "Ministry of Water Resources, River Development & Ganga Rejuvenation" in 2014 and became the National Authority of River Ganga, the largest river in the country. The Ministry formulates policies and plans for the water resource sector, providing support and technical guidance for projects like irrigation development, flood-control projects, development of ground water resources, and inter-state river management.

One of the biggest projects undertaken by the Ministry is rejuvenation of the River Ganga with the aim of abating pollution due to industries and sewage by 2017. Another challenging project undertaken by the Ministry is the Interlinking of Rivers Program for equitable distribution of water by enhancing the availability and supply of water in drought prone and rainfed areas.[15]

DISCUSSION QUESTIONS

- *Discuss the objectives and key functions of Tanzania's Rural Energy Agency, an autonomous body under the Ministry of Energy and Minerals.*
- *Environmental Management at the national level is managed by the Apex Environmental Ministries. In addition to the Environmental Ministries, various Line Ministries are focusing on Water, Energy and Agriculture. Do you see challenges in coordination between the Environment and Line Ministries? Are there any successful examples?*

BOX 2.4 EXAMPLES OF INSTITUTIONAL STRUCTURES OPERATING AT SUB-NATIONAL AND LOCAL LEVELS

LEVEL—SUB-NATIONAL; ENVIRONMENT MINISTRY

Secretariat of Environment, Sao Paulo

Established in 1986, the Secretariat of Environment in the State of Sao Paulo has the objective to protect the environment by promoting preservation, improvement, and restoration of essential elements of the environment.

The Secretariat has the responsibility of formulating and deploying plans and policies related to the environment and sustainable development. It coordinates planning efforts for natural resources and biodiversity conservation, environmental education, urban parks conservation, water management, urban afforestation, and climate change mitigation.[16]

In 2009, the State drafted the Climate Change Policy that the Secretariat is responsible for implementing. The key objectives of the policy were to reduce Greenhouse Gas (GHG) emissions, improve public transportation access and use, reduce waste landfilling, reduce deforestation, promote use of renewable fuels, encourage public participation and encourage research and dissemination of scientific and technological knowledge for protection from climate change.[17]

LEVEL: SUB-NATIONAL; LINE MINISTRY

State Water Board of California, USA

Sub-national governance extends to many levels of governance like the municipal, metropolitan, and resource agencies. In the cities of Hong Kong, Mexico City, Naples, Krakow, and California the drinking water supply is managed at the local level. For example, the Water Supplies Department under the Government of Hong Kong supplies potable water in Hong Kong.[18]

Let us look at the role and responsibilities of State Water Board of California. Established in 1967, the State Water Board has been working on water management projects such as operating and repairing existing sewer systems, building new wastewater treatment plants and cleanup of polluted ground water sources. The Board's tasks are to allocate water rights, resolve water right disputes, develop statewide water protection plans, establish water quality standards, and guide the nine Regional Water Quality Control Boards located in the major watersheds of the state.[19]

The Water Board introduced a Recycled Water Policy in 2009 (amended in 2013). The Policy included a mandate to increase the use of recycled water in California by 200,000 acre-feet per year (afy) by 2020 and by an additional 300,000 afy by 2030. To meet this objective, State Boards, Regional Water Boards, water purveyors, environmental community and publicly owned treatment works will work collaboratively. $1 billion was allocated in state and federal funds to meet the objectives of this policy.[20]

LEVEL: LOCAL

Sustainability Appraisals in UK Counties

Under European and UK National law, the UK Counties are vested with environmental management and sustainable development responsibilities. Counties are required by law to ensure their local development plans undergo Sustainability

Appraisal (SA). SA is a *decision aiding* tool that helps to ensure that spatial development plans are in line with the principles of sustainable development. This iterative assessment process involves integrating the environmental, social and economic concerns and priorities of the area, and then ensuring that they are integrated into planning policies to review/develop actions. Factoring in a broader range of perspectives, objectives, evidence and constraints of the local areas will benefit the decision-makers during the planning process.

Local consultation bodies and the public are involved in the SA process for their inputs by making the development plans and appraisal reports publicly available and through seminars, workshops and brainstorming discussions. It creates a transparency in the decision-making process of the local government bodies.[21]

One of the Counties' responsibilities is to carry out Strategic Environmental Assessments, Sustainability Appraisals, and Habitats Regulations Assessments for certain plans and projects. These assessments are used to help identify and evaluate the main environmental, social, and economic concerns that need to be addressed during developmental planning and also reflects global, national, regional, and local concerns. The areas of study include air quality, biodiversity, and green infrastructure, climate change mitigation and adaptation, flood risk assessment, community and well-being, waste, water, land use, housing, and transportation, amongst other things. The outcome of the assessments is identification of measures and changes to the plan to avoid or minimize negative effects and enhance positive effects.[22]

We will discuss the topic of Environmental Impact Assessment in Section 2.12.

LEVEL: LOCAL

Waste Management at Urban Local Bodies (ULBs), India

ULBs are municipal governing bodies in India that function as autonomous entities and are vested with multi-fold responsibilities. Some of the relevant environmental responsibilities include regulation of land-use, planning economic and social development, water supply public infrastructure development including buildings, roads and bridges, public health and sanitation, and public health and safety.[23]

Of the many responsibilities ULBs have, waste management is a critical one. Collecting, processing, transporting and disposing of municipal solid waste (MSW) are the responsibility of ULBs. The Municipal Solid Waste (Management & Handling) Rules require ULBs to segregate waste in multiple categories including organic waste, hazardous waste, recyclables and to ensure safe and scientific transportation, management, processing and disposal of municipal waste. However, given the complexity and quantum of the waste in the country, the ULBs face significant challenges for waste management and are usually supported by the central government, private, public, and the non-profit sector.[24]

DISCUSSION QUESTIONS

- *As one descends from national to sub-national and local levels, the focus of environmental management often gets more thematic depending on the priorities, for example, water, energy, or waste. There are also situations where local level institutions interact with multiple institutions at sub-national and national levels to come up with rounded or multi-dimensional action plans. Capacity building of local institutions in such cases becomes very critical. Develop a capacity development plan in this respect for Urban Local Bodies (ULBs).*

- *Discuss the case of Sustainability Appraisal plans and the requirements at the Counties in the UK and linkages with the National and EU levels. How are these requirements at the multiple levels harmonized? What are the challenges faced and how are these challenges addressed?*
- *There is often a need to achieve harmonization between the Ministries to achieve greater and sustainable impact and strike a congruence. Take the case of waste management at ULBs in India and discuss the harmonization with the interests of Ministry of Environment & Forests & Climate Change, Ministry of Urban Development and the Ministry of New and Renewable Energy.*

BOX 2.5 CASE OF MANILA BAY OIL SPILL CONTINGENCY PLANNING: A MULTI-LEVEL APPROACH[25]

Manila Bay is an important channel for business and trade in the Philippines. Over 30,000 vessels enter the Manila Bay per year to transport goods and cargo including oil. Due to the presence of oil refineries, transportation of oil as fuel or cargo, the risk of oil spills in this region is significantly high. In 2006, there was a major oil spill in the Philippines, where a commercial oil tanker spilled over 500,000 liters of oil in the gulf. The spill affected the communities, biodiversity and marine life in the area. Prior to this, 18 oil spills occurred between 1998 and 2004 in the Marina Bay.

Partnerships in Environmental Management for the Seas of East Asia (PEMSEA) is formed of partner countries such as Cambodia, China, Indonesia, Japan, Korea, Philippines, Korea, Singapore, Vietnam, Laos, and Timor Leste. PEMSEA, the Philippines Government and Coast Guard recognize the high-risk potential of oil spills. Their collective response was to create a mechanism for oil spill management within the Manila Bay. The response is called the Manila Bay Oil Spill Contingency Plan, which is a part of the Coast Guard (District) Oil Spill Contingency Plan that, in turn, is part of a larger National Oil Spill Contingency Plan (Figure 2.7).

The aim of the plan is to provide a timely and effective response program to address oil spills from ship-based sources through a coordinated mechanism among relevant agencies. The parties that need to prepare plans are oil refineries, terminals and depots, power plants and barges, manufacturing plants or similar establishments that use oil, shipping companies, ship yards, and Coast Guard Districts.

The three-tiered contingency plan classifies spills into three types based on the volume of oil spilt and proximity to the response center. The tiers determine the agencies responsible for action and containment of the spill. The 3-tiers and corresponding response agencies are included in the table below:

Tier	Volume Spilt	Response
I (Small spill)	Up to 10 m^3	Company or Ship Response Organization/District Response Organization
II (Medium spill)	Up to 1000 m^3	First Tier Response plus National Response Organization
III (Large spill)	>1000 m^3	All the national resources, with support of international organizations

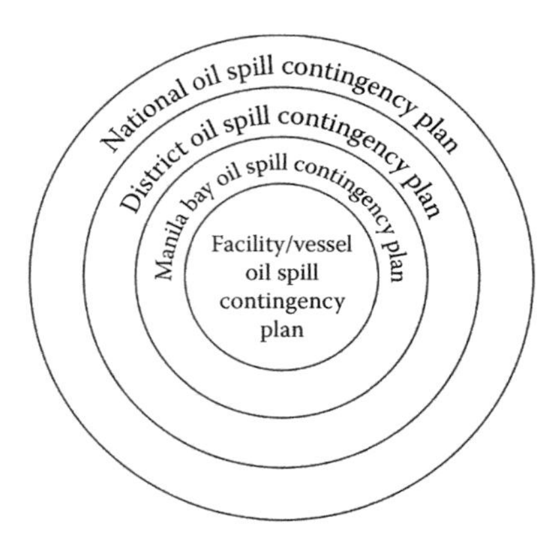

FIGURE 2.7
Illustration of a multi-level National Oil Spill Contingency Plan for Manila Bay. (From Manila Bay Oil Spill Contingency Plan, GEF/UNDP/IMO Regional Programme on Partnerships in Environmental Management for the Seas of East Asia (PEMSEA); p. 4, 2006. With permission, online source: http://www.pemsea.org/sites/default/files/manilabay-oilspill-plan.pdf.)

DISCUSSION QUESTIONS

- *Read the National Oil Spill Contingency Plan of Ghana at—http://epaoilandgas.org/national%20oil%20spill.html.*
- *Compare the Manila and Ghana oil spill contingency plan in terms of institutional arrangements.*

2.6 Inclusion of Environmental Rights in the Constitution

One of the first responses by countries was to reflect the environmental concerns and associated rights in the constitution. Accordingly, provisions for protection of the environment from pollution and reserving and guarding the resources for the future generations were enshrined in the national constitutions. These provisions were then translated into the relevant policies and laws. Figure 2.8 gives an overview of the typical process steps followed in a country with allocation of responsibilities to different institutions.

The original *frames* of most Constitutions lacked the foresight to visualize the acute environmental issues we face today. The first response of some countries was to introduce or amend Constitutional Frameworks suitably to include provisions for environmental protection and social justice.

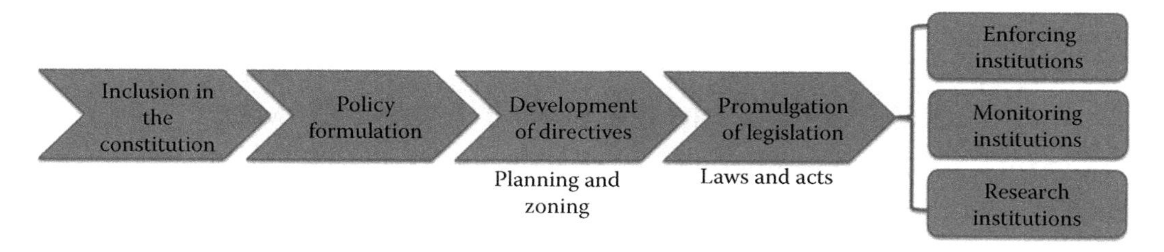

FIGURE 2.8
Typical constitutions, policy and regulation based response with institutional arrangements.

In some countries, the constitutional right to a healthy environment for all citizens became a guiding principle and trickled through the entire body of environmental legislation. Constitutional inclusion had a comprehensive impact on environmental policies and laws in countries such as Argentina, Portugal, Costa Rica, Brazil, Colombia, India, South Africa, and the Philippines.[26]

We present in Box 2.6 illustrations of relevant articles in various countries demonstrating constitutional inclusion in the interest of the environment.

2.7 Formulation of National Policies

Constitutional provisions guide and support the environmental policies and laws to execute its spirit, aspiration, or vision. A policy outlines a Ministry's goals and objectives that it aims to achieve and states the principles, strategies and appropriate institutional arrangements. A policy document is not a law but a guide to devise the needed legislative framework so that the policy and principles are implemented in practice.

Many national governments drafted policies to handle issues like protection of resources such as energy, water, land, and biodiversity, encourage sustainable consumption and production (SCP), and undertake cleanup of pollution hot spots.

Box 2.7 presents illustrations of environmental policies drafted in various countries on this basis.

2.8 Planning and Zoning Related Directives

One of the government responses to control pollution, deforestation, biodiversity depletion, and resource management was to demarcate zones and create buffers around the zones or areas of ecological or social significance to control and guide development without threatening the sustainability.

Zoning is used as an important national, regional, and urban planning and development tool in many countries. By creating zones, regulators aim to control land uses and the associated developmental activities. Zoning is typically applied for restricting development activities in ecologically sensitive areas like wetlands, national parks, forests, coastal areas, and areas of heritage value. Buffers are often set up around the zones to protect sensitive areas.

BOX 2.6 ILLUSTRATIONS OF RELEVANT ARTICLES IN VARIOUS COUNTRIES DEMONSTRATING CONSTITUTIONAL INCLUSION IN THE INTEREST OF THE ENVIRONMENT[27]

EXAMPLE 1: ARTICLES FROM GERMAN CONSTITUTION[28]

Article 72 (3) includes legislative provision for "Protection of nature and landscape management" and "Management of water resources"

Article 74 (17) includes legislative provision for "Promotion of agricultural production and forestry, ensuring the adequacy of food supply, the importation and exportation of agricultural and forestry products, deep-sea and coastal fishing, and preservation of the coasts."

Article 74 (24) includes legislative provision for "Waste disposal, air pollution control, and noise abatement."

EXAMPLE 2: ARTICLES FROM THE INDIAN CONSTITUTION[29]

The constitutional (forty-second Amendment) Act, 1976 incorporated two significant articles viz. Article 48-A and 51A (g).

Article 48-A of the constitution says, "the state shall endeavor to protect and improve the environment and to safeguard the forests and wild life of the country."

Article 51-A (g) says, It shall be duty of every citizen of India "to protect and improve the natural environment including forests, lakes, rivers and wild life and to have compassion for living creatures."

EXAMPLE 3: ARTICLES FROM THE CONSTITUTION OF CHINA, 1982[30]

The Chinese Constitution has provisions for Environment Protection and Health in the following Articles:

Article 21. The State develops medical and health services, promotes modern medicine and traditional Chinese medicine, encourages and supports the setting up of various medical and health facilities by the rural economic collectives, state enterprises and undertakings and neighborhood organizations, and promotes sanitation activities of a mass character, all to protect the people's health.

Article 26. The state protects and improves the living environment and the ecological environment, and prevents and controls pollution and other public hazards. The state organizes and encourages afforestation and the protection of forests.

EXAMPLE 4: CONSTITUTION OF THE UNITED REPUBLIC OF SOUTH AFRICA[31]

Article 24:

Everyone has the right:

1. To an environment that is not harmful to their health or well-being; and
2. To have the environment protected, for the benefit of present and future generations, through reasonable legislative and other measures that:

a. Prevent pollution and ecological degradation;
b. Promote conservation; and
c. Secure ecologically sustainable development and use of natural resources while promoting justifiable economic and social development.

Article 27:

Everyone has the right to have access to health care services, including reproductive health care; and sufficient food and water.

EXAMPLE 5: COLOMBIA'S CONSTITUTION OF 1991 WITH AMENDMENTS THROUGH 2005[32]

Article 78

Those who in the production and marketing of goods and services may jeopardize the health, safety, and adequate supply to consumers and users will be held responsible in accordance with the law.

Article 334

The general management of the economy is the responsibility of the State. By mandate of the law, the State will intervene in the exploitation of natural resources, land use, the production, distribution, use, and consumption of goods, and in the public and private services in order to rationalize the economy with the purpose of achieving an improved quality of life of its inhabitants, the equitable distribution of opportunities, and the benefits of development and conservation of a healthy environment.

Note that only parts of the articles relevant to the environmental challenges are included. Sections of the articles that are not pertinent to the challenges addressed in this book have been excluded.

DISCUSSION QUESTIONS

- *How important do you think it is to have constitutional provisions for the environment in any country and why? What may be the downstream impact of such provisions on policies and legislation? Give an example of such downstream linkages.*
- *Discuss the constitutional provisions for the interest on environment in a: South American country; European country; and countries in the Pacific Islands and make a comparison.*

Many international organizations played a key role to come up with *uniform* zoning policies to guide the national governments. Box 2.8 illustrates examples of ecological zoning recommended by a few international organizations as a *plan to protect* measure.

Box 2.9 illustrates case studies of zoning directives issued by the national governments.

BOX 2.7 ILLUSTRATIONS OF ENVIRONMENTAL POLICIES BY VARIOUS COUNTRIES

EXAMPLE 1: NATIONAL HEALTH POLICY, NIGERIA, 1988 (REVISED IN 2004)

The National Health Policy and Strategy was introduced in 1988 to provide a comprehensive health care system for Nigerians. The policy was amended in 2004 to factor in new health trends and challenges and improve the nation's health situation. Some of the key principles include providing access to quality and affordable healthcare to all Nigerians as their human right, equity in health care, provision of primary health care, providing cost-effective treatments for priority health issues, efficiency and accountability in the national health system, and promoting partnership and collaboration amongst various health care stakeholders. The health care goals, strategies and targets are heavily influenced by the Millennium Development Goals (MDGs). The Nigerian Health policy included targets like reduction in under-5 mortality rate by 66.66%, reduce maternal mortality rate by 75%, and to control or cease the spread of diseases like HIV/AIDS and malaria by 2015. The policy includes detailed guidelines on establishment of various health agencies, stakeholder involvement, resource development like education, training, technologies selection, quality control, and others to develop a robust and sustainable health policy.[33]

EXAMPLE 2: NATIONAL POLICY ON BIODIVERSITY, MALAYSIA, 1998 (REVISED IN 2015)

Malaysia's National Biodiversity Policy aims "to conserve its biological diversity and to ensure that its components are used in a sustainable manner for the continued progress and socio-economic development of the nation". The Policy's vision is "to transform the country into a center of excellence for research, conservation and utilization of tropical biodiversity".

Malaysia's diverse biodiversity elements include national parks, forests, wildlife sanctuaries, marine and coastal areas, mangroves, and endangered flora and fauna. These elements are under duress due to increasing population, introduction of invasive plant species, declines in fish stock, loss of forest cover due to poaching and illegal wildlife trade of species like turtles or rhinoceros.

This policy was first launched in 1998 and amended in 2015 and its policy statement and goals were amended. The 1998 Policy's objectives included optimizing the economic benefits through sustainable use of the nation's biodiversity elements; ensuring long-term food security for the nation; maintaining and improving environmental stability of ecological systems; preserving the unique biological heritage of the nation, and enhancing scientific and technological knowledge, and the educational, social, cultural, and aesthetic values of bio-diversity.[34]

In the revised 2015 policy statement were included five goals and the corresponding 17 targets to achieve the policy objective. The targets revolve around increasing public involvement and awareness, mainstreaming biodiversity conservation into national plans and policies, sustainable management of agriculture, forestry and fisheries, sustainable tourism, protection of vulnerable ecosystems like wetlands, coral reefs, sea grass beds, terrestrial and marine corridors, controlling poaching, illegal wildlife and plants trade, controlling introduction of alien species, increasing bio-diversity related knowledge and allocating adequate funds and resources for biodiversity conservation. The policy details actions to be taken to achieve these targets by 2025 along with a phase-wise plan. It also includes the roles and responsibilities of key stakeholders, sub-targets and indicators that will be used to measure the success of each of the 17 targets.[35]

EXAMPLE 3: NATIONAL POLICY FOR DISASTER PREPAREDNESS AND MANAGEMENT, REPUBLIC OF UGANDA, 2010

Uganda is inflicted by natural disasters like famine, drought, earthquakes and floods that disrupt human and economic activities. The Department of Relief, Disaster Preparedness and Management developed a policy to mitigate associated risks, increase preparedness, response and recovery systems to natural and human-inflicted disasters.

The Policy's goal is "to establish institutions and mechanisms that will reduce the vulnerability of people, livestock, plants and wildlife to disasters in Uganda." Some of the specific objectives are to establish institutions at national and local government levels for disaster preparedness and management, to train and equip these institutions to be prepared at all times for disaster management, to integrate disaster preparedness and management systems into developmental plans and activities, to promote research and technology in disaster risk reduction, to generate and disseminate information on early warning for disasters and hazard trend analysis, to promote public, private partnerships in Disaster Preparedness and Management and to create timely, co-ordinated and effective response systems at national, district, and local levels for disasters and emergency situations.[36]

EXAMPLE 4: NATIONAL WATER POLICY, INDIA, 2012[37]

Developed in 2012, India's National Water Policy is a comprehensive policy that addresses multiple areas that are critical elements for holistic water management. The policy recognizes that the issue of water management requires policies and governance at national, sub-national, and local levels due to unevenness in source availability, distribution, infrastructure, and demand in various parts of the country. The Policy follows the principles of equity, sustainability, and social justice in use and allocation of water. The Policy addresses the need for legislation for development of interstate rivers for interstate planning of land and water resources.

The Policy includes guidelines on optimization and awareness of the use of water and encourages community-based water management. Climate Change Adaptation strategies like water-efficient agriculture technology, and land-soil-water management are suggested. Other water management strategies like watershed management, recycling and reuse for demand management, use of water pricing as a tool to incentivize conservation, recycling and reuse, conservation of river corridors, water bodies and infrastructure, management, and preparing for water related disasters like flood and drought that commonly occurs in the country. The Policy recommends establishment of forums at the national level in the form of a Water Disputes Tribunal to resolve water related issues. Transboundary river management by open communications, data sharing with neighboring countries and negotiations about sharing and management of water is also incorporated as a guideline. The Policy recognized the importance of creating a national database (National Water Informatics Center) and promotion of research and training in the water sector.

This policy is a good example to demonstrate how holistic understanding of the cause and effects of the issues, long-term resilience, and efficient management was taken into account while developing a National Policy.

EXAMPLE 5: NATIONAL FERTILIZER POLICY, REPUBLIC OF RWANDA, 2014[38]

Agriculture is a key sector that contributes 34% to Rwanda's GDP. Rwanda experienced tremendous agricultural growth between 2004 and 2014. One of the key sector challenges is fertilizer availability, efficiency of usage, quality, and trade. To addresses these issues, the Ministry of Agriculture and Animal Resources, developed a fertilizer policy with the Vision "to have a functional and efficient private sector led fertilizer sector that is responsive of farmers' needs and the environment." The Policy's Mission is "to have a competitive and profitable fertilizer sector that ensures fertilizer access and affordability at farm gate in a timely manner creating acceptable fertilizer use by farmers for increased and sustainable agricultural productivity and farm incomes."

The specific objectives of the policy are to enable private sector activity of importing and distributing fertilizers (the policy recommends introducing appropriate incentives to garner private investments), a private sector friendly monitoring system to ensure quality standards of the fertilizers, enhancement of fertilizer usage through product diversity and technology, incentives in the form of subsidies, insurance to facilitate increased usage by farmers available to them at affordable prices. The Policy supports local production of fertilizers and research and development to improve efficiency and crop yields. The development of this Policy was supported by the Agriculture Working Group and incorporates gender and environmental concerns.

This is an example of an instance wherein the government has developed a policy keeping in mind the role of the private sector and the local priorities in order to successfully meet the policy objectives.

DISCUSSION QUESTIONS

- *Discuss the Integrated Climate and Energy Policy of Sweden, and its objectives and targets.*
- *Discuss the National Waste Policy of Australia, and its key areas and strategies.*

2.9 Eco-Cities, Eco-Industrial Parks and Eco-Towns

In Chapter 1 we discussed the challenges we are facing today due to intense urbanization. Globally, cities account for about 70% of CO_2 emissions. The bulk of the GHGs are generated due to building construction and operations, urban transportation, and energy generation and usage sectors. Reducing CO_2 emissions in urban areas will require reduction in energy consumption of buildings during construction and operation and maintenance; encouraging use of low carbon forms of transportation such as public transportation,

BOX 2.8 ILLUSTRATIONS OF ECOLOGICAL ZONING RECOMENDED BY INTERNATIONAL ORGANIZATIONS

EXAMPLE 1: PROTECTED AREAS RECOMMENDED BY THE INTERNATIONAL UNION FOR THE CONSERVATION OF NATURE[39]

The International Union for the Conservation of Nature (IUCN) has developed a system to classify protected areas into seven categories based on the specific management objectives of the protected areas. IUCN defines protected areas as "a geographical space, recognized, dedicated and managed, through legal or other effective means, to achieve the long-term conservation of nature with associated ecosystem services and cultural values."[40]

IUCN helps countries and communities manage areas of ecological, economic, or social significance like national parks, wilderness areas, natural reserves, or community conserved areas that provide protection from harsh weather or natural disasters and is a source of food, water, and medicines.

IUCN developed seven protected area categories that are:

- Ia: Strict Nature Reserve
- Ib: Wilderness Area
- II: National Park
- III: Natural Monument or Feature
- IV: Habitat/Species Management Area
- V: Protected Landscape/Seascape
- VI: Protected area with sustainable use of natural resources

To learn more about the above seven categories visit: https://www.iucn.org/theme/protected-areas/about/protected-areas-categories

National governments and international bodies like the UN recognize these categories as a global standard to define and document protected areas under their jurisdiction and to incorporate them into government legislation for conservation.

EXAMPLE 2: IMPORTANT BIRD AND BIODIVERSITY AREAS BY BIRDLIFE INTERNATIONAL

An Important Bird and Biodiversity Area (IBA) is an area identified, documented, and protected by the organization BirdLife International with the objective of conserving areas on earth that are of great significance for conservation of the world's birds. Since the 1970s, BirdLife International has identified over 12,000 IBAs through a network of 120 BirdLife Partners, one partner per country or territory. Birds are effective indicators of a wider biodiversity; hence, the protection of bird sites will translate into protection of other animal and plant species in the region.[41]

Of the 1200 plus IBAs, BirdLife International identifies and prioritizes the more critical or threatened areas under the IBAs in Danger Initiative. About 422 IBAs have been identified as critically threatened due to threats like unsustainable agriculture or aquaculture, human intrusions, residential or commercial development, pollution, climate change, introduction of invasive species, energy production or mining, transportation and service corridors, and geological events (see Figure 2.9).

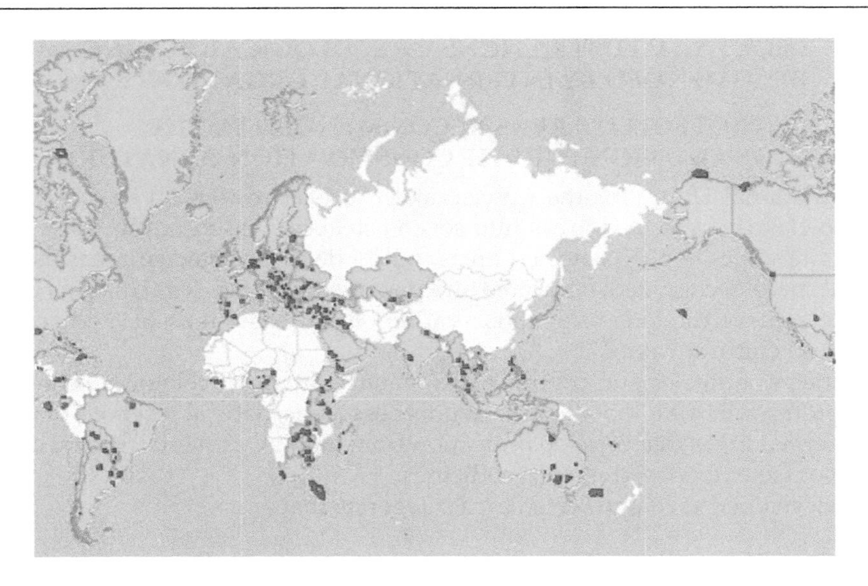

FIGURE 2.9
World Map indicating Important Bird and Biodiversity Areas under risk ("IBAs in Danger"). (Updated from BirdLife International, 2014, Important Bird and Biodiversity Areas in Danger—priority sites for immediate action. Available at http://www.birdlife.org/worldwide/programmes/sites-habitats-ibas-and-kbas.)

BirdLife is working with its Partners to protect these areas through awareness campaigns, advocacy or legal ways to control unsustainable development, or take effective measures to protect IBAs.[42] BirdLife Australia launched a Facebook page to provide *IBA Champions* a platform to collaborate with the community and interested stakeholders towards IBA protection.[43]

DISCUSSION QUESTIONS

- *Discuss WWF's Global 200 project designed to conserve endangered ecoregions. Pick an endangered ecoregion in the Global 200 list and discuss its ecological importance and conservation strategies.*
- *Give examples where zoning has worked and where it has failed. Discuss why.*

pooling or efficient vehicles and fuels; and energy production. However, achieving such goals will require huge investment and new paradigms of planning and rehabilitating the cities.

Existing cities and its infrastructure need to adapt to the environmental and social changes as they grow. The bus rapid transit system in Guangzhou, People's Republic of China, for example, is integrated with the city's metro system and other non-motorized means of transport. The city planning and zoning also facilitates this mixed use of transportation. The urban design and transit service facilitates a dense yet pedestrian-friendly environment which allows easy access to services and for workers or employees and lowers carbon emissions.

A city's ecological footprint is measured in hectares per person and gives an idea about the cities' resource use intensity per person. In rural China, the average footprint is 1.6,

BOX 2.9 ILLUSTRATIONS OF ZONING DIRECTIVES BY NATIONAL GOVERNMENTS

ENVIRONMENTALLY SENSITIVE AREAS, AUSTRALIA

Environmentally Sensitive Areas (ESAs) were defined under the Environmental Protection Act, 1986 in Australia to protect national parks, state forests, heritage sites, wetlands, and other endangered areas. The specific ESA categories include areas declared as a World Heritage Property or of heritage value under different Commonwealth Acts, defined wetlands and areas within 50 meters of the wetland, areas covered by vegetation within 50 meters of rare flora, areas that contain an endangered ecological community, the areas covered by the lakes, wetlands, and native vegetation as covered under various national and regional policies.[44]

THE TAJ TRAPEZIUM, INDIA

Taj Trapezium Zone (TTZ) is a defined area of 10,400 sq. km around the Taj Mahal to protect the monument from pollution.

Foundries, refineries and chemical and hazardous industries located in the vicinity of the monument caused major damage to it. Sulphur dioxide from power plants, refineries, and industries combined with oxygen and moisture caused acid rains that corroded the gleaming white marble structure of Taj Mahal causing it to yellow and blacken. The court defined the Trapezium area, which covers around 40 protected monuments, including three world heritage sites. The industries within the area were required to have pollution control measures in place or shut down operations if they failed to do so. The industries that did not comply with the air pollution control measures within TTZ, either had to relocated outside the area or switch coal-based operations to alternate fuels like natural gas or propane.[45,46]

LOW EMISSION ZONE, LONDON

Administered by Executive Agency Transport for London and Greater London Authority, the Low Emission Zone (LEZ) came into operation in 2008. LEZ was introduced to incentivize polluting heavy diesel commercial vehicles in London to switch to cleaner fuels. It covers most of Greater London. If vehicles entering the LEZ do not conform with the emission standards, they have to pay a daily charge if they wish to enter the zone. Lorries, buses, and coaches had to meet the Euro IV emissions standard while large vans, minibuses and specialist vehicles had to meet the Euro 3 standard.[47] This pollution charge was expected to motivate vehicle users to switch to cleaner fuels or install emission control filters.

Vehicles that did not meet the emission standards had the option to switch to pure gas, could fit an approved filter to their exhaust or upgrade their vehicle to a gas or electric vehicle or pay the charge of 100–200 British Pounds per day.[48]

PHILIPPINES ENVIRONMENTALLY CRITICAL AREAS (ECA) AND ENVIRONMENTALLY CRITICAL PROJECTS (ECP)[49,50]

Environmentally Critical Areas (ECAs) are those areas that are ecologically, socially, or geologically sensitive, frequently exposed to hazards or areas that are historically interesting. The Philippines Presidential Proclamation defines 12 Classes of ECAs, which include:

- National parks, watershed and wildlife preserves, and sanctuaries declared by law
- Areas marked as potential tourist spots
- Habitats of endangered or threatened species indigenous to the Philippines
- Areas of unique historic, archaeological, or scientific interest
- Areas traditionally occupied by indigenous people and cultural communities
- Areas frequently affected by natural calamities like floods, typhoons, volcanic activity
- Areas with critical (steep) slopes
- Areas classified as prime agricultural lands
- Aquifer recharge areas
- Water bodies used for domestic supply or support of fish and wildlife
- Mangrove areas supporting critical ecological functions or livelihoods
- Coral reefs with critical ecological functions

ECAs are demarcated to protect them from environmentally detrimental developmental activities. Any development project that is located in the ECAs is subjected to an Environment Impact Assessment (EIA) process. Some ECAs have defined buffer zones to ensure that the adjacent land is free from intensive land use to protect the ECAs.

Similarly, the Philippines Government defines Environmentally Critical Projects (ECPs) that may have critical impacts on the environment and need to undergo an EIA. ECPs include projects in industries like metal, iron, steel, petroleum, mining, quarrying, forestry, fishery projects or infrastructure projects like dams, power plants, reclamation projects, roads, and bridges.

COASTAL REGULATION ZONE, INDIA[51]

In 1991, the Ministry of Environment and Forests (MoEF) issued a notification on Coastal Regulation Zone (CRZ) in order to protect and conserve the coastal areas and ecosystems in India. The main objectives of defining the CRZ was to protect the livelihood of the fishing communities and local communities living in the coastal areas, to conserve and protect coastal stretches from natural hazards like tsunamis and from rising sea levels as a result of global warming.

There are four zonal classifications under this notification: CRZ I, II, III, and IV. CRZ I areas are ecologically sensitive areas like mangroves, corals, sand dunes, national parks, wildlife habitats, marine parks, forests, and habitats for endangered species. CRZ II are areas that are developed up to or close to the shoreline and that fall within the municipal jurisdiction. CRZ III areas include areas that are relatively undisturbed and that include rural and urban areas that are not sustainably developed. CRZ IV areas are aquatic areas from low tide line up to territorial limits.

The Notification defines what activities are permissible and not permissible in the defined CRZs.

FLOOD ZONES IN THE UNITED KINGDOM[52]

The National Planning Policy Framework set out by the UK Government defines planning policies for England for development of local and neighborhood plans. One of the key aspects of the framework is the definition of flood zones to avoid development in areas that are at a risk of flooding. The defined zones and their allowed land uses are:

Zone 1

The areas in this zone have a low probability, less than 0.1% annual probability of river or sea flooding. All uses of land are permissible in this zone. Developments over a hectare in area need to undergo a flood risk assessment process.

Zone 2

This zone is comprised of lands that have medium probability, between 0.1% and 1% annual probability of river flooding or between 0.5% and 0.1% annual probability of sea flooding in any year. All infrastructures except ones that are not highly vulnerable to flooding like basement dwellings, utilities, and public services that need to be operated during floods can be developed in this zone. All developments in this Zone need to undergo flood risk assessment.

Zone 3a

This zone is comprised of land that has a high probability of 1% or greater annual probability of river flooding or a 0.5% or greater annual probability of flooding from the sea in any year. Water-compatible and less vulnerable infrastructure like navigation facilities, water and sewage pumping stations, ship building activities and others can be developed in this zone and need to undergo a flood risk assessment process.

Zone 3b

This zone is comprised of land which would flood with an annual probability of 5% or greater in any year, or is designed to flood in an extreme (0.1%) flood could be considered as a Zone 3b area. The land in this zone is where water *has* to flow or be stored in times of flood. Only the water-compatible uses and the essential infrastructure like utilities, electricity generating power stations, primary substations, water treatment works, and wind turbines, which are safe for users in flood plains and do not undergo severe loss of floodplain storage, can be located in this zone.

 **DISCUSSION QUESTIONS**

- *Compare the concept of ESA in Australia with ECAs in the Philippines.*
- *Discuss the impact of London's Low Emission Zone on mitigating vehicular emissions.*
- *Discuss the four area classifications (CRZ I, II, III & IV) under the Coastal Regulation Zone of India. Do you see any overlaps or ambiguities?*

in Shanghai, it is 7 while in a typical city in the United States the footprint is 9.7. These footprints are significantly higher than an ideal footprint of about 1.8 hectares per person, which is considered sustainable.

Governments started responding to this challenge by conceptualizing Eco-Cities that were designed to utilize minimal energy, water, conserve available resources including harnessing urban biodiversity and release minimal wastes, emissions and residues through reuse, recycle and recovery. Apart from the green field cities where these principles were applied, urban revival or transformational projects were also undertaken, especially for Tier-2 and Tier-3 cities. Different governments set different themes and guidelines for eco-cities. Some focus on green technologies, on information, communications and technologies (ICT) based smart systems, green infrastructure or on social inclusivity. Accordingly, programs like GrEEN cities (promoted by Asian Development Bank), ECO^2Cities (conceived by the World Bank) and Sustainable Cities (UN HABITAT) were launched in partnership with the national governments. More recently, programs like solar cities, smart cities, and climate resilient cities have come about focusing on reducing GHG emissions and the vulnerability due to climate change.

According to United Nations Development Program (UNDP) and Renmin University of China, an eco-city is defined as "a city that provides an acceptable standard of living for its human occupants without depleting the ecosystems and biochemical cycles on which it depends." Eco-cities are developed using three concepts:

- Eco-industry
- Eco-scape
- Eco-culture

The concept of Eco-industry revolves around promoting industry symbiosis and metabolism, life-cycle production, resource conservation, and use of renewable energy. Eco-scape focuses on developing a sustainable built environment, open spaces for public and ecosystem conservation, and maximizing accessibility, while minimizing resource use and urban problems. Eco-culture is about understanding the relation between humans and nature, environmental ethics in order to enhance people's contribution to maintaining a high-quality urban ecosystem, which provides long-term sustenance to generations.[53] The idea is to make public services and infrastructure more user and environment-friendly, especially with regard to basic amenities like air and water pollution management, wastewater, and waste management.

We present in Box 2.10 examples of Tianjin Eco-city and the Solar City of Barcelona. Box 2.11 illustrates the impact of Information, Communication, and Technology (ICT) initiatives in a remote village in India.

In planning, the Eco-cities *process* is equally important as much as the *projects* or the outcomes. We present in Box 2.12 the process followed in developing an Eco-City Action Plan for the cities of Sangli-Miraj-Kupwad (SMK) in the State of Maharashtra in India.

Similar to the concepts like Eco-cities, the idea of developing Eco-Industrial Parks (EIPs) emerged. The EIPs were based on the concept of industrial symbiosis, wherein industries or businesses work in a collaborative manner for mutual benefit on the social, economic, and environmental aspects. The companies in the EIP shared infrastructure, resources, energy, water or material waste that could be used as an input for another industry. EIPs were developed on specific industrial themes like Resource Recovery EIP, Agricultural EIP, Renewable Energy EIP, or Petrochemical EIP. EIP were initiated by governments, DFIs and private businesses.[61] We present in Box 2.13 the case of an EIP in Hong Kong.

BOX 2.10 CASE STUDIES ON ECO-CITIES AND SOLAR CITIES[54]

TIANJIN ECO-CITY[55]

Tianjin is the third largest city of the People's Republic of China (PRC). Tianjin as an Eco-city was undertaken as a collaborative project between PRC and Singapore. The Sino–Singapore Tanjin Eco-City (SSTEC) was planned on non-arable salt land located in the Tianjin Binhai New Area. By 2020, the city of 34.2 square kilometer area is to be built to house 350,000 people. A mixed land-use city plan is developed to include housing, service-oriented, high-technology, and environment related businesses and industries to create 190,000 jobs. CNY25.5 billion ($3.8 billion) was invested to develop public infrastructure and facilities for the entire SSTEC area. Construction began in September 2008.

Designed on the principles of sustainability, the eco-city is intended to be an energy- and resource-efficient and a low carbon eco-city that takes into account economic viability and elements to promote social harmony amongst the inhabitants. SSTEC's transport plan integrates with land use, includes a higher floor area ratio allocation in the areas near metro stations, thus balancing the population densities with those at the Tianjin city center. Further, the mixed land-use plan will reduce the need for workers to commute from outside the SSTEC area. SSTEC will follow green building standards and promote use of renewable energy. The plan includes provision of affordable public housing equivalent to 20% of the city's total housing provision.

Government and private sectors of both countries are involved in the project and share expertise and experience in areas like urban planning, environmental protection, resource conservation, water and waste management, and sustainable environmental and social development.

SOLAR CITY, BARCELONA, SPAIN[56,57]

The city of Barcelona is amongst the more energy efficient global cities with a low per capita GHG emission level. It is the first European city to have a Solar Thermal Ordinance (STO) in place in 2000. Under the STO, all private and public new buildings, renovated buildings, and re-purposed buildings were required to supply 60% of their running hot water using solar thermal systems. Under the STO, the surface of solar thermal in Barcelona increased from 1.1 sqm per 1,000 inhabitants in 2,000 to 59 sqm per 1,000 inhabitants in December 2010. Other Spanish cities like Seville, Madrid, and Pamplona followed Barcelona's lead. In 2002, Barcelona adopted a Barcelona Energy Improvement Plan to reduce energy consumption and improve air quality through demand management, technology improvements and increased public awareness. In 2004, the city built Europe's largest solar array at the time at Forum Esplanade. Under this plan, the city continues to expand its solar PV portfolio on public infrastructure like public buildings, schools, and city buses.

DISCUSSION QUESTIONS

- *Building brand new ecocities has not been a very successful experiment. The much talked about Eco-city of Masdar will probably not happen due to the financial crisis. The project that was expected to be completed by 2015 may be delayed by at least*

10 years.[58] The Tianjin Eco-city in China described in this Box is under implementation. The experience of Phase I has, however, shown that it has been difficult to get occupiers. Every effort is needed to provide incentives and attractive packages, despite the promise to live in a sustainable world.[59] Discuss why building of brand new eco-cities does not seem to work.

- *A thematic approach on the other hand seems more promising. The Solar city of Barcelona is a good example. Look for more such examples, especially those in the United States where cities are operated on 100% renewable energy.*

Ministry of Economy Trade and Industry (METI) and the Ministry of Environment (MoE) launched Japan's Eco-town program jointly to promote regional economic development and environmental sustainability. The objective of the program was to create resource-recycling socio-economic systems by driving local industries towards recycling and suppressing the generation of waste. To meet this objective, the initiative promoted local innovation and entrepreneurship with focus on integrated waste management.

Grants up to 50% of the project cost were awarded to the Eco-town projects. A key criterion was to create an industry cluster that aimed at zero emissions using wastes and by-products from one industry as useful resources in another. This initiative necessitated

BOX 2.11 WARANA *WIRED VILLAGE* PROJECT[60]

India's Prime Minister's Office Information Technology (IT) Task Force initiated the Warana *Wired Village* project in 1998 to pilot the use of Information and Community Technology (ICT) solutions for rural and socio-economic development in undeveloped regions of India. Using ICT solutions, advanced information and services were provided to 70 villages around the Warana district with such information in the native language about growing crops, market rates, and government programs to increase employment opportunities. This information and knowledge helped farmers increase the efficiency and productivity of the sugar cane agriculture, which was managed through a co-operative. Facilitation booths were set up to provide agricultural, medical, and educational information to villagers through the Internet via the National Informatics Center Network. For further support, distance education programs were delivered at both primary and higher educational institutes. A Geographic Information System (GIS) was established leading to greater transparency in administration especially in matters related to land. The project catered to local needs such as revenue records, health cards, credit cards, and agricultural products prices in both national as well as international markets.

DISCUSSION QUESTIONS

- *Discuss the potential and challenges of such an innovation government project/initiative.*
- *Do you think that such initiatives will help in faster and wider capacity building and establish data collection mechanisms for faster reform?*

BOX 2.12 SANGLI-MIRAJ KUPWAD ECO-CITY ACTION PLAN PROJECT

The SMK Eco-City Action Plan Project was conceived in two distinct phases. In the first phase, SMK Eco-City Action Plan has been prepared in consultation with stakeholders through a collaborative process. This phase includes identification of key stakeholders, opening a dialogue, evolving environmental policy in the form of a Charter for SMK, developing project concepts and outline of programs estimating investments required and recommending institutional arrangements.

The second phase was visualized as the implementation phase in which the projects and programs proposed under the SMK Eco-City Action Plan were to be implemented by building institutional capacities, forging partnerships, mobilizing funds and resources, and promoting Public Private Partnerships.

The Eco-City Action Plan was developed based on the Environmental Management Systems (EMS) approach. Figure 2.10 illustrates how the milestones were planned to achieve the objectives in a systematic manner. The Eco-City Action Plan was developed with the participatory approach as illustrated in Figure 2.11.

Workshops, public meetings, one-to-one meetings with experts and NGOs, and interviews with key stakeholders were conducted to draw the various issues and opportunities in SMK. In order to involve stakeholders at a decision-making or expert level, a Core Committee of key stakeholders was formed. Public meetings were organized twice; at the start and at the time of disclosure of the Draft Report. The feedback received from the public was incorporated in the Final Report.

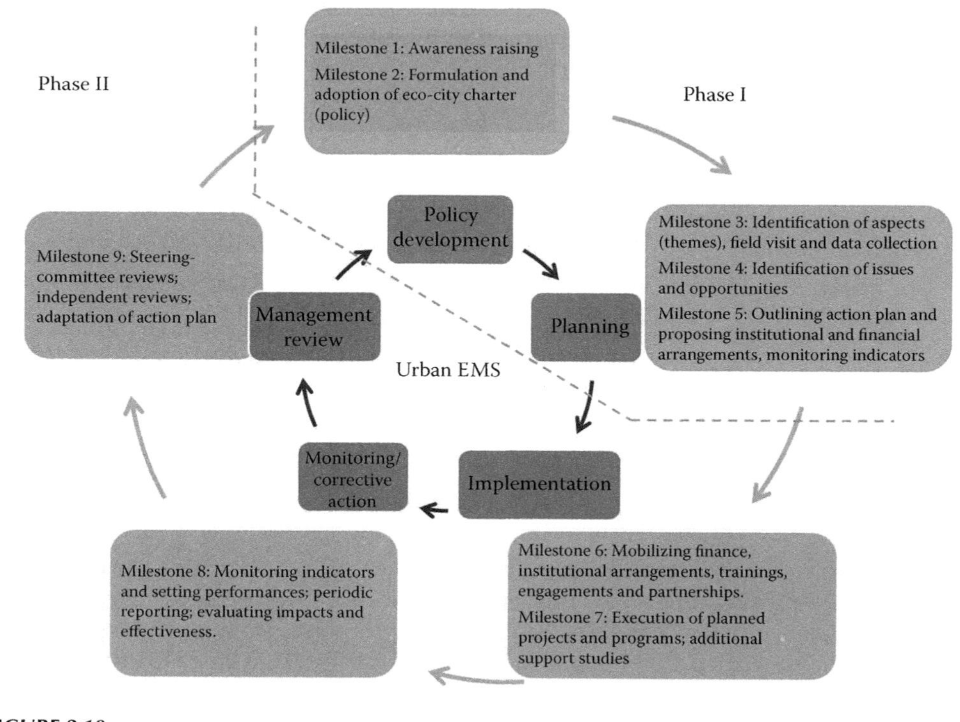

FIGURE 2.10
EMS approach for planning SMK's Eco-City Action Plan.

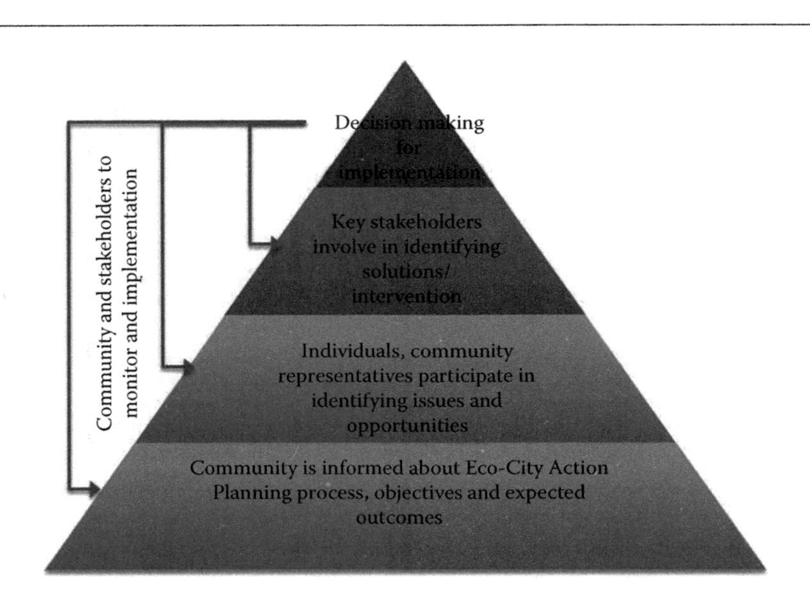

FIGURE 2.11
Participatory approach followed in the SML Eco-City Action Plan.

To promote the concept of the Eco-City, share various outputs, stimulate networking, and provide a mechanism for response, a website for SMK Eco-City was developed. The website called EkoVoices had several interactive features like mapping, opinion polls, and discussion groups.

DISCUSSION QUESTIONS

- *Look for examples similar to SMK where a multi-level participatory process was used to develop an Eco-City Action Plan. How does one build financial resources for implementation?*
- *To what extent should private sector participation be sought? What could be the elements of risks?*
- *Who should be the right agency to implement the Eco-City Action Plan? Should this be an independent Special Purpose Vehicle (SPV) considering the weak institutional capacities of the Urban Local Bodies (ULBs)?*

adoption of resource productivity, innovation in environmental technologies, and instilled competition in regional markets.

With the help of Government subsidies, over 62 eco-town facilities have been built that either deal with products covered by the various recycling laws like packaging, home appliances, vehicles, construction material or food wastes, process domestic wastes or deal with difficult-to-recycle wastes. Box 2.14 presents a case of an Eco-town in Japan.

Fueled by government policies, subsidies, and private funding, the program launch was an initial success. However, its sustainability has been quite challenging. Some of

BOX 2.13 ECO-INDUSTRIAL PARK IN HONG KONG

In Hong Kong, the Environment Protection Department set-up an EcoPark that was tailor-made in 2007 for the recycling businesses. At the EcoPark, 13 plots were created for recycling of waste cooking oil, metals, wood, electrical and electronic equipment, plastics, batteries, construction and demolition waste, glass, rubber tyres, food and PCBs. Figure 2.12 shows the situation of EcoPark between Jan 2017 to April 2017.

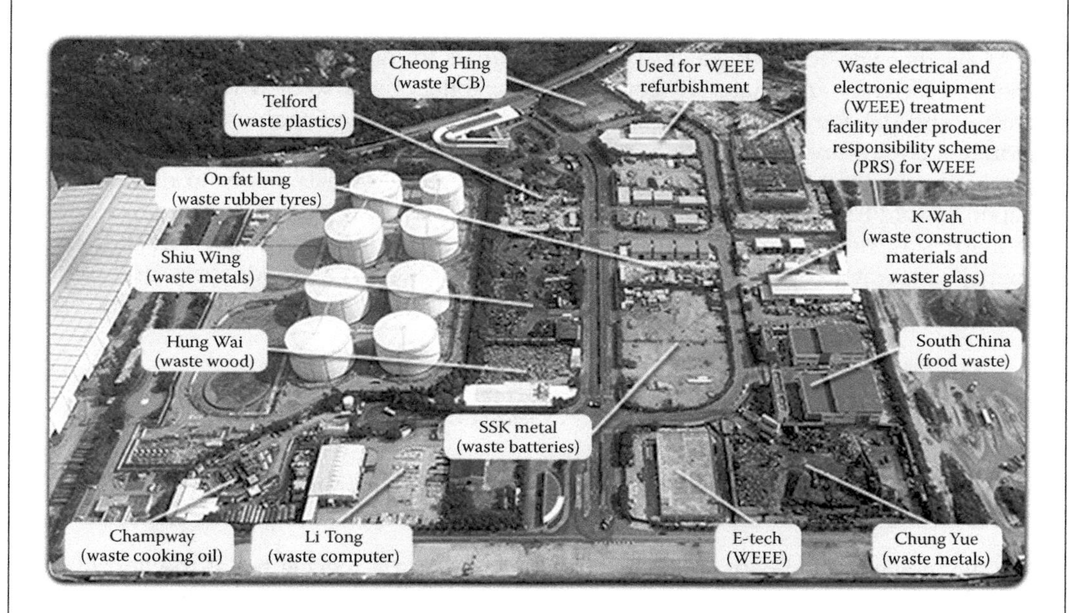

FIGURE 2.12
Configuration of Hong Kong's EcoPark for Recycling (as of April 2017) (From Environmental Protection Department, The Government of the Hong Kong Special Administrative Region, Urban Property Management Limited. With permission, online Source: http://www.ecopark.com.hk/en/about.aspx.)

This government run EcoPark helped its tenants with set-up, commissioning and operations and encourages adoption of advanced technologies and recycling processes.167 And tenants of EcoPark are subject to change upon termination or expiry of their leases.

DISCUSSION QUESTIONS

- *Do you think that setting up Eco-Parks for recycling on an exclusive basis is a good idea?*
- *Do you see recycling units established for every industrial estate/park as a part of the infrastructure? Such units are also called Material Recovery Facilities (MRFs).*
- *Research on MRFs at industrial estates and compare the pros and cons with Eco-industrial Parks that are set up for recycling on an exclusive basis. Give examples.*

BOX 2.14 ECO-TOWNS: CASE OF INDUSTRIAL AND URBAN SYMBIOSIS IN THE RECYCLING SECTOR IN JAPAN[62]

Kitakyushu Eco-town has the most extensive range of recycling and environmental industries in Japan. Some of the initiatives undertaken included increasing public awareness to improve collection of raw material (waste), use of high-quality material for new equipment production and diversion of low-grade material as fuel in the cement and steel industries. Innovations were encouraged to reduce equipment disassembly time. For this, industrial collaborations and academic research were key.

At another Eco-Town in Kawasaki, 15 companies formed an association to improve energy efficiency and material recycling. They collaborated to deal with waste, both within and outside the association, developed individual environmental policies and stringent environmental targets. Some of the interesting projects included, use of waste plastics as fuel, as construction material, hard to recycle waste paper to make toilet paper, use of PET bottles to produce new PET bottles, use of waste material from cement industries as fuel, and recycling of home appliances and fluorescent tubes. In addition, energy and water conservation projects were undertaken by the industries, and these projects were promoted amongst citizens and schools through outreach programs.

 **DISCUSSION QUESTIONS**

- *Discuss the role of the Academic and Research City and Practical Research Area in the Kitakyushu Eco-town model.*
- *What is the difference between Eco-Towns and Eco-Industrial Parks?*

the challenges included consumer's resistance to paying a recycle fee at the end of their product's useful life, quality and supply of raw materials, and adverse impact of the collaborative approach on innovation.

Despite the challenges, the Eco-Town initiative highlights the importance of:

- Collaboration between government, industries and society
- Need for a policy framework which includes legislation and economic incentives
- Academic or research activities for technological and process innovations in the environmental industry
- Need for focus on energy, water and material conservation as part of integrated waste management
- Engaging community and consumers is critical for success and sustainability of businesses

2.10 Common Environmental Infrastructure

Governments take up building of environmental infrastructure for the common interests of a targeted group of users. This is often referred to as Common Environmental Infrastructure (CEI). CEI is generally developed specifically for pollution management in an industrial area or industrial cluster (e.g., Common Effluent Treatment Plants or CETPs) or for collective

BOX 2.15 RAINWATER HARVESTING IN JAPAN

The practice of rainwater harvesting is thriving in Japan, both for public and private utilities. In Tokyo, at the community level, *Rojison*, a rainwater harvesting utilization facility, has been set up by the local residents in the Mukojima district. The rainwater collected from the Rojison is then used by private homes for watering gardens, firefighting, and as drinking water during emergencies.[63]

In 2014, the Act to Advance the Utilization of Rainwater was passed, under which municipalities are required to take maximum efforts to set and work towards rainwater utilization targets. Moreover, the national government provided financial support through subsidy programs. In 2011, 179 out of 209 municipalities were providing subsidies for rainwater storage. Residents who wanted to install rainwater tanks with a capacity of less than 1,000 liters received a subsidy for half of the cost of the tank, which includes installation costs from the city.[64] The Japanese organization, People for Rainwater, provides information and guidance on how to build your own rainwater tanks, with easily available products such as plastic buckets.

DISCUSSION QUESTIONS

- *In India, Bangalore Water Supply and Sewerage Board (BWSSB) set up Sir M. Vishweshwarayya's "Rain Water Harvesting Theme Park" on 1.2 acres of land. This park demonstrates 26 different types of Rain Water Harvesting models along with the water conservation tips at inside and outside the main building. Write a note on this Park and look for similar such examples across the world.*
- *Visit http://bwssb.gov.in/bwssbuat/content/rain-water-harvesting-0 for more details.*
- *Read the working paper prepared by K. S. Umamani and S. Manasi on the Rainwater Harvesting Initiative in Bangalore City: Problems and Prospects (see http://www.isec.ac.in/WP%20302%20-%20Umamani%20and%20Manasi.pdf).*

management of residues (e.g., common solid waste and biomedical waste management facilities). CEIs also include recovery/reuse of waste lead, metals (such as chrome from tanneries and metals from electroplating industries) and used oil. More recently, waste-to-energy plants and Combined Heat and Power (CHP) facilities also require the establishment of CEI.

Recognizing the need for a collective pollution control and treatment mechanism, the central and state governments have introduced a number of subsidies for pollution control equipment and treatment installations. In India, a scheme on Common Effluent Treatment Plants (CETP) operates to build as well as upgrade the common effluent management facilities. The CETP scheme in India focuses on clusters of small-scale industries as these industrial units do not have the financial resources and technical capacities to install individual Effluent Treatment Plants (ETPs). The central and state governments subsidize 25% of the total CETP development costs while 30% of the required funds are secured through loans from financial institutions, and the remaining 20% is contributed by the participating industries on the principle of *polluter pays* (Figure 2.13). There are nearly 200 CETPs operating in India today under such as the Credit Linked Capital Subsidy Scheme. The concept of CETPs is now followed in Sri Lanka, Bangladesh, and Vietnam.

The CETPs are gradually evolving to more sophisticated reuse and recovery systems (refer to Figure 2.14) where the effluents are recycled to the industries or recoveries are

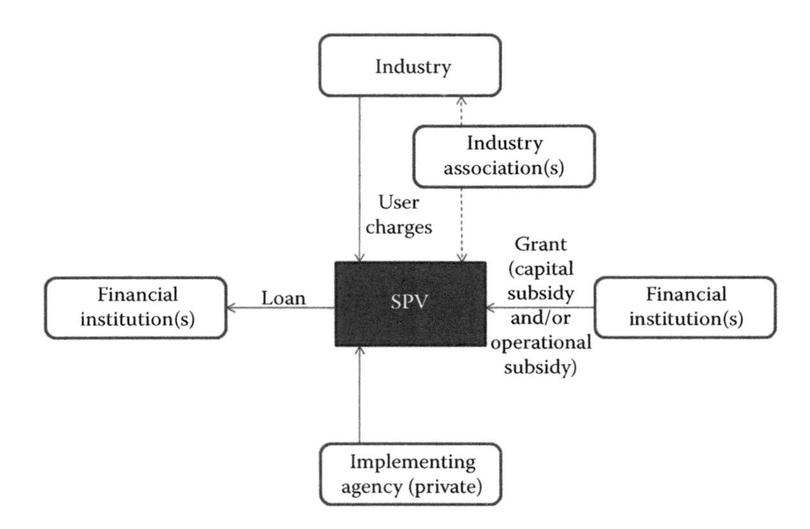

FIGURE 2.13
A typical Institutional Framework for CEI.

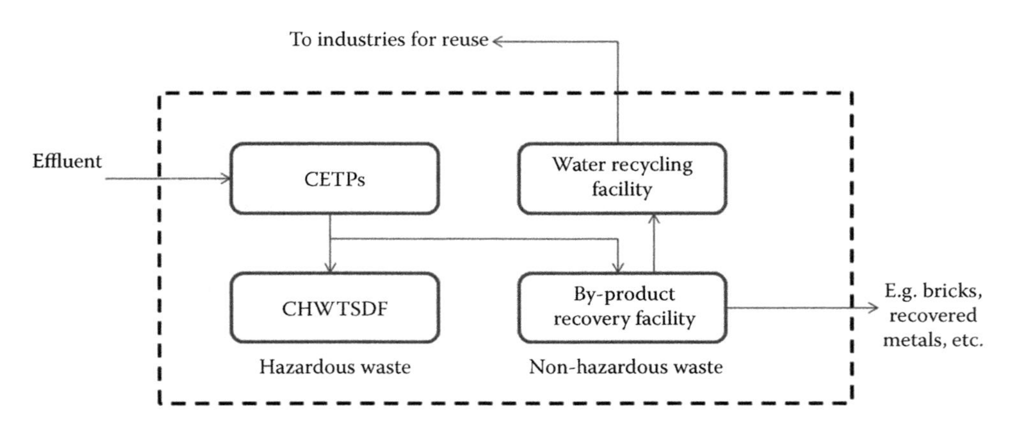

FIGURE 2.14
Gradual evolution of CETPs from stand-alone to more sophisticated reuse and recovery systems.

made of valuable chemicals. CETPs are now part of a more holistic treatment-recovery-reuse solution comprised of add-ons such as a Common Hazardous Waste Treatment, Storage and Disposal Facility, a By-Product Recovery Facility (for non-hazardous wastes), and a Water Recycling Facility (for water reuse in industries). The concept of recovery systems can also address harvesting of rain water especially for the interest of communities. Box 2.15 illustrates Rain Water Harvesting in Japan on this bases.

2.11 Laws & Regulations

Laws are legal instruments framed to execute objectives and goals stated in the Constitution and Government Policy. Laws set out procedures, standards and relevant documentation that may be necessary to comply with environmental policies. If a law is not followed, then those responsible for non-compliance can be charged or penalized and prosecuted in court. Business operations may be discontinued, supply of water or energy withheld, and products can be banned from markets. In some countries, if the regulator or the enforcer of law does not take action in time, then the stakeholders or the affected entities can move to special courts and seek justice.

Laws to control pollution and limit use of natural resources have long existed. Human-caused environmental disasters like the 1969 Santa Barbara Oil Spill, the Bhopal Gas Tragedy in India in 1984, and the Exxon Valdez oil spill in Alaska in 1989 prompted countries to sit up, formulate, and enforce what we call *Environmental Laws*.

Figure 2.15 and Table 2.1 depict how the governments of Japan and Bangladesh are following a progressive approach to widen as well as tighten waste related regulation guided by the national policies.

Such a progressive approach is recommended as it gives opportunities to assess the effectiveness in the interim and accordingly take appropriate corrective measures. Further, in

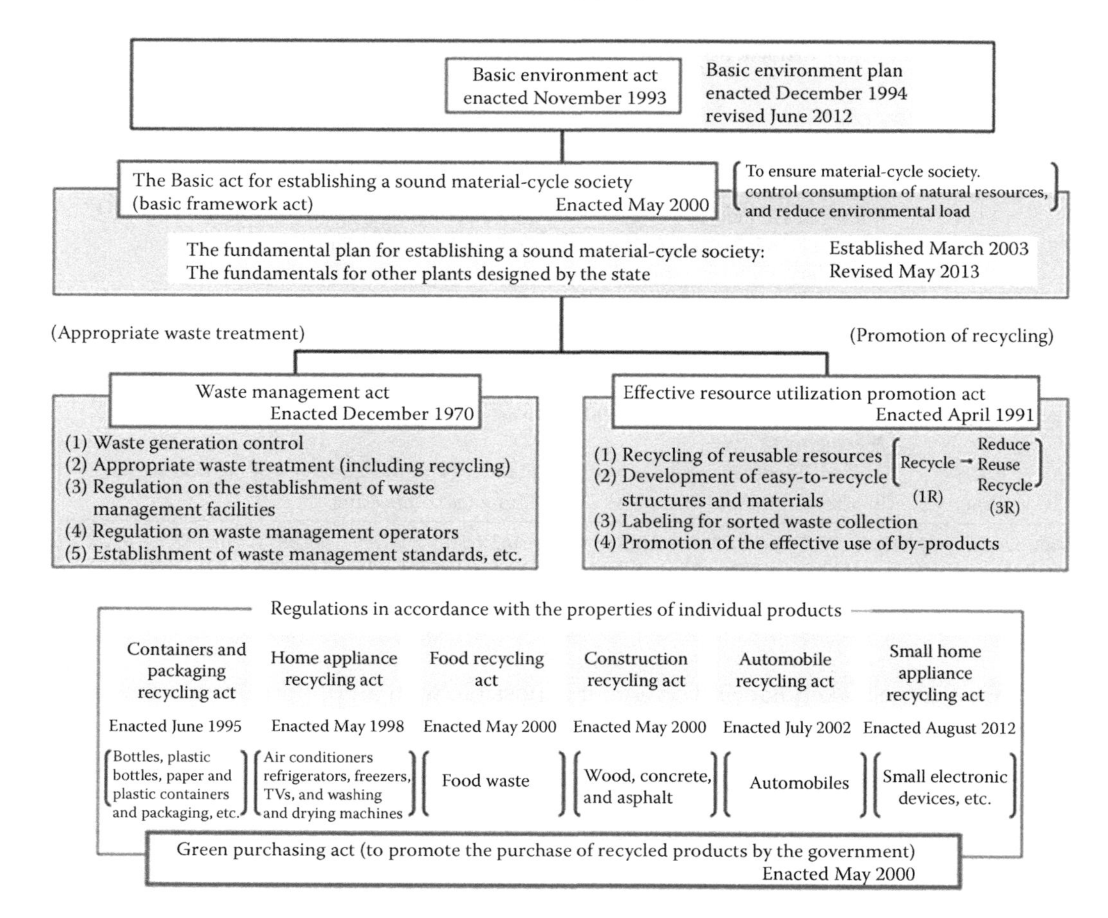

FIGURE 2.15
History waste management related legislation in Japan. (From Waste Management and Recycling Department Policy Planning Division, Office of Sound Material-Cycle Society, History and Current State of Waste Management in Japan, Ministry of the Environment, p. 17, 2014. With permission, online source: https://www.env.go.jp/en/recycle/smcs/attach/hcswm.pdf.)

TABLE 2.1

Development of a National Policy Framework in Bangladesh to Support Sustainable Waste Management[a]

Date	Type	Name	Summary of Issues Relevant to Waste
1995	Action Plan	National Environmental Management Action Plan	Waste Management Hierarchy promoted
1998	Policy	National Policy for Water and Sanitation	Recycle waste; organic waste to be used for compost and biogas.
1998	Policy	Urban Management Policy Statement	Privatization of services and for slum dwellers to get sanitation and solid waste disposal.
2004	Other	Dhaka Declaration on Waste Management by South Asian Association of Regional Cooperation countries	Countries agree to encourage NGOs and private companies to establish community-based composting, segregation of waste at source, separate collection and resource recovery from wastes, especially focusing on composting.
2005	Rules	Draft National Solid Waste Management Handling Rule	Waste management hierarchy incorporated.
2005	Strategy	Poverty Reduction Strategy Paper	Environmental management systems promoted with focus on waste segregation at source along with the waste management hierarchy.
2005	Strategy	National Sanitation Strategy	Achieve 100% sanitation coverage by 2010 with an emphasis on recovery and recycling.
2005	Action Plan	Dhaka Environmental Management Plan	Recycling promoted as are environmental management systems for industry.
2005	Action Plan	Solid Waste Action Plan for Eight Secondary Towns in Bangladesh	Promotes the waste management hierarchy.
2006	Policy	Draft National Urban Policy	Clean Development Mechanism and recycling emphasized.
2006	Act	Fertilizer Act	Promotes composting and subsequently standards set.
2006	Rules	Lead-Acid Battery Recycling and Management Rules	Waste management hierarchy incorporated.
2008	Rules	Medical Waste Management Rules	Standards for measurement of medical waste.
2008	Other	Circular to Promote Compost	Promotes composting.

[a] Adapted from National 3R Strategy Development, *National 3R Strategy Development: A progress report on seven countries in Asia from 2005 to 2009*. United Nations Centre for Regional Development, United Nations Environmental Programme/Regional Resource Centre in Asia and the Pacific and Institute for Global Environmental Strategies.

this process more stakeholder involvement is possible with an opportunity to strengthen the institutions responsible for enforcement.

It may be worth comparing the case of Japan and that of Bangladesh regarding management of wastes and research on the outcomes as well as the challenges faced.

2.12 Environmental Impact Assessment

Environmental Impact Assessment (EIA) is a requirement that came about due to the thinking of the *precautionary principle* and *do no harm* to the environment. EIA was conceived as a proactive tool to ensure that the projects in their construction and operational activities produced the least negative impacts possible, and that the residual impacts were communicated to the stakeholders and mitigated by appropriate environmental management plans.

EIA helps to take measures to mitigate the anticipated impacts/risks by improving the concept, design, implementation and operation during the project life cycle. Impact represents a Change (Positive, Neutral or Zero, Negative). It is generally understood by its extent and magnitude or the severity. Risk is a probability of impact indicating its likely magnitude considering scenarios.

Conducting an EIA adds value, improves developmental effectiveness, sustainability, and secures acceptance of the project by the stakeholders through the processes of consultation and participation. EIA brings in a transparency in the governance. From the regulatory as well as investment perspective, EIA is, thus, an important decision-making tool.

The first country which promulgated a legislation on EIA on a national scale was the United States. The National Environmental Protection Act (NEPA) laid the foundation of the EIA process requiring certain projects to undergo an environmental examination. The NEPA also stated the reporting requirements for obtaining approval of the responsible administrative body.

The initiation of NEPA in the United States led to ripples in other countries notably in Canada, Australia, the Netherlands, and the United Kingdom. Each of these countries, based on experience in the United States and priorities and context of their own countries, came up with EIA related policies and legislation.

The EIA process at project level largely involved the following eight steps:

- Screening (does the project require EIA?)
- Scoping (what issues and impacts should the EIA address?)
- Baseline studies (establish the environmental baseline)
- Alternatives (consider the different approaches) and arriving at Preferred Option
- Impact prediction and Assessment (forecast the environmental impacts)
- Preparation of Environmental Management Plan (EMP), Disaster Management Plan (DMP) and Resettlement & Rehabilitation Plan (RRP) as appropriate
- Monitoring and Evaluation (M&E)
- Resource Mobilization and Institutional Capacity Building to implement EMP/DMP/RRP and M&E

Most countries followed the criteria of project screening based on project type, size, and location. Only certain types of projects required EIA and most did not including schools, parks, renewable energy projects like solar, wind, and even metros (!) on the argument that these projects are intrinsically environment friendly. Projects required deeper or rigorous examination if the project was proximal to a sensitive location such as a reserved forest or a sanctuary or a turtle breeding ground. Many countries, therefore, delineated environmentally critical or sensitive zones and the buffer distances that should be maintained. Depending on the sensitivity and complexity of the project, EIA was categorized and conducted at two levels—Rapid EIA or Initial Environmental Examination (IEE) and Comprehensive or Detailed EIA.

Deciding the threshold on project size is however a complex exercise and rather a subjective matter. In Hawaii for example, hotels having less than 100 rooms were not required to perform EIA. Consequently, after the legislation, most hotels were built with 99 rooms. Thresholds on requirement of EIA have always been abused.

In the last three decades, and especially so in the developing countries, EIA has been followed on the project level, and the instrument is primarily used to identify mitigation plans prior to the approvals. Experience has shown that limiting EIA to projects alone is not going to help in the interest of sustainability, and consideration will need to be given to regional level impacts where multiple projects contribute to impacts cumulatively. In other words, EIA must look at the regional situation and development scenarios to recognize the indirect/induced and cumulative impacts and ensure that the carrying capacity of the region is not unduly exceeded.

Figures 2.16 and 2.17 show the importance of introducing EIA to higher levels as appropriate to the categories to address impacts/risks in a comprehensive manner and ensuring development effectiveness over a long run. Category A projects have higher sensitivity to the environment, followed by categories B and C. It may be observed that the category C projects could be handled through best practices to contain the direct impacts. Category B projects require both project specific mitigations as well as consideration of the cumulative and indirect impacts. These impacts may be best addressed through cumulative, regional, and programmatic EIAs depending on the scale and spatial extent of the projects. Category A projects require both project specific and programmatic approaches with consideration of cumulative and regional scale impacts. In some category A projects, where there is a basket of projects with details not known a priori, Strategic Environmental Assessment (SEA) may be relevant especially to address the induced impacts. Here, SEA provides an overarching framework to guide the project level EIAs and in this process, ensure the development effectiveness.

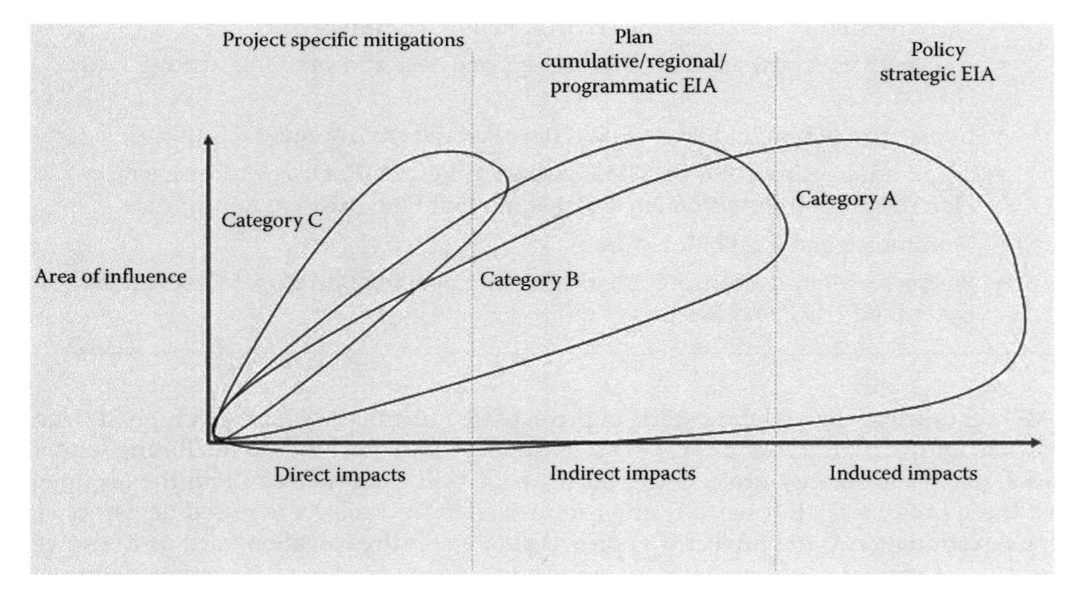

FIGURE 2.16
Relationships between EIA categories, impacts and levels/types of EIA.

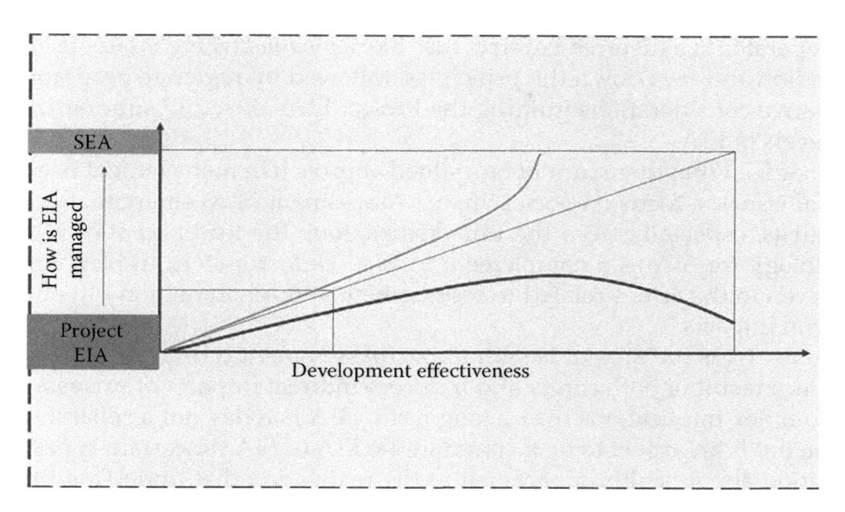

FIGURE 2.17
Project EIAs when guided by SEA ensure development effectiveness over long run.

It is not surprising, therefore, that many countries enhanced their EIA legislation to address requirements of regional and cumulative EIAs especially when dealing with area wide development projects. Examples are industrial estates/parks, network of transport corridors, urban and peri-urban development, etc. Here, apart from mitigation plans, planning as well as policy related measures had to be evolved. Box 2.16 describes SEA.

BOX 2.16 STRATEGIC ENVIRONMENTAL ASSESSMENT

Strategic environmental assessment (SEA) addresses environmental, social and possibly other sustainability aspects while formulating or modifying policies, plans and programs. SEA is a structured and participative process of multiple stakeholders operating at multiple levels.

At a sector level, SEA has been applied, for example, for transport, irrigation and agricultural sectors. Application of SEA has helped in introducing sector wide reforms interfacing with environmental and social considerations. Another variant that emerged in this process is the programmatic EIA that is applicable to projects at a *program level* or when projects are replicated on a regional scale. While the project by itself may not require EIA or an Environmental Clearance, the same project when replicated or up scaled can have potential to cause adverse impacts on a regional scale. Programmatic EIA brought out the idea of developing mitigation plans in the form of best practices. These best practices get then embedded in the very project design.

 **DISCUSSION QUESTION**

- *If Environmental and Social Management Plans (ESMP) are the outcomes of Project limited EIAs, then what is the outcome of a SEA? Compile examples that illustrate such outcomes underscoring benefits of the SEA approach.*

EIA thus operates in a hierarchical structure like a *pyramid* where Strategic EIA gives the overall direction and lays down the principles, followed by regional/programmatic EIAs with cumulative considerations guiding the Project EIAs. Box 2.17 summarizes different types and levels of EIA.

As it was realized that there cannot be a siloed approach in management of environmental and social issues, a focus on Social Impact Assessment (SIA) emerged. Social development specialists, especially from the universities, took the lead and started working on the methodology for SIA as a complement to EIA. Unfortunately, in SIA, emphasis was primarily given to the issues related to resettlement and rehabilitation rather than on the induced social impacts.

Very close to SIA is the area of Health Impact Assessment (HIA). Impact on health was considered as a result of both direct and induced/indirect impacts of unsustainable practices with complex implications over a long term. HIA is today not a separate mandatory requirement, but is an aspect to be factored in the EIA or SIA for certain types of projects.

Consideration about health impacts led to the realization that understanding of health impacts is rather complex to track the path from pollution release, contamination of resources and their consumption leading to phenomenon such as bio-accumulation. There are several uncertainties about such an assessment and our understanding is still evolving. This led to the discipline of Environmental Risk Assessment (ERA) focusing on the nexus between health and the eco-systems.

The next wave after ERA was to focus on biodiversity. Biodiversity Impact Assessment is today a field by itself where the EIA process is followed and adapted, for example,

BOX 2.17 DIFFERENT TYPES AND LEVELS OF ENVIRONMENTAL IMPACT ASSESSMENT[65]

- *Strategic EIA*: Strategic EIA refers to systematic analysis of the environmental impacts of development policies, plans, programs and other proposed strategic actions at a broader level.
- *Regional EIA*: Regional EIA addresses the environmental impacts of regional development plans. It integrates environmental concerns into development planning for a geographic region, typically at the sub-country level. For this, the cumulative effects of all projects within a region should be taken into account.
- *Sectoral EIA*: Sectoral EIA is conducted in the context of both regional or sectoral level planning. Once EIA is carried out at a sector level, project-level EIA requirements get guided.
- *Project level EIA*: Project level EIA refers to the specific developmental activity and assesses the impacts/risks that it exerts on the receiving environment to come up with an appropriate environmental and social management plan.

DISCUSSION QUESTIONS

- *Are there any overlaps between the above four types of EIA? Draw a Venn diagram to show the linkages as well as the overlaps.*
- *Are there any examples where at least three out of the four types of EIAs are applied for a development project/program?*

Environmental Management Plan becomes the Biodiversity Protection and Conservation Plan.

Given the expanse of EIA from project level to policy and planning levels, integration with social development, and health and safety related aspects, the next logical step was to wrap all these perspectives into one crucible, that is, Sustainability.

Application of sustainability frameworks at the national level EIA has been a major extension of EIA in the last decade. The idea was not just management of risks but to identify and leverage opportunities while staying within the *limits of growth*. Sustainability integration or sustainability based appraisal provided EA a new dimension and a role beyond permitting or having an Environmental Clearance. Sustainability Appraisal (SA) essentially integrated economic, environmental, and social considerations. Integrated Assessment (IA) is synonymous with Sustainability Appraisal.

SA has been widely practiced for planning local area development in the United Kingdom, and it has been made mandatory that such appraisals are carried out every year and reported. Few private sector equity investors have also adopted sustainability appraisal frameworks. Some of the DFIs such as the World Bank, IFC, and DFID have already set up sustainability based/driven environmental and social assessment requirements.

We must remember that EIA is essentially a generic tool that links activities with environmental components. It, therefore, has a place in the establishment of an Environmental Management System (EMS) of ISO 14001. In these systems, EIA is used to analyze project activities, associated aspects, and their influence on the environmental component so as to check whether the impacts are in compliance or whether they pose a threat or risk to human health and eco-systems. We introduce the key elements of ISO 14001 in Chapter 4. Apart from record keeping and actioning, ISO 14001 can be used to cast the Environmental Management Plans (EMPs) to ensure continual improvement as well as maintaining a focus on the prevention.

As the understanding on the environmental impacts of making, packaging, distributing, and servicing products increased, the tool of Life Cycle Assessment (LCA) emerged. LCA uses the core principles of EIA to predict, assess, and manage the adverse impacts, influencing thereby the product design, material sourcing, and product use and management of rejects/residues. The framework of EIA can also be used for the sustainability assessment of technologies (SAT). SAT considers economic, environmental, and social perspectives across the life cycle incorporating stakeholder consultation. We will discuss the key elements of LCA and SAT in Chapter 4.

EIA thus provided a generic framework to address manufacturing systems and services following once again the *precautionary* and *do no harm* principles. This power of EIA is not recognized by most as the understanding of EIA is limited simply to obtain Environmental Clearance.

More recently, the dimension of climate change has been added. In specific, the Asian Development Bank (ADB) has come up with climate proofing and its integration in the environmental assessment with several sectoral guidelines.

2.13 Integrating Climate Change Considerations in Environmental Impact Assessment

Planners and regulators today are not addressing the impacts of Climate Change (CC) adequately in future development plans. This is not just the case with the developing countries but with the developed countries as well. There are very few examples available where you see CC is reflected in policies, plans, and project designs and especially in the process of EIA.

The level of progress in integrating CC considerations in EIA varies considerably. Countries like the Netherlands, Canada, and Australia have been the pioneers in implementing incorporation of CC in EIA. While the Netherlands includes CC through a Strategic Environmental Assessment (SEA), Canada and Australia have taken the route towards CC integration through the project level EIAs. The European Commission, in its directive on the assessment of the effects of certain public and private projects on the environment, aims to reflect CC-related concerns.

Although CC related concerns and understanding are growing, incorporation of CC in the EIA process has not seen an acceptance as expected. Project developers in countries like Canada, a pioneer in this area, believe that not much climate related information is available to analyze the impacts of climate change on the projects. Besides, data availability and expertise on CC modeling is still an issue.

Figure 2.18 shows possible integration of CC related considerations specific to adaption across the project cycle. It may be observed that integration of CC helps in reducing the adverse impacts as well as risks or vulnerabilities. This example can be easily expanded to include aspects on the mitigation, especially in the development of environmental management plans, for example, use of clean technologies and clean fuels or maximum use of renewable energy.

The International Association of Impact Assessment surveyed the Australian CC-EIA system from the point of view of EIA practitioners. In all, 63 respondents were drawn

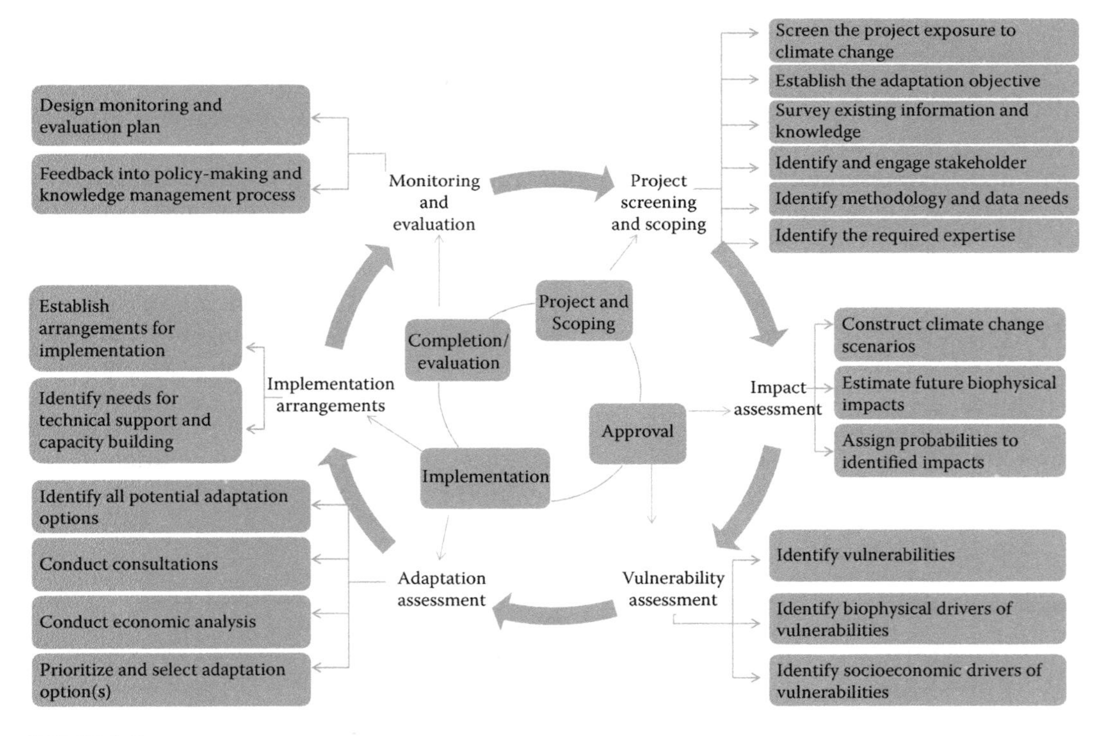

FIGURE 2.18
Possible integration of climate change related considerations focusing on adaption across the project cycle.

from across the country. It was found that majority practitioners believed that CC is highly relevant in EIA and SEA. In addition, the survey suggested that project EIAs cannot take the lead in incorporating CC EIA. CC considerations must start or originate from SEAs.

Major barriers to a project EIA being able to address CC were ranked as follows:

1. Lack of government policy and incentives to address CC in EIA
2. Lack of political and agency will to address climate change.
3. EIA scoping does not address CC, that is, which projects need to address CC
4. Lack of expertise and appropriate EIA tools

Let us understand the complexity of the issue.

CC considerations in EIA typically result in mitigation and adaptation plans. The adaptation plans need to be developed at a regional level, often beyond the boundaries of an individual project. For designing and implementing adaptation related plans, a simultaneous consideration for multiple projects is required to assess the cumulative impacts over the region. Public consultations need to be used as an important milestone to link the SEA, Regional EIA, and Project level EIAs.

The entry point for developing an adaptation plan is thus at a strategic level, where tools such as Regional EIA (REIA), SEA and Cumulative Impact Assessment (CIA) need to be used.

Unfortunately, in most nations in the world these three extensions of Project EIA have not been legislated. If we want to address CC in EIA, then we will require a major reform in the EIA system.

The mitigation plans, on the other hand, are generally project-limited and influence the project design and operations. Here, aspects such as energy mix, water use and conservation, afforestation, and erosion control need to be examined. Many of these aspects get addressed in the preparation of a Project focused Environmental Management Plan (EMP).

To address abnormal and emergent situations, however, the Project level EMPs need to be accompanied by the Disaster Management Plan (DMP). Once CC considerations are included, adaptation and mitigation elements get factored and the DMP assumes a form of a Disaster Risk Reduction Plan (DRRP). This DRRP needs to address both onsite and offsite risks.

Management of onsite risks becomes a part of the Project EIA, while the management of offsite risks needs to be integrated with regional DRRP. Both EMP and DRRP need to abide by the framework of the REIA and SEA with clear institutional and cost sharing arrangements. Again, DRRP needs to be synchronized with the adaptation related plans at the regional level—especially on matters related to policy, plans, and supporting commonly shared infrastructure. Relations between SEA/REIA, Project level EIA, EMP and DRRP in the context of CC integration are shown in Figure 2.19.

Project EIAs are generally processed by state and national level environmental regulatory authorities. Separate departments/ministries operate for management of disaster related risks. Often, there are no linkages occurring between these institutions. SEA with a focus on CC may be used to ensure mainstreaming of CC in the project and regional EIAs and more critically to ensure coordination between key institutions and the project developer/sponsor.

Key stakeholders in the CC integration will be National/Regional Planning agencies, Environmental and Disaster Management Agencies, and the Project Proponent. Table 2.2 lists roles and responsibilities of key stakeholder institutions in the conduct of SEA, Regional EIA, and Project EIAs.

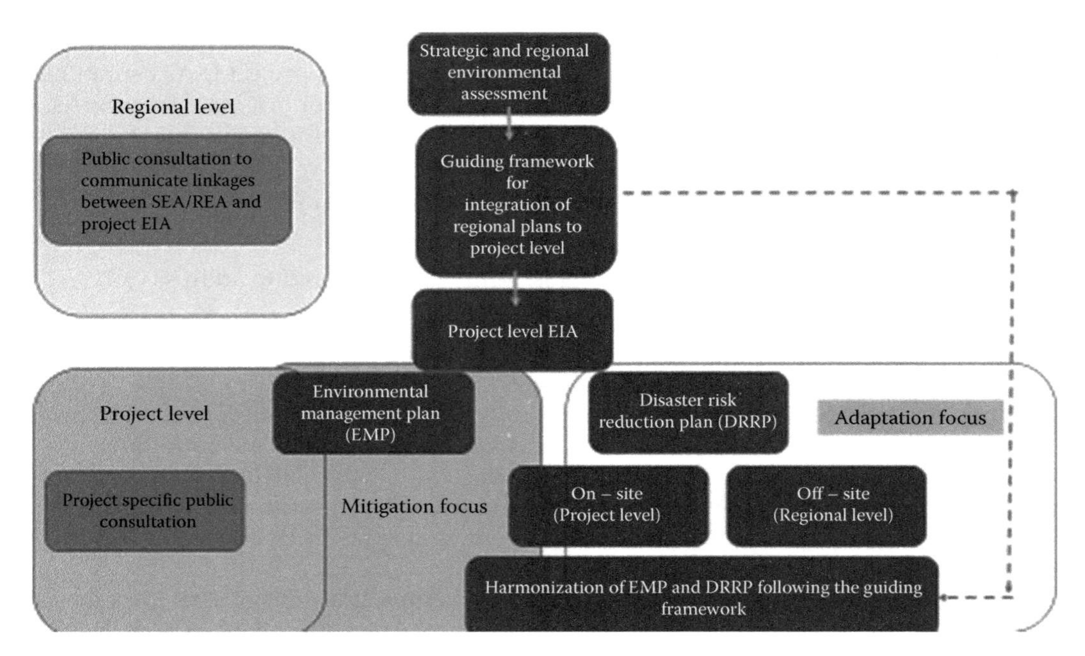

FIGURE 2.19
Integration of CC consideration in EIA.

Many countries have set up CC cells. These cells need to undertake required coordination supported by CC related research organization that holds the required databases and expertise on CC related modelling. Figure 2.20 shows the role of various institutions for coordination.

REA and SEA clearly assume an important role to ensure harmonization between Project level EMP and DRRP with the CC adaptation plans at the regional level. Cumulative Impact Assessment (CIA) will remain the key. Institutional coordination with cost sharing will be important in the implementation of the CC related recommendations. Involvement of the stakeholders is necessary to appreciate the concerns of the CC, especially its economic, social, and environmental implications. Capacity building of the planners, regulators and professionals is also required. Finally, pilots should be implemented to demonstrate how CC in EIA could be mainstreamed. Based on the experience of the pilot, the EIA related legislation may be suitably amended. We will need to develop screening and categorization to define which projects or regions will need CC considerations based on the vulnerability atlas, type, and scale of projects development.

2.14 Environmental Standards, Progression and Expanse

Environmental standards became part of environmental policies and laws to provide quantitative guidelines and set standards for emissions and ambient as well as work space environment. Environmental standards served as guidelines and tools for organizations to manage their processes and environmental responsibilities to ensure compliance.

TABLE 2.2

Roles and Responsibilities of Key Stakeholder Institutions

Activity	Planning Institutions Engaged with Development and Development Controls	Environmental and Allied Regulators Involved in Environmental Clearance	Project Proponent
SEA/REIA			
Baseline data of climate parameters like rainfall, temperature, Hydrological maps, infrastructure mapping, natural resource maps	◆	◆	
Future projections of climate at regional level	◆	◆	
Probable CC related impacts/risks at regional level	◆	◆	
Strategic/Regional Environmental assessment incorporating CC	◆		
Consultation with authorities and stakeholders	◆	◆	◆
Development of Guiding Framework and Operational Principles for Integration of adaptation and mitigation in the development plans and policies	◆	◆	
Prepare response mechanism plans for disaster risk reduction at regional level	◆	◆	
Monitoring effectiveness of the plan in terms of mitigation and adaptation	◆	◆	
Project EIA			
Impact of climate change on project/program			◆
Mitigation measures			◆
EMP			◆
DRRP	◆		
EMP, DRRP Integration with outcomes of REA/SEA	◆	◆	◆
Stakeholder consultation	◆	◆	◆

Setting of environmental standards has been a scientific as well as a consultative process. Protection of health and the ecosystems has remained the focus. As a next step, technical feasibility and economic viability were considered for setting minimum national standards. Figure 2.21 shows the standards set at various stages of operations in the pulp and paper production process over a period of time.

In the early stages, the Paper and Pulp industry was asked to meet effluent standards such as for Biochemical Oxygen Demand (BOD). When the effluents were treated to meet this objective, it led to generation of sludges at the effluent treatment plants that had to be regulated. The next level of intervention demanded changes in the processes and chemicals such as discouraging use of chlorine in bleaching due to the formation of AOX (Adsorbable Organic Halogens). As resources started becoming scare, norms were set up for consumption of water, putting a cap.

As understanding of the life cycle impacts improved, paper and paper products became the next point of regulation. Many eco-labels emerged that asked for the eco-friendliness based on criteria on chemicals used, resources consumed, biodegradability, and recyclability. The standards and regulations thus got evolved encompassing the Life Cycle.

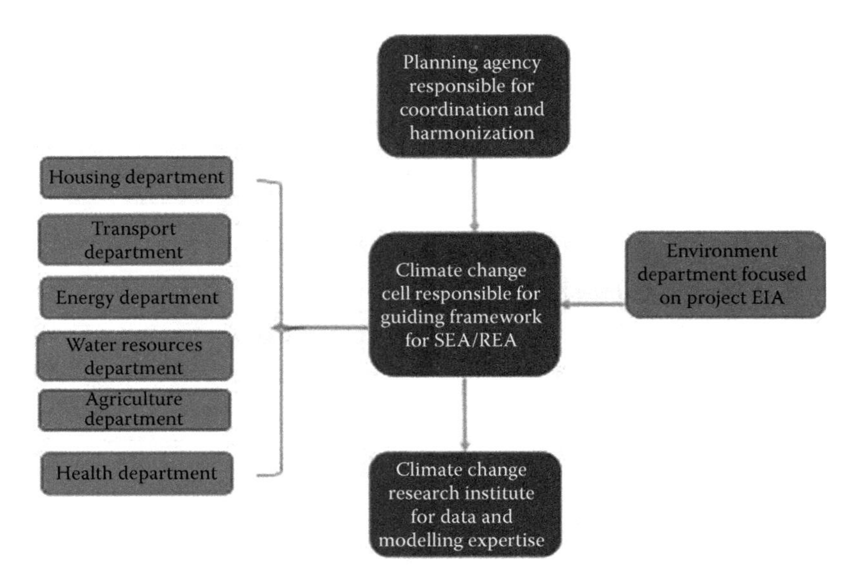

FIGURE 2.20
Institutional arrangements for mainstreaming CC considerations in EIA.

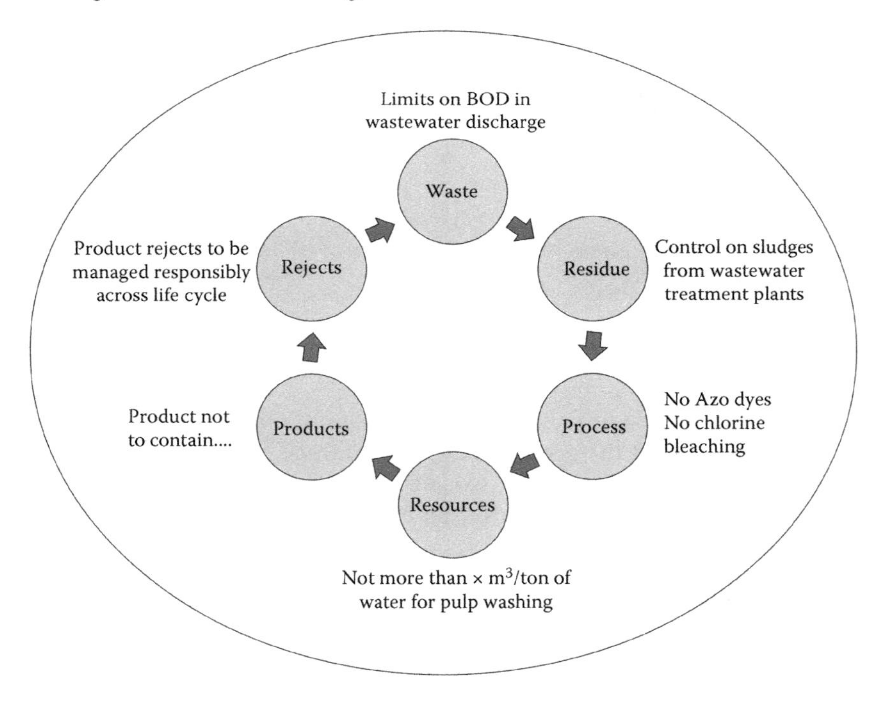

FIGURE 2.21
Evolution of environmental standards across life cycle—Illustration of Paper and Pulp Industry.

Standards on emissions of vehicles has been another example where targets are set in advance and the automobile industry keeps adapting/improving the technologies to minimize and control the vehicular emissions. We present in Boxes 2.18 and 2.19 examples of changing emission standards and ambient air quality standards, respectively.

BOX 2.18 EXAMPLES OF CHANGING ENVIRONMENTAL STANDARDS

EUROPEAN EMISSION STANDARDS OR EURO STANDARDS

Euro standards define the exhaust emission standards for automobiles. The Euro standards were introduced through a series of European Union Directives. Six standards were introduced between 1994 and 2014, wherein the standards for emissions of carbon monoxide (CO), particulate matter (PM), nitrogen oxide (NO_x), total hydrocarbon (THC), and non-methane hydrocarbons (NMHC) are set. The emission standards have become increasing difficult to comply with over the years. The standards were introduced for various vehicle types like heavy-duty vehicles, passenger cars and motorcycles and different fuel types like diesel and petrol. Figure 2.22 below shows the Euro standards I–VI have become stringent over the years for diesel operated passenger cars.[66]

Various countries have adopted the Euro Standards for road vehicles. The plot below shows the timeline of adoption of Euro Standards by Asian countries when compared to its adoption in EU (Figure 2.23).[67]

DISCUSSION QUESTION

- *Tightening of the emission standards requires innovations at the end for automobile manufacturers. One of the strategies to reduce emissions has been making the automobile lighter in weight by making use of plastic-based materials. At the end of life of the vehicles, it may be difficult to manage the plastic-based waste materials. Discuss how to address this dichotomy and how this challenge is addressed by some of the world's leading automobile manufacturers.*

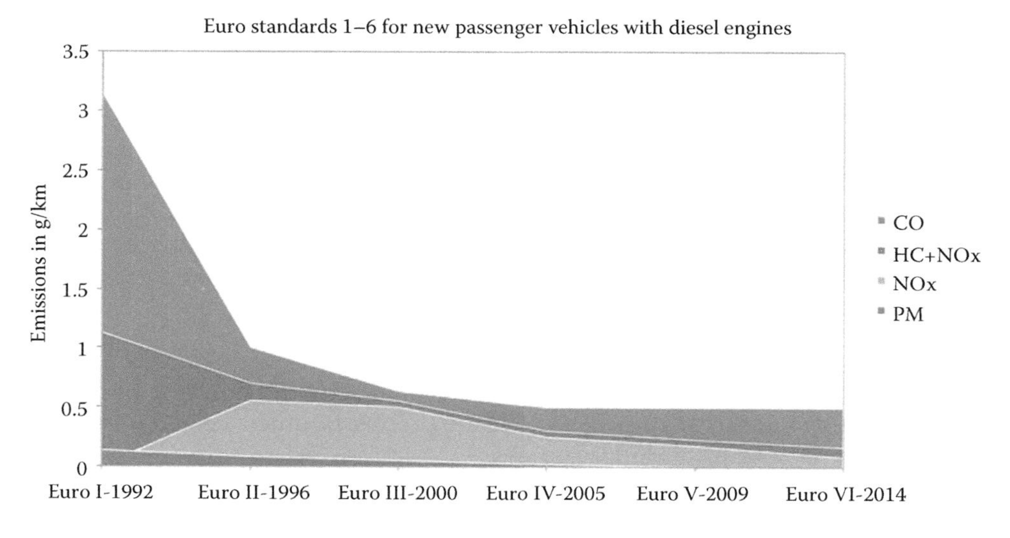

FIGURE 2.22
Euro standards 1–6 for diesel operated passenger vehicles. (Created using data from EU: Cars and Light Trucks, Emission Standards, DieselNet, 2016, online source: https://www.dieselnet.com/standards/eu/ld.php.)

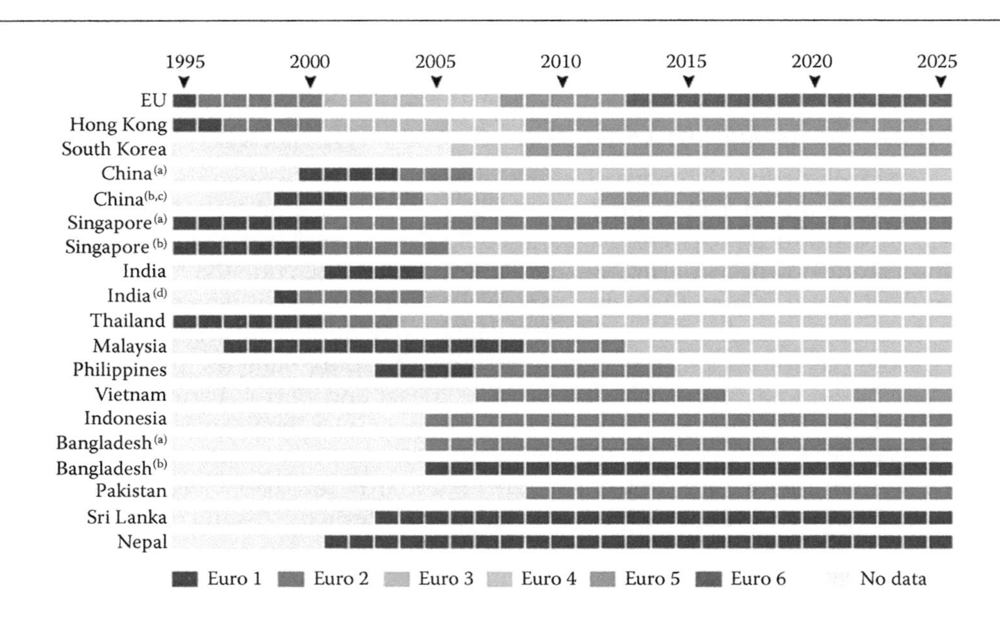

FIGURE 2.23
Adoption of EU EURO 1–6 standards for road vehicles in Asian countries. (Taken from: Adoption of the EU Euro Emissions standards for road vehicles in Asian Countries, European Environment Agency, 2016, online source: http://www.eea.europa.eu/data-and-maps/figures/number-of-international-environmental-agreements-adopted-1.) (a) For gasoline (petrol) vehicles (b) For diesel vehicles (c) For entire country (d) For the following cities—Delhi, Mumbai, Kolkata, Chennai, Hyderabad, Bangalore, Lucknow, Kanpur, Agra, Surat, Ahmedabad, Pune, and Sholapur.

BOX 2.19 NATIONAL ENVIRONMENTAL STANDARDS FOR AIR QUALITY, NEW ZEALAND, 2004[68]

The Ministry for the Environment of Canada issued the National Environmental Standards for Air Quality (NES) in 2004 to regulate the release of harmful pollutants. Released as part of the Resource Management Act 1991, the NES aimed at protecting the environment and providing the minimum level of health of all New Zealanders.

The pollutants regulated were carbon monoxide, sulfur dioxide, nitrogen dioxide, ozone, and particulate matter (PM) 10. The NES for air quality is comprised of 14 separate but interlinked standards.

These include:

- Seven standards for dioxins and toxics that ban activities emitting harmful pollutants to the air such as open burning of tires, bitumen, coated wire, oil, landfill fires, and hazardous wastes
- Five standards and codes to keep the ambient air quality safe and healthy
- One standard for the design of new wood burners installed in urban areas to curtail emissions of air. Emission limits in terms of fine particle emitted per kg of wood burnt and the minimum thermal efficiency rating required was specified.

- One standard for landfills with a capacity of over 1 million tons of refuse to collect greenhouse gas emissions and destroy it by flaring to avoid uncontrolled release into the environment

AIR QUALITY STANDARDS, CHINA, 1996 AND 2012[69,70]

The Ministry of Environmental Protection commenced regulating ambient air quality in China in 1982 and limits were set for emissions of Total Suspended Particulates (TSP), sulfur dioxide (SO_2), nitrogen dioxide (NO_2), lead (Pb), and benzo(a)pyrene (BaP). In 1996, the standards were expanded to include particulate matter 10 μm or less (PM_{10}), carbon monoxide (CO), ozone (O_3), and fluoride (F). The standards were amended again in 2000 and then made more stringent in 2012. In 2012, the standards were expanded to include particulate matter 2.5 μm or less ($PM_{2.5}$) and nitrogen oxide (NO_x). Table 2.3 includes a comparative assessment of

TABLE 2.3

Progression of Air Pollution Standards in China[a]

Pollutant	Collecting Time	Standards Set in 1996 Limit			Standards Set in 2012 Limit		Unit
		Class 1	*Class 2*	*Class 3*	*Class 1*	*Class 2*	
SO_2	Annual	20	60	100	20	60	μg/m³
	24 hours	50	150	250	50	150	
	Hourly	150	500	700	150	500	
NO_2	Annual	40	80	80	40	40	μg/m³
	24 hours	80	120	120	80	80	
	Hourly	120	240	240	200	200	
O_3	Daily, 8-hr max	–	–	–	100	160	μg/m³
	Hourly	160	200	200	160	200	
PM_{10}	Annual	40	100	150	40	70	μg/m³
	24 hours	50	150	250	50	150	
$PM_{2.5}$	Annual		–		15	35	μg/m³
	24 hours		–		35	75	
TSP	Annual	80	200	300	80	200	μg/m³
	24 hours	120	300	500	120	300	
Pb	Seasonal		1.5		1	1	μg/m³
	Annual		1		0.5	0.5	
Benzopyrene (BaP)	Annual		–		0.001	0.001	μg/m³
	24 hours		0.01		0.0025	0.0025	
NO_x	Annual		–		50	50	μg/m³
	24 hours		–		100	100	
	Hourly		–		250	250	
F	24 hours		7			–	
	Hourly		20				
	Monthly	1.8		3			
CO	24 hours	4	4	6	4	4	mg/m³
	Hourly	10	10	20	10	10	

[a] Based on China Air Quality Standards, TransportPolicy.net, 2014, online source: http://transportpolicy.net/index.php?title=China:_Air_Quality_Standards

the standards set in 1996 and in 2012. The Standards were defined for three classes (Class I, II and III) in 1996, which was reduced to two classes (Class I and II) in 2012 based on the geographical locations of targeted areas. Class 1 standards applied to special regions such as national parks; Class 2 standards applied to all other areas, including urban and industrial areas. Class 3 applied to special industrial areas in 1996 standards. The Ministry has a phase-wise implementation plan targeting key cities and capitals in the initial years followed by nation-wide implementation by 2016.

DISCUSSION QUESTION

- *Example of expansion of standards on air quality in New Zealand shows how a regulator develops the regulations to address key emissions as well as sensitive areas. The example of China shows how the government expands the parameters, tightens them, and makes the standards relevant to the type of land use. Examine the air quality related regulations in your country from these perspectives.*

Environmental Standards are continuously changing to keep up with new research and discoveries. The United States Clean Air Act of 1970 was the first legislation on controlling air pollution and setting limits to emissions from stationary and mobile sources like industries and vehicles, respectively. The Act has been continuously modified to include hazardous pollutants, air pollution problems from interstate transport of pollutants, acid deposition control, and industrial monitoring of volatile organic compounds. In a landmark case of Massachusetts vs US, EPA established the regulation of greenhouse gases including carbon dioxide, methane, nitrous oxide, and fluorinated gases under the United States Clean Air Act. In this case, Massachusetts and other states petitioned that EPA should regulate emissions of carbon dioxide and other gases that contribute to global warming (climate change) from new motor vehicles as they meet the definition of air pollutants.[71] Box 2.20 illustrates the CO_2 standard introduced in the United States for the new power plants.

BOX 2.20 CARBON DIOXIDE STANDARDS FOR NEW POWER PLANTS, US, 2015[72]

In 2015, US EPA set standards to limit CO_2 emissions from new, modified, and reconstructed power plants. The standards define separate emission limits for natural gas and coal fired plants. New natural gas plants have to meet an emission limit of 1,000 pounds (454 kgs) of carbon dioxde per megawatt-hour of electricity produced while a new coal fired plants has an emission limit of 1,400 pounds (635 kgs) of carbon dioxde per megawatt-hour of electricity produced. The difference in emission limits for both types of plants illustrates the technological considerations to be made while developing the standards.

To comply with these standards, natural gas plants need to employ an efficient generation technology like the combined cycle technology, which can help them operate at a thermal efficiency of 60%, as opposed to 30%–35% efficiency achievable through older technologies. Coal plants will have to explore and employ the carbon capture and storage technology wherein carbon dioxide emissions are captured, transported, and stored in underground geological formations. This technology makes it possible to capture 90% of the CO_2 emissions that makes possible an emission rate of 500 pounds (227 kgs) of carbon dioxde per megawatt-hour of electricity produced.

DISCUSSION QUESTION

- *Check whether you find any other country that is regulating the emissions of CO_2, especially on power plants. Do you think that this regulation will be effective?*

2.15 Managing Indoor Air Quality

Indoor air is considered safe by many. When outdoors, we encounter pollution from vehicles and industries and these emissions worry us. No wonder, the statistics reported on air quality are generally on outdoor air or ambient air.

We tend to believe that staying indoors shields us from outdoor air pollution. We feel safe in enclosed spaces like offices, homes, malls and theatres, air conditioned cars, buses, and railway coaches. Unfortunately, the situation is to the contrary. Indoor Air Quality (IAQ) can be just as bad as or even worse than ambient or outdoor air quality.

There are a number of reasons why IAQ can be of concern. Emissions emanate from cooking, cigarette smoking, burning of scented sticks (agarbattis) and mosquito repellent coils, paint on the wall, seals from furniture, coatings from fabrics and carpets that release complex pollutants. The pollutants released include particulates, formaldehyde (HCHO), radon, tolune, ammonia, benzene, 2-furaldehyde, benzyl alcohol, monocyclic monoterpenes, dichloromethane, ethylhexyl phthalate, just to name a few. Most of these *micropollutants* are carcinogenic and difficult, as well as expensive, to monitor. Many times, naturally ventilated buildings are designed poorly and buildings with forced ventilation or enclosed with air conditioners, do not maintain adequate ventilation or cycles of air exchange. As a result, the micropollutants that are released indoors get accumulated—sometimes overshooting the acceptable values.

Acceptable IAQ is defined by the American Society of Heating, Refrigeration and Air Conditioning Engineers (ASHRAE) in ANSI/ASHRAE standard 62.1 as: "Air in which there are no known contaminants at harmful concentrations and with which a substantial majority (usually 80%) of the people exposed do not express dissatisfaction."

Traditionally, IAQ has been associated with Sick Building Syndrome (SBS). The World Health Organization (WHO) compiled common reported symptoms into what was defined as SBS. These symptoms included: eye, nose, and throat irritation; sensation of dry mucous membranes; dry, itching, and red skin; headaches and mental fatigue; high frequency of airway infections and cough; hoarseness and wheezing; nausea and dizziness; and unspecific hypersensitivity. Most of us are facing these problems today and frequently so. We have to keep visiting doctors and chemists.

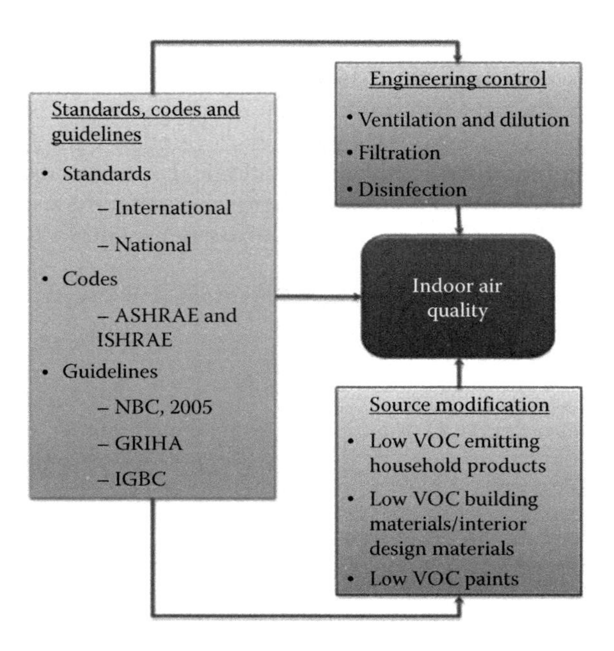

FIGURE 2.24
Need for a strategic approach to manage indoor air quality.

China set standards for IAQ as early as 1976. Korea has IAQ standards even for metro buses! IAQ standards in Japan and Germany cover most of the micro-pollutants listed above. India does not have an indoor air quality standard. A Society for Indoor Environment has just been formed in India.

But is just setting of IAQ standards going to be enough? How can we enforce such standards?

The approach will have to be multipronged, focusing on prevention and control. We will need to bring key stakeholders such as architects, paint manufacturers, furniture makers, HVAC manufacturers, green building rating agencies, medical professionals, etc., together. A combination of control, reduction (modification) with standards/guidelines may work. See Figure 2.24 developed in the context of India.

2.16 Phasing Out of Substances and Technologies

Technology, product and material innovations take place every day. Environmental research and monitoring studies across the globe reveal the impacts of the extensively used products, their components, and related emissions on the environment. This newly acquired knowledge equips governments to assess which products or materials should be taken out of the market or phased out following the precautionary and preventive principles. Box 2.21 shows an illustration of the 33/50 program in the United States.

Researchers F. Sherwood Rowland and Mario Molina discovered the adverse impacts of chlorofluorocarbons (CFCs) on the ozone layer in the upper atmosphere in 1973. Further research and monitoring of ozone at the stratosphere led to the ban of CFCs in the United States in 1978 followed by a ban in Canada, Norway, and Sweden. Finally, discussions

BOX 2.21 THE 33/50 PROGRAM OF PHASE-OUT IN THE UNITED STATES[73]

The 33/50 Program is one of US EPA's voluntary programs developed as part of US EPA's Pollution Prevention Strategy. The program aimed at reducing generation of chemical waste streams from the industrial sector by substitution with less harmful chemicals and/or modifying processes.

Launched in 1991, the program was designed to reduce releases of 17 high priority chemicals by 33% at the end of 1992 and by 50% at the end of 1995. The 17 chemicals accounted for 25% of the toxic industrial releases as monitored in the Toxic Releases Inventory (TRI) report. Industries were encouraged to use less toxic substitutes, reformulate products, and redesign production processes to achieve source reduction rather than cater to end-of-pipe clean-up.

Over 1000 companies reduced their emissions by 1994 leading to the program achieving its 50% reduction goal one year ahead of schedule.[74]

 **DISCUSSION QUESTIONS**

- *Discuss the funding mechanism of the Montreal Protocol and Multilateral Funds and their role in the success of the Montreal Protocol.*
- *List the 17 chemicals targeted under the 33/50 Program and identify the ozone depleting substance in the list.*
- *Discuss the success factors of the 33/50 Program.*

and concerns at the global level resulted in the Montreal Protocol in 1987.[75] The Montreal Protocol was ratified by 191 countries that committed to reducing the production and use of ozone depleting substances (ODS). By year 2010, almost all parties complied with the phase-out of CFCs, halons, carbon tetrachloride, methyl chloroform, and cholorobromomethane, leading to a 98% phase-out when compared to historic levels. Global production of CFC and halons ended in 2010. HCFCs mainly comprised of the 2%, and the phase-out of this transitional ODS was accelerated in 2007 giving way to less harmful substitutes like hydrofluorocarbons, hydrocarbons, and natural refrigerants.[76]

2.17 Green Public Procurement

Green Public Procurement (GPP) may be simply defined as "Public procurement for a better environment." GPP is formally defined as "a process whereby public authorities seek to procure goods, services and works with a reduced environmental impact throughout their life cycle when compared to goods, services and works with the same primary function that would otherwise be procured."

GPP is fundamentally a voluntary instrument, but it can be legislated. Japan already has a law on GPP. In 2000, in South Africa, Department of Environment Affairs adopted a Preferential Procurement Policy under the "Preferential Procurement Policy Framework Act, 2000." In China, from January 2007, provincial and central governments have made a list of environment friendly products certified by China Certification Committee for

Environmental Labelling and these products have to mandatorily meet environmental protection and energy saving standards. In Mexico, the 2007–2012 National Development Plan brought in sustainability criteria in the procurement policy followed by a procurement law. The law recognized that all wood and furniture procurement by public agencies requires a certificate highlighting its legal origin and paper procured by public agencies will need to have 50% recycled content.

Public authorities are major consumers in Europe. They spend approximately 1.8 trillion euro annually (2015 statistics), representing around 14% of the EU's gross domestic product. By using their purchasing power to choose goods and services with lower impacts on the environment, consumers in Europe can make an important contribution to Sustainable Consumption and Production (SCP).

The European Union (EU) adopted two directives on February 26, 2014. Today many of the EU countries have transposed these directives or rules into national laws. The new rules are driven by goals that include environmental protection, social responsibility, innovation, combating climate change, employment, public health, and other social and environmental considerations.

Importantly, these directives support innovation partnerships where a contracting authority wishes to purchase goods or services, which are not currently available on the market. The authority may establish an innovation partnership with one or more partners allowing research and development (R&D), piloting and subsequent purchase of a new product, service, or work. The procedure for establishing an innovation partnership is set out in Article 31 of Directive 2014/24/EU. Furthermore, these procurement directives allow for preliminary market consultation with suppliers in order to get advice, which may be used in the preparation of the procedure.

Green purchasing is thus about influencing the market. By promoting and using GPP, public authorities can provide industry with real incentives for developing green materials, technologies and products. GPP is, therefore, a strong stimulus for eco-innovation.

GPP has great environmental benefits too. In Brazil for instance, procurement of recycled paper notebooks in middle and high schools has helped in saving 8 million liters of water, 1,766 tons of waste, 241 kg of organohalogen compounds from procurement of 17,97,866 high school and 19,94,149 middle school *green* kits.

To be effective, GPP requires the inclusion of clear and verifiable environmental criteria for products and services in the public procurement process. Several countries in the world have developed guidance in this area, in the form of national GPP criteria.

Key goods of focus for greening could include ceiling fans, refrigerators, air conditioners, motors where emphasis could be on energy efficiency and, hence, reduction in GHG emissions and life cycle costs. To promote GPP, we need to make a strong case for GPP that makes an economic, an environmental, and a social force.

The Ministry of Finance in Mauritius piloted GPP. The main objective was to develop a Framework for Sustainable or Green Public Procurement that will ensure that procurement decisions take the following key factors into account when evaluating goods and services:

- *Economic*: The need to achieve better value for money with the financial resources available.
- *Environmental*: The product, service or work requirements should include environmental performances following environmentally friendly production methods,

higher energy efficiency as well as maximum use of renewable energy, lower generation of waste and emissions and avoiding use of non-biodegradable and toxic substances.

- *Social*: The reduction of poverty and inequality by promoting security and social inclusion; improving working conditions and employee welfare; and promoting gender balance.

An Action Plan was developed on the basis of the above and was approved by the Cabinet of Ministers in December 2011. A workshop was organized with a view to develop sustainability criteria for five products as an initial phase. These products included paper, ICT equipment, office furniture, passenger cars, detergents, and cleaning materials. A second workshop was held to train procurement officers and, hence, facilitate implementation of GPP. The participants included both procurement officers and suppliers. Model bid documents were then prepared after training and consultation.

Today, 30% of the GDP of India is spent on public procurement. Given the massive size of public spending, the public sector in India can be a prime driver towards sustainable production and consumption and can create environmental and economic benefits. Unfortunately, in India, GPP is still in its infancy. Some public sector entities and government departments have started internalizing environmental and energy efficiency criteria in their procurement decisions. The challenge of making GPP as a common practice still remains. Box 2.22 illustrates the efforts taken towards GPP by the Ministry of Railways in India.

BOX 2.22 GPP AND THE MINISTRY OF RAILWAYS IN INDIA

Indian Railways (IR) has already taken steps in this direction by specifying minimum 3-star energy rating during procurement. These requirements could be heightened gradually as the market matures.

Introduction of biodegradable water bottles, biodegradable paper cups, use of leaf plates, etc., are examples of greening of IR. IR has already installed Bio-toilets for efficient waste disposal and resource recovery. Paperless e-ticketing has been successfully introduced leading to significant reduction in the environmental footprints. Water is now recycled after washing the wagons and solar energy producing plants and self-sustaining hydro-electric and bio-diesel plants at vacant railway lands are getting commissioned.

IR is using Compressed Natural Gas (CNG) and bio-diesel in its fleet of multiple diesel units. Test runs for dual fuel mode compatible units have been done, which have demonstrated positive results.

DISCUSSION QUESTION

- *Indian Railways have taken several green initiatives described above without formally launching a Green Public Procurement Policy. Is formulation of a GPP policy necessary? What are the advantages?*

2.18 Market Based Instruments

Policymakers can adopt multiple approaches for environmental management. The traditional approach as discussed previously is the command-and-control approach (CAC), where regulators enact a law, to enforce environmentally sound practices. The CAC approach involves the setting of technical standards to protect and improve the environmental quality.

Market based Economic Instruments (EIs), on the other hand, encompass a range of economic tools to incentivize the polluters to reduce or eliminate environmental pollution. EIs range from emission charges, fees, taxes, marketable or tradable permits to deposit-refund systems, performance bonds, liability payments, green funds, rebates, and subsidies. Unlike most CACs, EIs operate at a decentralized level. Under most scenarios, EIs operate on the *Polluter Pays* principle, which shifts the costs and responsibilities, associated with pollution back on to the polluter.

Box 2.23 presents how the world's first *Polluter Pays* policy in Colombia succeeded and paved the path for similar responses from other regions in the world.

Due to their economic implications, EIs typically are more efficient than CACs, which rely on mandated technologies and/or pollution reduction targets applied universally across polluters.[78] The reason is evident in that businesses and industries respond to economic incentives due to pressures from markets and shareholders.

BOX 2.23 HOW COLOMBIA SUCCESSFULLY FOUGHT WATER POLLUTION: ILLUSTRATION OF THE POLLUTER PAYS MODEL

By the early 1990s, the rivers in the State of Antioquia in Colombia were heavily polluted as a result of economic and industrial development. The rivers, frothing with chemicals and detergents, ran yellow or electric blue in color due to cement and textile industry activities. Attempts like environmental policies, fines, and factory shut downs proved to be unsuccessful. In 1997, the government responded by introducing economic incentive-based regulations to control industrial emissions and effluents. All polluters including towns, factories, and farms had to pay per unit of organic pollution they discharged into the waterways of the Antioquia district. This economic implication to businesses had immediate impacts as business as usual practices became more expensive to the company. Industries and even municipal authorities that traditionally released untreated organic matter into waters, built wastewater treatment plants worth millions of dollars. There was a significant decrease in organic wastes and suspended solids in the region's watersheds. This success story inspired other regions in the country and across the globe to introduce such economic instruments to regulate pollution.[77]

DISCUSSION QUESTIONS

- *Give another example where the Polluter Pays Principle was successfully implemented for environmental management.*
- *Discuss the Polluter Pays Principle with the help of an example.*

2.18.1 Economic Instruments

Box 2.24 illustrates a few economic instruments and financial support received.

BOX 2.24 ILLUSTRATIONS OF A FEW ECONOMIC INSTRUMENTS AND FINANCIAL SUPPORT

SUPERFUND PROGRAM, USA, 1980

The Superfund program funds remediation of some of the nation's most contaminated sites and responds to environmental emergencies like oil spills and natural disasters. The program was created under a law, Comprehensive Environmental Response, Compensation, and Liability Act (CERCLA). The law created a tax on the chemical and petroleum industries that goes to a trust fund for cleanup of abandoned or uncontrolled hazardous waste sites. The law also makes the polluter liable for releasing hazardous wastes at the sites.[79]

Of the many success stories, a cleanup that played an instrumental role in passage of the program is the Cleanup of the Valley of Drums. Under the program 4,200 waste-containing drums (Valley of Drums) were cleared from a site in Kentucky that were contaminating the site due to leakage and run-off from the drum contents like oil, solvents, and hazardous substances.[80]

RENEWABLE ENERGY CERTIFICATES (OR TRADABLE RENEWABLE CERTIFICATES)

A Renewable Energy Certificate (REC) mechanism is a market-based instrument to promote renewable energy and facilitate organizational compliance with renewable energy purchase obligations (RPOs[81]). RECs are tradable and non-tangible energy commodities that can be sold and purchased by customers who do not have access to green and clean energy through their utility provider or the option to switch power suppliers. One REC is equivalent to 1 MWh of renewable energy. Purchase of REC is deemed as purchase of power generated from renewable sources. RECs have been used extensively in many countries, such as Australia, Japan, the United States, Netherlands, Denmark, India, and the UK.

GREEN TECHNOLOGY FINANCING SCHEME, MALAYSIA, 2010

Malaysia recognizes that Green Technology will be a driver for sustainable development and climate change adaptation and mitigation. They launched the National Green Technology Policy in 2009 to promote use of Green Technology, that is, development and application of products, equipment, and systems that conserve natural resources and minimize negative impact of development activities. To incentivize adoption and utilization of Green Technology, they allocated funds of USD 86 million. The Scheme offers 60% guarantee of the financing amount and a rebate of 2% on the interest/profit rate charged by the financial institutions. The program also provides its proponents access to private and commercial financial institutions looking to make green investments.[82]

REPUBLIC OF KOREA'S GREEN CREDIT CARD

In 2011, the Korean Ministry of Environment introduced a green credit card scheme to encourage consumers to adopt more environmentally friendly lifestyle patterns by incentivizing them with tangible economic rewards. The consumers could use a green credit card to accumulate green points for achieving savings on the use of utilities like tap water, electricity, and gas heating, for using public transport, or purchasing eco-friendly products certified with eco-labels, such as Korea Eco-Label and Carbon Label. Forty-nine private companies including manufacturers, retailers, and coffee shops were looped in to facilitate the scheme. Green points are also awarded when people bring their own cups to the coffee shop and return used cellular phones to providers.

The accumulated green points can be used like cash to purchase eco-friendly products, such as hybrid cars or efficient light bulbs and services at hotels, restaurants, and theaters.

This scheme enabled users to track their carbon emissions and consequent improvements upon adopting a greener lifestyle.[83]

Similarly, in 2009 Seoul introduced the Eco-mileage incentive program to incentivize households and businesses to reduce their consumption of resources such as electricity, water, and natural gas. As an incentive, they would accumulate points that could be used for purchases of eco-friendly products, or for use of public transportation or at authorized merchant stores. The users could track their savings and points earned online. Between 2009 and 2014, the program had saved 377,000 tons of oil equivalents, which amounts to an annual operation of 471,000 cars.[84]

MUNICIPAL GREEN BONDS, USA

Green Bonds are bonds issued to fund environmentally beneficial projects such as renewable energy, green infrastructure, clean water, waste management, and eco-friendly transportation. In 2013, the Commonwealth of Massachusetts issued $100 million in Green Bonds to fund environmental projects. Since this initiative, the national Green Bond issuance has increased to $2.5 billion in 2014 and $1.3 billion in 2015.[85]

In 2015, the New York State Environment Agency issued $376.5 million of green bonds for water projects. The proceeds were distributed to local governments, state public authorities, and specified private entities to finance or refinance clean water and drinking water projects under the State's water pollution program.[86]

Green Bonds will be discussed in detail in Chapter 3.

ENVIRONMENTAL TAXES IN EU

The United Nations System of Environmental-Economic Accounting (UN SEEA), a global statistical standard, defines environmental tax as "a tax whose tax base is a physical unit (or a proxy of it) of something that has a proven, specific, negative impact on the environment." The primary objective of introducing environmental tax is to encourage environmentally positive behavior change.

In the European Economic Area (EEA) member countries (28), the environmental taxes are levied on use of:

- *Energy*: for transport, stationary and greenhouse gases
- *Transport* (excluding fuel)—sale, import or use of vehicles, road use for passenger cars and commercial vehicles
- *Pollution*: air emissions, effluent discharge to water, waste management, pesticides and fertilizer use
- *Resource use*: water abstraction and extraction of raw materials

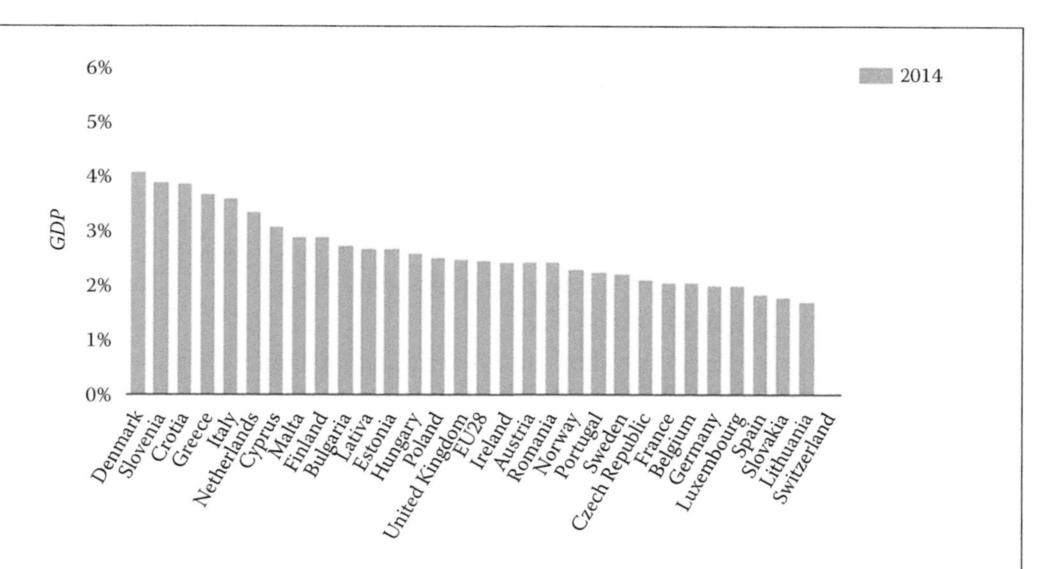

FIGURE 2.25
Environmental tax revenues as a percent of GDP in EU member states, Norway and Switzerland. From Environmental taxation and EU environmental policies, European Environment Agency, No. 17, p. 18, 2016. With permission, online source: http://efaep.org/sites/enep/files/EEA%20Environmental_taxation_and_EU_policies.pdf.)

EU uses environmental taxes as enabling instruments to achieve economic, social, and environmental objectives. Some countries have viewed these taxes as a source of national income. Environmental taxes in EU have witnessed a growth in real terms and as a percentage of the country's GDP. Figure 2.25 shows environmental tax revenues as a percentage of GDP for EU member countries, Norway, and Switzerland in 2014.[87,88]

DISCUSSION QUESTIONS

- *Discuss two successful economic instruments used to promote sustainable:*
  - *Agriculture*
  - *Waste management*

2.18.2 Information Based Instruments

The Indonesian national pollution control agency, BAPEDAL, launched an unconventional but effective environmental disclosure program called PROPER—Program for Pollution Control, Evaluation and Rating. This program utilizes motivating factors like public disclosure of environmental performance, environmental awards, and reputational incentives to incentivize improvement in environmental performance of polluting industries. BAPEDAL launched the initiative in view of poor and ineffective enforcement and compliance of pollution standards across industries. Box 2.25 provides the details.

BOX 2.25 PROPER IN INDONESIA[89,90,91,92]

Under the PROPER program, each participating (polluting) industry was rated for its environmental performance based on five colors or categories (Figure 2.26):

- *Gold* for near zero emissions level
- *Green* for exceeding national compliance standards
- *Blue* for adhering to national compliance standards
- *Red* for falling short of compliance
- *Black* for those companies that caused serious damage to the environment and have made no effort towards compliance

Environmental Performance was achieved based on implementation of environmental management systems, hazardous and nonhazardous waste management, application of 3R (reduce, reuse and recycle), water and air pollution control, energy efficiency, natural resource conservation, and social responsibility for the community.

Upon the launch of PROPER and initial rating, the poor performing industries (with black and red rating) made serious efforts to improve their environmental performance in the next one year. Only 18 months after full disclosure, PROPER program resulted in 40% pollution reduction in the pilot group of 187 industries. During the financial crisis period of 1997–2001 and due to governmental re-organization; however, the PROPER program went dormant. The program was resuscitated in 2001 and was back in operation once again in 2004 where performance of 85 companies was announced to the public. By 2010 the total number of companies in the program increased to 690.

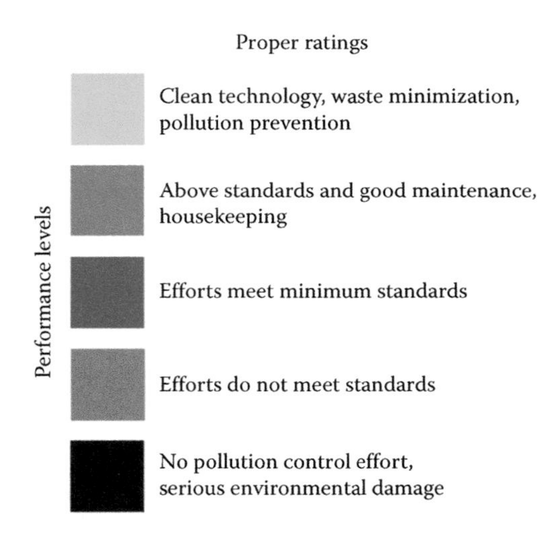

FIGURE 2.26
Performance ratings in the PROPER program. (Based on Development Research Group, Greening Industry: New Markets for Communities, Markets and Governments, *A World Bank Research Report*, p. 65, 1999, Oxford University Press, online source: http://documents.worldbank.org/curated/en/421701468772781985/310436360_20050007024038/additional/multi-page.pdf.)

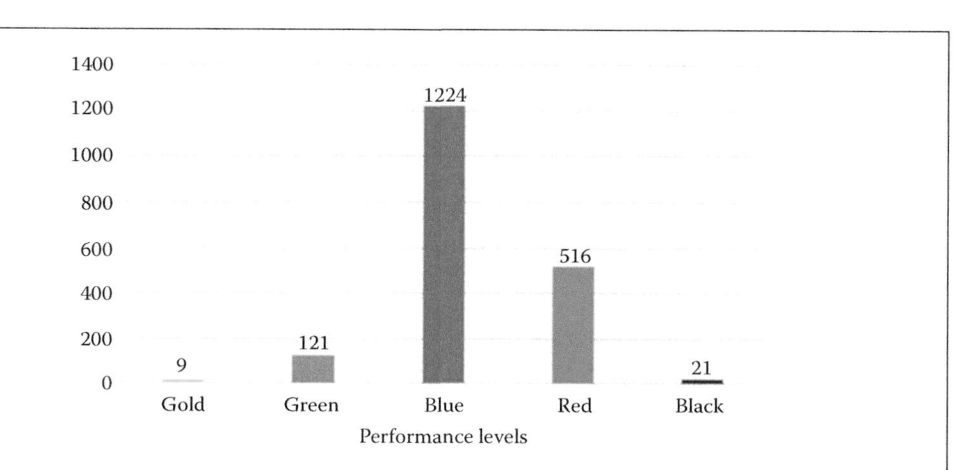

FIGURE 2.27
Performance assessment of companies under the PROPER program. (Based on the data from Widayati, N., Environmental Performance Rating Program, PROPER, Ministry of Environment and Finance, Jakarta, Indonesia, slide 9/21, online source: http://www.aecen.org/sites/default/files/panel_2_nety_widayati_indonesia.pdf.)

STATUS AND IMPACT OF PROPER, 2015[93]

The performance of the 1,908 companies in the program in 2015 was (Figure 2.27):

In quantitative terms, the impact until 2015 has been energy cut by 26 million GJ, 488 million cubic meters of water savings, a reduction of 11 million tons of non-hazardous and 2.4 million tons of hazardous waste, and investment of USD 88.2 million.

The PROPER model was adopted by countries like the Philippines (Ecowatch) and Mexico (Public Environmental Performance Index).

DISCUSSION QUESTION

- *Compare the PROPER with Ecowatch and the Public Environmental Performance Index. What have been the relative successes and the challenges faced?*

2.19 Environmental Courts and Tribunals

Principle 10 of the Earth Summit in 1992 states that "environmental issues are best handled with participation of all concerned citizens, by providing them appropriate access to information and an opportunity to participate in decision making processes. Effective access to judicial and administrative proceedings, including redress and remedy" was a key part of this Principle.[98]

Countries recognized the significance of making institutional changes to provide accessible environmental justice to people, especially in emerging markets like India and China, where trade and development issues tend to clash with environmental imperatives. It is interesting to note that countries like Brazil, Finland, and India have Constitutional provisions for environmental justice. Box 2.26 provides information on EU Energy Labels for communication to consumers. Box 2.27 illustrates communication on air quality; and Box 2.28 provides for illustrations of Environmental courts and tribunals.

BOX 2.26 EU ENERGY LABEL FOR COMMUNICATION TO CONSUMERS

The European Union issued a Directive in 2010 to indicate energy and resource consumption information of products by using labeling or standard product information. The objective of this Directive was to inform consumer choices while purchasing household electrical appliances and help them select energy products with lower energy and resource consumption.[94]

As a result of this Directive, an EU Energy Label was developed to serve as a uniform label providing energy efficiency based ratings for products. The Label was introduced for air conditioners, dishwashers, refrigerators, washing machines, heaters, lighting fixtures, boilers, televisions, and vacuum cleaners.

The label rates products from A-G with dark green (Rating A) being most efficient and red (Rating G) being least efficient. In the past, the topmost rating was always an *A* but now the rating has expanded to A+, A++ or an A+++. The label provides information on total energy consumption, water consumption, and noise levels for washing machines, and screen size for televisions.[95]

DISCUSSION QUESTIONS

- *What parameters are considered in the assigning of a label in the EU? Explain with examples.*
- *Compile the EU label nomenclature and requirements for televisions, refrigerators, and vacuum cleaners. Compare these with your country's national standards.*
- *Are rating A goods more expensive? Why?*

BOX 2.27 COMMUNICATING AIR QUALITY

AIR QUALITY INDEX IN CHINA[96]

The Air Quality standards and monitoring were used to develop an Air Quality Index (AQI), a scale to monitor the air quality, inform the public, and help them limit their exposure to air pollution. The publicly available AQI could help citizens plan their outdoor activities and limit them during a period of increased levels of air pollution.

The Chinese AQI is indicated by six parameters: SO_2, NO_2, CO, PM_{10} and $PM_{2.5}$ are measured in daily average and by O_3 that is measured in 1 hour and 8 hour intervals. Using six color classifications, the AQI results are used to indicate real-time air quality levels. The six classifications of the index are excellent, good, light polluted, moderately polluted, heavily polluted, and severely polluted.

ADB'S CLEAN AIR SCORECARD

The Clean Air Scorecard was developed by ADB through the Clean Air Initiative for Asian Cities Center (CAI-Asia). It gives an overall clean air score ranging from zero, the lowest grade, to a maximum of 100. It comprises three individual inducers: the Air Pollution and Health Index, the Clean Air Management Capacity Index, and the

Clean Air Policies and Actions Index. The score thus captures institutional capacity and policy related aspects. A Microsoft Excel tool has been developed for the computation of the scores.

The Clean Air Scorecard has been piloted in the cities of Bangkok, Thailand, Jakarta, Indonesia, Manila, the Philippines, and in various cities in China.[97]

DISCUSSION QUESTIONS

- *Give examples where information on air quality is used to set alarms for elderly people and school children or for directing real time traffic diversion.*
- *The air quality index provides a single number and information on individual pollutants is lost in the amalgamation process. How can we address this issue?*

BOX 2.28 ILLUSTRATIONS OF ENVIRONMENTAL COURTS AND TRIBUNALS

PARLIAMENTARY COMMISSIONER FOR FUTURE GENERATIONS, HUNGARY

Office of the Parliamentary Commissioner for Future Generations of Hungary started operating in 2008 after the Ombudsman Act was passed in 2007. The Office consists of 19 lawyers, two economists, one engineer, two biologists, a climate change expert, and a medical doctor. The Office serves as a consultative body to the Parliament regarding environmental legislation. It may initiate proceedings at the Constitutional Court or intervene in court litigations in the interest of future generations and the enforcement of the right to a healthy environment. The Office also resolves environmental conflict cases.

One example of such a case is when the Commissioner intervened in a spatial development plan that did not factor in environmental aspects. A Budapest municipality was planning an increase in the number of residential units in an already crowded area, thus impacting the noise pollution, air pollution levels, and also decreasing the ratio of green open areas. The Commissioner concluded that the plan was not in line with the principles of sustainable development. As a result, the municipality did not pass the spatial plan and conducted an impact assessment in line with the Commissioner's recommendations.[99]

NATIONAL GREEN TRIBUNAL, INDIA[100,101]

The Government of India established the National Green Tribunal (NGT) under the National Green Tribunal Act of 2010. The Act draws inspiration from Article 21 of the Indian Constitution that states that Indian citizens have a right to a healthy environment. NGT was established for "effective and expeditious disposal of cases relating to environmental protection and conservation of forests and other natural resources including enforcement of any legal right relating to environment and giving relief and compensation for damages to persons and property."

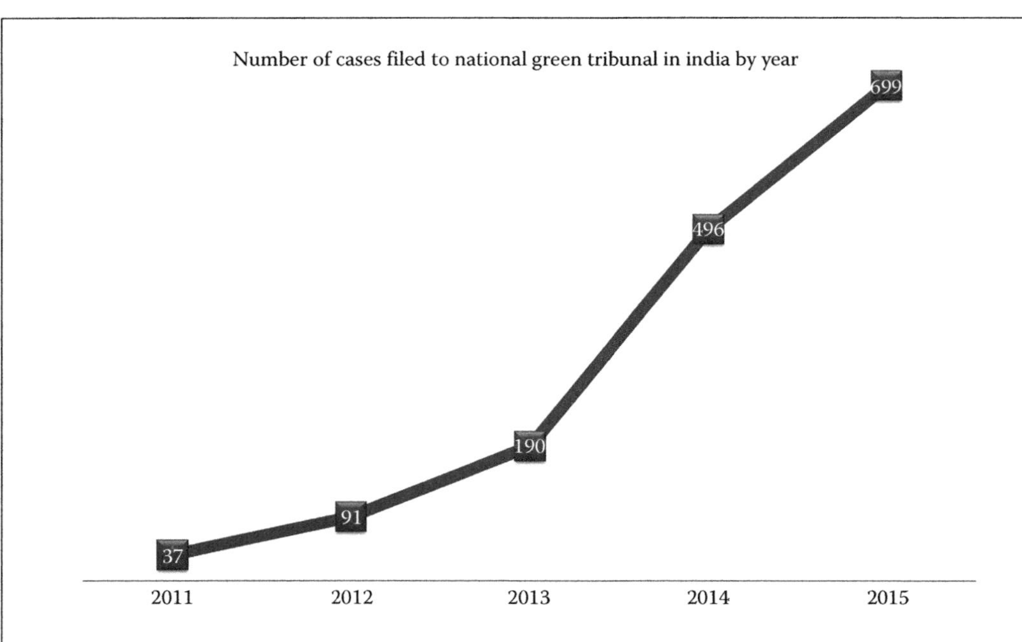

FIGURE 2.28
Number of cases filed to National Green Tribunal in India by year. (Based on data from the Indian Environmental Portal website [2014–2015] and National Green Tribunal and Environmental Justice in India [2011–2013]. Compiled by Mamata Mandan [under the supervision of Dr Prasad Modak] (2016). Online source: http://www.indiaenvironmentportal.org.in/ and http://www.greentribunal.gov.in/.)

NGT is guided by the principles of natural justice and has a mandate to dispose appeals or applications within 6 months of their filing. This will ensure expedited environmental justice, which is critical in India where cases at times are unsettled for decades. Figure 2.28 shows the numbers of cases filed to NGT in India between 2011–2015.

DISCUSSION QUESTIONS

- *Recommended Read: Greening Justice—Creating and Improving Environmental Courts and Tribunals by George Pring and Catherine Pring*
- *Discuss the Amazon-Cargill case study and the role of the environmental courts in containing the adverse environmental impact of Cargill's soybean cultivation in the region.*

There were concerns regarding the ability of general, non-specialized court systems to handle environmental and land use issues affecting development and future sustainability. This led to the creation of Environmental Courts and Tribunals (ECTs). ECTs are judicial bodies and tribunals that specialize in resolving environmental, resource development, land use, and climate change disputes.

One of the early examples of environmental courts was the establishment of a Water Court in Finland and Sweden in 1918, which solely dealt with cases revolving around protection and allocation of the country's water resources. In a study called *Greening Justice: Creating and Improving Environmental Courts and Tribunals* over 350 specialized ECTs were identified in 41 countries.

Endnotes

1. For further reading: Griger, A., *Only One Earth: Stockholm and the Beginning of Modern Environmental Diplomacy,* Environment & Society Portal, Arcadia, no. 12, Rachel Carson Center for Environment and Society, 2012, online source: http://www.environmentandsociety.org/arcadia/only-one-earth-stockholm-and-beginning-modern-environmental-diplomacy
2. Taken from: Recommended Binational Phosphorus Targets—Recommended Binational Phosphorus Targets to Combat Lake Erie Algal Blooms, EPA: United States Environmental Protection Agency, 2016, online source: https://www.epa.gov/glwqa/recommended-binational-phosphorus-targets
3. For further reading: Franklin, C.A., Burnett, R. T., Paolini, R.J.P. and Raizenne, M.E., *Health Risks from Acid Rain: A Canadian Perspective,* Environmental Health Perspectives, Ontario, Vol. 63, pp. 155–168, 1985, online source: http://www.ncbi.nlm.nih.gov/pmc/articles/PMC1568495/pdf/envhper00446-0153.pdf
4. For further reading: Schneider, J.C., Significance of acid rain to Michigan lakes and their fisheries, Michigan Department of Natural Resources Fisheries Division, *Fisheries Technical Report* No. 83(1), 1993, online source: http://quod.lib.umich.edu/f/fishery/5020303.0001.001?rgn=main;view=fulltext
5. Based on: 2016 Progress Report of the Parties, Pursuant to the Canada-United States Great Lakes Water Quality Agreement, 2016, online source: https://binational.net/wp-content/uploads/2016/09/PRP-160927-EN.pdf
6. For further reading: Great Lakes Air Deposition Program, Great Lakes Fact Sheet, 2008, online source: http://gis.glin.net/prdb/docs/how-glc-glad-fs-final.pdf
7. Based on: Duties and Restrictions, Environment Guide, Environment Foundation, 2014, online source: http://www.environmentguide.org.nz/rma/duties-restrictions-and-existing-uses/
8. For further reading: Palmer, Sir G., Ruminations on the problems with the Resource Management Act, *Keynote Address to the Local Government Environmental Compliance Conference,* Auckland, 2015, online source: https://www.planning.org.nz/Attachment?Action=Download&Attachment_id=3538
9. For further reading: Partnership in Environmental Governance, Swedish Environmental Protection Agency, online source: http://www.swedishepa.se/Documents/publikationer/978-91-620-8495-0.pdf?pid=4176
10. For further reading: Data, databases and applications, Swedish Environmental Protection Agency, online source: http://www.swedishepa.se/State-of-the-environment/Data-databases-and-applications/
11. Based on: Ministry of Agriculture, Land and Fisheries, Government of the Republic of Trinidad and Tobago, 2013, online source: http://www.agriculture.gov.tt/
12. Based on: Ministry of Energy and Minerals, The United Republic of Tanzania, 2017, online source: https://mem.go.tz/background-history/
13. Based on: Roles and Functions, Tanzania Minerals Audit Agency (TMAA), 2017, online source: http://www.tmaa.go.tz/tmaa/about/category/roles_and_functions
14. Based on: Mining Cadastre Portal, United Republic of Tanzania Ministry of Energy & Minerals, online source: https://portal.mem.go.tz/site/CustomHtml.aspx?PageID=d7f3f61d-4689-4280-a59a-b865f002dd60
15. For Further Reading: Interlinking of Rivers, Ministry of Water Resources, River Development & Ganga Rejuvenation, Government of India, 2016, online source: http://wrmin.nic.in/forms/list.aspx?lid=1279
16. Based on: The Secretariat, Governo do Estado de Sao Paulo, Environment System of Sao Paulo, online source: http://www.ambiente.sp.gov.br/en/the-secretariat/
17. For Further Reading: World Resources Institute, A Compilation of Green Economy Policies, Programs, and Initiatives from Around the World, *The Green Economy in Practice: Interactive Workshop 1,* February 11, 2011, online source: http://pdf.wri.org/green_economy_compilation_2011-02.pdf

18. Based on: Water Supplies Department, The Government of the Hong Kong Special Administrative Region, 2016, online source: http://www.wsd.gov.hk/en/about_us/our_vision_mission_and_values/index.html
19. Based on: California Environmental Protection Agency, History of the Water Boards, State Water Resources Control Board, 2012, online source: http://www.swrcb.ca.gov/about_us/water_boards_structure/history.shtml
20. Based on: Policy for Water Quality Control for Recycled Water (Recycled Water Policy), State Water Resources Control Board, 2013, online source: http://www.waterboards.ca.gov/water_issues/programs/water_recycling_policy/docs/rwp_revtoc.pdf
21. Based on: Guidance on Sustainability Appraisal, Prepared for Hertfordshire County Council, 2007, online source; http://www.hertfordshire.gov.uk/infobase/docs/pdfstore/ssustappguidance0907.pdf
22. Based on: The Role of Sustainability Appraisal Sustainability Appraisal, Planning Advisory Service, Local Government Association, 2014, online source: http://www.pas.gov.uk/chapter-6-the-role-of-sustainability-appraisal
23. For further reading: Anand, V., Functions of the Urban Local Bodies, Financing Small Cities, IFMR, online source: http://financingcities.ifmr.co.in/blog/2012/12/17/functions-of-the-urban-local-bodies-ulbs/
24. For further reading: Urban Management Consulting Private Limited (UMC) and Centre for Environment Education (CEE), Urban Solid Waste Management in Indian Cities, National Institute of Urban Affairs, 2015, online source: http://smartnet.niua.org/sites/default/files/resources/NIUA-PEARL%20Good%20Practices%20SWM.pdf
25. Based on: The Manila Bay Oil Spill Contingency Plan, GEF/UNDP/IMO Regional Program on Partnerships in Environmental Management for the Seas of East Asia (PEMSEA); p. 4, 2006, online source: http://www.pemsea.org/sites/default/files/manilabay-oilspill-plan.pdf
26. For further reading: Boyd, D.R., The Status of Constitutional Protection for the Environment in Other Nations, David Suzuki Foundation, 2014, online source: http://davidsuzuki.org/publications/2014/whitepapers/DSF%20White%20Paper%204.pdf
27. Note: The citations in this box have been taken from the Constitutions providing a reference. The citations are however only illustrative and not exhaustive.
28. Taken from: Basic Law for the Federal Republic of Germany, 2012, online source: https://www.bundestag.de/blob/284870/ce0d03414872b427e57fccb703634dcd/basic_law-data.pdf
29. Taken from: Dr. Ambedkar, B.R. et al., The Constitution of India, 1950, online source: https://india.gov.in/sites/upload_files/npi/files/coi_part_full.pdf
30. Taken from: Constitution of The People's Republic of China, 1982, online source: http://china.usc.edu/constitution-peoples-republic-china-1982
31. Taken from: Constitution of the Republic of South Africa No. 108 of 1996, online source: http://www.gov.za/sites/www.gov.za/files/images/a108-96.pdf
32. Taken from: Constitute Project, Colombia's Constitution of 1991 with Amendments through 2005, online source: https://www.constituteproject.org/constitution/Colombia_2005.pdf
33. Based on: Revised National Health Policy, *Federal Ministry of Health*, Nigeria, 2004, online source: https://www.advancingpartners.org/sites/default/files/cadres/policies/revisednationalhealthpolicydocument.pdf
34. Based on: Malaysia's National Policy on Biological Diversity, Ministry of Science, Environment and Technology, 1998, online source: http://www.chm.frim.gov.my/getattachment/85c78522-fe93-4ca7-aaf3-454bef892a7e/NBP.pdf.aspx
35. Based on: National Policy on Biological Diversity 2015–2025, *Ministry of Natural Resources and Environment*, Malaysia, 2015, online source: http://www.nre.gov.my/ms-my/Documents/PengumumanNRE/NPBD%202015_2020.pdf
36. Based on: Directorate of Relief, Disaster Preparedness and Refugees, Office of the Prime Minister, The National Policy for Disaster Preparedness and Management, Republic of Uganda, 2010, online source: http://www.ifrc.org/docs/IDRL/Disaster%20Policy%20for%20Uganda.pdf

37. Based on: Ministry of Water Resources, National Water Policy, Government of India, 2012, online source: http://wrmin.nic.in/writereaddata/NationalWaterPolicy/NWP2012Eng6495132651.pdf
38. Based on: The Ministry of Agriculture and Animal Resources, National Fertilizer Policy, Republic of Rwanda, 2014, online source: http://www.minagri.gov.rw/fileadmin/user_upload/documents/STRAT.PLC/FERTILIZER_POLICY_-FINAL.pdf
39. Based on: Protected Areas Categories, IUCN, 2017, online source: https://www.iucn.org/theme/protected-areas/about/protected-areas-categories
40. Taken from: Protected Areas: Delivering the Promise of Sydney, IUCN, 2017, online source: https://www.iucn.org/theme/protected-areas/about
41. Based on: Important Bird and Biodiversity Areas (IBAs), BirdLife International, 2017, online source: http://www.birdlife.org/worldwide/programmes/sites-habitats-ibas
42. For further reading: Heath, M., IBAs in Danger—Urgent action needed, BirdLife International, 2013, online source: http://www.birdlife.org/worldwide/news/ibas-danger-%E2%80%93-urgent-action-needed
43. For further reading: Important Bird and Biodiversity Areas are in Danger, BirdLife Australia, online source: http://birdlife.org.au/ibas-in-danger
44. Based on: Department of Environment Regulation, Clearing Regulation Factsheet: Environmentally Sensitive Areas- Environmental Protection Act 1986, Government of Western Australia, 2014, online source; https://www.der.wa.gov.au/images/documents/your-environment/native-vegetation/Fact_sheets/fs24-clearing-regs_ESAs.pdf
45. Based on: This Case is Commonly Referred to As Taj Trapezium Case Law Essay, Essays, UK, 2013, online source: http://www.lawteacher.net/free-law-essays/european-law/this-case-is-commonly-referred-to-as-taj-trapezium-case-law-essay.php
46. Based on: M.C. Mehta v. Union of India, WP 13381/1984 (1996.12.30) (Taj Trapezium Case), Environmental Law Alliance Worldwide, 2015, online source: https://www.elaw.org/content/india-mc-mehta-v-union-india-wp-133811984-19961230-taj-trapezium-case
47. Based on: New London Low Emission Zone (LEZ) standards now in operation, Transport for London, 2011, online source: http://urbanaccessregulations.eu/images/stories/pdf_files/London%20PN%20new%20standards%20now%20in%20operation.pdf
48. Based on: Low Emission Zone, Transport for London, online source: https://tfl.gov.uk/modes/driving/low-emission-zone
49. Taken from: The President of the Philippines, Proclamation No. 2146, s. 1981, Official Gazette, 1981, online source: http://www.gov.ph/1981/12/14/proclamation-no-2146-s-1981/
50. Based on: Environmental Management: Environmentally Critical Areas/Project, *GIS Cookbook for LGUs*, 2008, online source: http://www.cookbook.hlurb.gov.ph/4-08-08-environmental-management-environmentally-critical-areas-project
51. Based on: Government of India, Frequently Asked Questions on the Coastal Regulation Zone Notification, 2011 and Island Protection Zone Notification, Ministry of Environment, Forest and Climate Change, 2011, online source: http://www.moef.nic.in/downloads/public-information/FAQ-CRZ.pdf
52. Based on: Department for Communities and Local Government, Technical Guidance to the National Planning Policy Framework, 2012, online source: https://www.gov.uk/government/uploads/system/uploads/attachment_data/file/6000/2115548.pdf
53. Based on: Sustainable Urban Development in the People's Republic of China Eco-City Development—A New and Sustainable Way Forward? ADB Urban Innovations and Best Practices, online source: https://www.adb.org/sites/default/files/publication/27868/urbandev-prc-nov2010-ecocity.pdf
54. For further reading: Lindfield, M. and Steinberg, F., Urban Development Series: Green Cities, ADB, 2012, online source: https://www.adb.org/sites/default/files/publication/30059/green-cities.pdf
55. Based on: Sino-Singapore Tianjin Eco-city: A model for sustainable development, Government of Singapore, 2016, online source: http://www.tianjinecocity.gov.sg/

56. Based on: Brief case study on local climate and energy actions #14, Barcelona, Spain—A new energy model to tackle climate change, LG Action, online source: http://www.iclei-europe.org/fileadmin/templates/iclei-europe/files/content/Membership/MUTS/Barcelona/LG_Action_case_Barcelona_EN_final.pdf
57. Based on: Barcelona Solar Energy, WWF Global, 2012, online source: http://wwf.panda.org/?204380/Barcelona-solar-energy
58. Based on: Goldenberg, S., Masdar's zero-carbon dream could become world's first green ghost town, The Guardian, 2016, online source: https://www.theguardian.com/environment/2016/feb/16/masdars-zero-carbon-dream-could-become-worlds-first-green-ghost-town
59. Based on: Kaiman, J., China's 'eco-cities': Empty of hospitals, shopping centres and people, 2014, The Guardian, online source: https://www.theguardian.com/cities/2014/apr/14/china-tianjin-eco-city-empty-hospitals-people
60. Based on: Warana Wired Village Project, Warana Co-operative Continuous Energy, online source: http://waranapower.com/wired-village.html
61. Based on: Eco-Industrial Parks (EIP), Indigo Development: Creating Systems solutions for sustainable development through industrial ecology, 2006, online source: http://www.indigodev.com/Ecoparks.html
62. Based on: Higuchi, K. and Norton, M., Japan's eco-towns and innovation clusters: Synergy towards sustainability, *Global Environment, Environment & Science Portal*, Vol. 1, pp. 224–43, 2008, online source: http://www.environmentandsociety.org/mml/higuchi-kazukiyo-and-michael-g-norton-japans-eco-towns-and-innovation-clusters-synergy-towards
63. Based on: Examples of Rainwater Harvesting and Utilisation Around the World: Tokyo, Japan, UNEP, An Environmentally Sound Approach for Sustainable Urban Water Management: An Introductory Guide for Decision-Makers, Rainwater Harvesting and Utilisation, 2002, online source: http://www.unep.or.jp/ietc/publications/urban/urbanenv-2/9.asp
64. Based on: Let's Use Rainwater! Recent Trends in Rainwater use in Japan, Japan for Sustainability, 2014, online source: http://www.japanfs.org/en/news/archives/news_id035023.html
65. Based on: Standard Terms of Reference (TOR) for EIA/EMP Report for Projects/Activities Requiring Environment Clearance Under EIA Notification, 2006, Ministry of Environment, Forest and Climate Change, India, 2015, online source: http://www.moef.gov.in/sites/default/files/final%20Booklet.pdf
66. Taken from: EU: Cars and Light Trucks, Emission Standards, *DieselNet*, 2016, online source: https://www.dieselnet.com/standards/eu/ld.php
67. Taken from: Adoption of the EU Euro Emissions standards for road vehicles in Asian Countries, European Environment Agency, 2016, online source: http://www.eea.europa.eu/data-and-maps/figures/number-of-international-environmental-agreements-adopted-1
68. Based on: Resource Management (National Environmental Standards for Air Quality) Regulations 2004, New Zealand Legislation, Parliamentary Counsel Office, online source: http://www.legislation.govt.nz/regulation/public/2004/0309/latest/whole.html#DLM1850305
69. Based on: China: Air Quality Standards, TransportPolicy.net, 2014, online source: http://transportpolicy.net/index.php?title=China:_Air_Quality_Standards
70. Based on: Appraisal Center for Environment & Engineering—Ministry of Environmental Protection, Ambient Air Quality Standards, National Standards of the People's Republic of China, Ministry of Environmental Protection of the People's Repulic of China, General Administration of Quality Supervision, Inspection and Quarantine of the People's Republic of China, 2012, online source: http://www.china-eia.com/en/policiesregulations/technicalstandards/9152.htm
71. Based on: ELP and Clean Air Act Regulation of Greenhouse Gases, Harvard: Environmental Law Program—Emmett Clinic, Policy Initiative, online source: http://environment.law.harvard.edu/issues/the-clean-air-act-and-greenhouse-gases/
72. Based on: EPA Regulation of Greenhouse Gas Emissions from New Power Plants, C2ES: Center for Climate and Energy Solutions, online source: http://www.c2es.org/federal/executive/epa/ghg-standards-for-new-power-plants

73. Based on: Martin, G., Global climate deal to phase out fast-growing pollutant. What happens next? news@Northeastern, Society & Culture, 2016, online source: http://www.northeastern.edu/news/2016/10/global-climate-deal-to-phase-out-fast-growing-pollutant-what-happens-next/
74. Based on: Initiative: 33/50, CSD Major Groups, CSD Follow-up: Multi-Stakeholder Review of Voluntary Initiatives and Agreements for Industry, online source: https://sustainabledevelopment.un.org/content/dsd/dsd_aofw_mg/mg_VIA/viaprofiles_33_50.htm
75. Based on: Molina, M., Chlorofluorocarbons (CFCs), and Ozone Depletion, Research and Development of the U.S. Department of Energy, 2016, online source: https://www.osti.gov/accomplishments/molina.html
76. Based on: Montreal Protocol—Achievements to Date and Challenges Ahead, Ozone Secretariat, UNEP, 2017, online source: http://ozone.unep.org/en/focus/montreal-protocol-achievements-date-and-challenges-ahead
77. For further reading: Ambrus, S., Colombia Tries a New Way to Fight Water Pollution … and It Works; Eco Americas, Colombia, World Bank, 2000, online source: http://siteresources.worldbank.org/INTRES/Resources/469232-1321568702932/Greening_ColombiaprogramArticle.pdf
78. Based on: The Use of Economic Instruments in Environmental Policy: Opportunities and Challenges, United Nations Publication, 2004, online source: https://www.cbd.int/financial/doc/several-several-unep.pdf
79. Based on: Superfund: CERCLA Overview, EPA: United States Environmental Protection Agency, 2016, online source: https://www.epa.gov/superfund/superfund-cercla-overview
80. Based on: Valley of the Drums, Bullitt County History, 2016, online source: http://bullittcounty history.org/bchistory/valleydrum.html
81. Based on: Renewable Purchase Obligation (RPO) are targets for distribution companies to purchase certain percentage of their total power requirement from renewable energy sources.
82. Based on: Green Technology Financing Scheme: Empowering Green Businesses, Green Technology Financing Scheme (GTFS), Malaysian Green Technology Corporation, 2016, online source: https://www.gtfs.my/news/green-technology-financing-scheme-gtfs
83. For further reading: Low Carbon Green Growth Roadmap for Asia and the Pacific—Case Study: Republic of Korea's green credit card, UNESCAP: United Nations Economic and Social Commission for Asia and the Pacific, 2012, online source: http://www.unescap.org/sites/default/files/32.%20CS-Republic-of-Korea-green-credit-card.pdf
84. Based on: Eco-mileage: A citizen's participation programme for protecting the environment, C40 Cities: Case study, 2014, online source: http://www.c40.org/case_studies/eco-mileage-a-citizen-s-participation-programme-for-protecting-the-environment
85. Based on: Green City Bonds, How to Issue a Green Muni Bond, The Green Muni Bonds Playbook, Green City Bonds Coalition, NRDC, 2016, online source: https://www.nrdc.org/sites/default/files/greencitybonds-ib.pdf
86. For further reading, Kidney, S., India's 1st green corporate bond from CLP (INR 6 bn) + gossip/Nordic Investm Bank EUR500m/New York Env Fund $376.5m for water/Colarado State Uni $42.1m for green buildings/unlabeled climate bond for offshore wind + Ulrik Ross + more, Climate Bonds Initiative, 2015, online source: https://www.climatebonds.net/2015/09/india%E2%80%99s-1st-green-corporate-bond-clp-inr6bn-gossip-nordic-investm-bank-eur500m-new-york-env
87. Based on: Environmental taxation and EU environmental policies, European Environmental Agency.
88. Based on: Designing tax systems for a green economy transition, European Environment Agency, 2016, online source: http://www.eea.europa.eu/highlights/designing-tax-systems-for-a
89. For further reading: Afsah, S., Garcia, J.H. and Sterner, T., The Institutional History of Indonesia's Environmental Rating and Public Disclosure Program (PROPER), Frederick A. and Barbara, M.- Erb Institute: University of Michigan, 2011, online source: http://erb.umich.edu/News-and-Events/news-events-docs/11-12/eco-labels2011/JorgeGarciaLopez.pdf

90. For further reading: Kanungo, P. and Moreno, M., Indonesia's Program for Pollution Control, Evaluation, and Rating (PROPER), Empowerment case studies, World Bank Case, 2003, online source: http://siteresources.worldbank.org/INTEMPOWERMENT/Resources/14825_Indonesia_Proper-web.pdf
91. Taken from: Development Research Group, Greening Industry: New Markets for Communities, Markets and Governments, *A World Bank Research Report*, p. 65, 1999, Oxford University Press, online source: http://documents.worldbank.org/curated/en/421701468772781985/310436360_20050007024038/additional/multi-page.pdf
92. For further reading: Khoirunissa, I., Napitupulu, W. and Dwikorianto, T., Gold "PROPER" Achievement of Environmental & Social Management, *Proceedings World Geothermal Congress*, Australia, pp. 19–25, 2015, online source: https://pangea.stanford.edu/ERE/db/WGC/papers/WGC/2015/02003.pdf
93. Based on: Indonesia's Green Industrial Estates & Best Practices, Solidiance, 2015, online source: http://www.slideshare.net/dduhamel/indonesias-green-industrial-estates-and-best-practices
94. Based on: Directive 2010/30/EU of the European Parliament and of the Council of May 19, 2010 on the indication by labelling and standard product information of the consumption of energy and other resources by energy-related products, EUR-Lex, 2010, online source: http://eur-lex.europa.eu/legal-content/EN/TXT/?uri=CELEX:32010L0030
95. Based on: Commission Delegated Regulation(EU) No 812/2013 supplementing Directive 2010/30/EU of the European Parliament and of the Council with regard to the energy labelling of water heaters, hot water storage tanks and packages of water heater and solar device, EUR-Lex, 2013, online source: http://eur-lex.europa.eu/legal-content/EN/TXT/?uri=CELEX%3A32013R0812
96. For further reading: Gao, F., Evaluation of the Chinese New Air Quality Index: Based on Comparison with the US AQI system and the WHO AQGs, Novia: University of Applied Sciences, 2013, online source: https://publications.theseus.fi/bitstream/handle/10024/65044/Gao_Fanyu.pdf?sequence=1
97. For further reading: Joshi, V., Fu, L. and Ajero, M., Clean Air Scorecard Helps Clear the Air in the People's Republic of China, ADB Knowledge Showcases, Issue 49, 2013, online source: https://www.adb.org/sites/default/files/publication/31133/clean-air-scorecard-peoples-republic-china.pdf
98. For further reading: Report of the United Nations Conference on the Human Environment, UNEP, Stockholm, June 5–16, 1972, online source: http://www.unep.org/documents.multilingual/default.asp?documentid=78&articleid=1163
99. For further reading: Ambrusne, E.T., The parliamentary commissioner for future generations of Hungary and his impact, *Intergenerational Justice Review*, Vol. 10(1), pp. 21, 2010.
100. Based on: Government of India, National Green Tribunal (NGT), Ministry of Environment, Forest and Climate Change—Government of India, 2016, online source: http://envfor.nic.in/rules-regulations/national-green-tribunal-ngt
101. Based on: The Gazette of India, The National Green Tribunal Act, Ministry of Law and Justice, 2010, online source: http://www.moef.nic.in/downloads/public-information/NGT-fin.pdf

3

Financial Institutions

3.1 Introduction

In this chapter, we will look at the role of the finance and banking sector in fostering sustainable investments. This sector is one of the most influential sectors influencing infrastructure, business practices, products and services markets, and social development. We will discuss the different types of financing institutions like the Development Finance Institutions (DFIs) (also called development banks) and Private Sector Finance Institutions (PSFIs) that over time have evolved their business policies and frameworks to address the environmental and social impacts of their investments and associated activities. This chapter traces some of the early and more recent national and international responses by the financial sector and financial mechanisms and instruments such as stock market indices and green bonds.

Developing countries have often sought *Development Finance* to spur growth. Development Financing can be classified into three types, each of which plays a role in meeting public and private finance requirements.

- Budgets allocated by national governments.
- Assistance obtained in the form of concessional loans from DFIs or grants from foundations (like the Bill & Melinda Gates Foundation or Gordon and Betty Moore Foundation that are mostly targeted towards growth and social objectives).
- Non-concessional loans from DFIs or private banks or investors, typically used for infrastructure or other revenue generating projects like telecommunications, energy generation or e-commerce in the form of equity or commercial loans.

3.2 Development Financing Institutions

DFIs have been set up to support private and public sector development at national, regional, and global scales.

DFIs can be of various types—national, regional, bilateral or multilateral. Every DFI has its own set of social, environmental, and economic development objectives in either the public or private or both sectors. Most DFIs follow their own funding policies and frameworks, though there have been significant efforts in recent times to harmonize the policies and monitoring frameworks.

Bilateral DFIs are independent financing institutions governed by a single national government while Multilateral DFIs are governed by a group of member countries. Some examples of Bilateral DFIs are Netherlands Development Finance Company (FMO), Development Bank of Austria (OeEB), Belgian Investment Company for Developing Countries (BIO), Denmark's Investment Fund for Developing Countries (IFU), Finnish Fund for Industrial Cooperation Ltd. (FINNFUND), and German Development Bank (KfW). Examples of Multilateral DFIs include the World Bank, International Finance Corporation (IFC), Asian Development Bank (ADB), Inter-American Development Bank (IDB), European Investment Bank (EIB), and others.[1]

The World Bank is one of the largest and the most influential international DFIs that has been providing financial and technical assistance to developing countries across the globe. Headquartered in the United States, the World Bank is made up of 189 member countries compared to 193 member countries of the United Nations. World Bank is a co-operative, which is headed by a Board of Governors made up of its member countries. World Bank provides low-interest loans, zero to low-interest credits and grants to developing countries.[2]

ADB is a regional DFI headquartered in the Philippines with 67 member countries of which 48 are from Asia. ADB was established with the objective of fostering development in the poorest countries in the Asian region.[3] IDB is another regional DFI that caters to the Latin American and Caribbean region. KfW is a DFI owned by the German government that promotes development in Germany, Europe, and developing countries. FMO is amongst the largest bilateral DFIs and plays a crucial role in providing credit in the form of higher risk loans, equity and risk guarantee instruments to the private sector in developing countries. FMO focuses on sectors like financial institutions (intermediaries), energy, and agribusiness.[4]

DISCUSSION QUESTIONS

- *List three major international DFIs that in your opinion do impactful work in your region. Present statistics on the financial assistance provided to say ten of the beneficiary member countries and make a plot between percentage financial assistance to their GDPs and the Human Development Index (HDI). Make observations on this basis.*
- *There are DFIs operating at the national level as well, for example, National Agricultural Bank for Rural Development (NABARD) in India and the National Bank of Egypt. Describe two major DFIs of national mandate in your country. Check whether these DFIs work with the international DFIs listed above and describe some of the major national programs or schemes that are jointly supported over the last ten years.*

These DFIs generally operate on an independent basis but sometimes partner with other DFIs to meet their objectives and borrower's needs.

Though traditionally DFIs were set up to support development in the public sector, many DFIs have set up subsidiaries to support development in and through the private sector. IFC is a member of the World Bank Group (WBG) and is the world's largest development institution that supports the private sector in developing countries.[5] Similarly, ADB's Private Sector Operations Department (PSOD) invests in privately held and state-sponsored companies across a wide range of industry sectors like infrastructure, financial services, clean energy, and agribusiness throughout Asia.[6]

The sector-wise investments made by the DFI's are of two types: concessional financing and non-concessional financing. In 2012, the major portion of finances were distributed/ invested towards Economic infrastructure which represented 43% and 47%, for concessional financing and non-concessional financing, respectively.

Social infrastructure (25% and 20%) and production sectors (14% and 15%) while a minor portion (9% and 17%) of the finances went to sectors like Humanitarian aid, community aid, etc.[81]

3.3 Private Sector Finance Institutions

PSFIs are private entities like banks and hedge funds that are owned by shareholders and not by the government. PSFIs are usually both business and development oriented but getting a good economic return from the investments is the principal focus.

Some of the leading PSFIs across the world include JP Morgan, UBS, Credit Suisse, Citigroup Inc., and HSBC, that focus on sustainability and social initiatives. For example, Citigroup is committed to international agreements like Equator Principles, UNEP Financial Initiative and Roundtable on Sustainable Palm Oil. Citigroup has an environmental and social policy framework and invests in alternative energy and green bonds. JPMorgan Chase Bank has a Global Environmental and Social Risk Management team that implements the objectives of its environmental and social policy framework.

3.4 Evolution of Environmental and Social Safeguards at Development Financial Institutions

In the initial period, most DFIs did not have an environmental and social policy in place. Meeting the national and local requirements of compliance was the principal criteria during appraisal.

As it became evident that development effectiveness could be sustainable only when a comprehensive environmental and social policy and framework was applied, DFIs like the World Bank evolved Operational Policies (OPs) and Bank Procedures (BPs). These OPs and BPs were applied in addition to the requirements of the law of the land and influenced the project concept, design, and implementation as well as operations.

We will use the World Bank as an example to gain a general understanding of the evolution of environmental and social policies and the operational framework.

The World Bank was established in 1944 to help reconstruct European countries that were destroyed during World War II. Till the 1960s, World Bank funded infrastructure projects like dams, irrigation systems, and roads. In the 1970s, World Bank shifted its focus towards poverty alleviation and issues like food production, population, education, health, and nutrition in developing countries.[7]

The World Bank consists of five principle institutions—International Bank for Reconstruction and Development (IBRD), International Development Association (IDA),

International Finance Corporation (IFC), Multilateral Guarantee Agency (MGA) and International Centre for the Settlement of Investment Disputes (ICSID). IBRD is the original World Bank institution. IBRD provides assistance to middle-income and credit-worthy low-income countries while IDA gives credit-free loans and grants to the poorest countries. IFC provides assistance to only the private sector, MGA promotes foreign direct investments in developing countries and ICSID provides advisory conciliation and arbitration of investment disputes.[8]

World Bank projects faced numerous economic challenges and faced criticism due to their adverse social and environmental impact. In the 1970s and 1980s, many projects sponsored by the World Bank became controversies.

- In Indonesia, a World Bank funded project caused millions of non-Javanese people to resettle by force in India. The World Bank invested in the Narmada dam projects that displaced 300,000 people. Subsequently, the World Bank withdrew from this project.
- In Brazil, the World Bank supported a 1500 km long highway development project through the Amazon. It caused land conflicts, widespread deforestation, and triggered disease, epidemics, and crime.

Due to such controversies, civil society organizations and the media began to scrutinize and criticize the World Bank's investments and its social and environmental impact. Environmental and social safeguards were then developed to put in place binding environmental and social policies designed to protect the people and the environment.

In 1980, the World Bank adopted a policy on involuntary resettlement to protect the population displaced due to World Bank funded projects. In 1981, in anticipation of the controversial project in Brazil, World Bank adopted the indigenous people's policy, to protect the interests and rights of indigenous people in the project regions. Though these policies were not termed as *safeguards* at the time, they were a first attempt to addressing social and environmental concerns associated with World Bank projects.

Rising concerns and pressure from civil society organizations and donor governments like the United States and European governments led to the establishment of an Environmental Department in 1987 at the World Bank. Over the next decade, initiatives like mandatory Environmental Impact Assessment of projects prior to the Board's approval, creation of World Bank Inspection Panel for affected communities to register complaints and formalization of Safeguard policies were introduced.[9] Figure 3.1 provides an overview of this evolution.

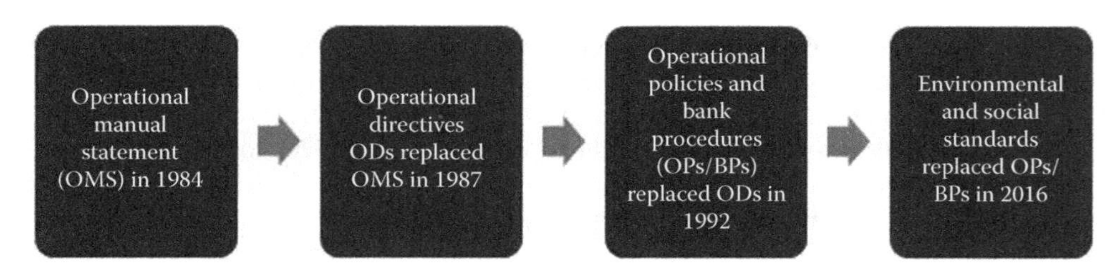

FIGURE 3.1
Evolution of safeguards frameworks at the World Bank. (Credit: Environmental Management Centre LLP, Mumbai.)

Based on an Independent Evaluation Group (IEG) report in 2010, the World Bank revised the existing safeguards policies to increase environmental and social coverage and harmonization across the five WBGs. The feedback from the report and consultations with peers and borrowers resulted in the development of the World Bank's new Environmental and Social Framework (ESF) that replaced the OPs/BPs. The ESF involved important advances in areas such as transparency, non-discrimination, social inclusion, public participation, and accountability—including expanded roles for grievance redress mechanisms. The framework's vision statement references social inclusion and human rights. Enhancing borrowing countries' capacities and management systems is a central development goal set by the World Bank and its shareholders. Mechanisms supporting this goal were included in the ESF.

Taking a cue from the World Bank, other DFIs like ADB and IDB introduced safeguards of their own. Inter-agency workshops were held to disseminate knowledge and share experiences. Some DFIs went beyond the original World Bank's policies and introduced mandates for Social Impact Assessment and Labor Rights, which are now addressed in the new World Bank ESF.

For instance, in 2003 ADB adopted a mechanism for communities adversely affected by ADB-financed projects to report violations of ADB's policies and safeguards, express their grievances and seek solutions or justice. In 2009, the ADB introduced a new Safeguard Policy Statement (SPS) combining the three separate policies—Indigenous People (1998), an Involuntary Resettlement Policy (1995) and an Environment Policy (2002). See Table 3.1 for an overview of safeguard policies of various DFIs.

Figure 3.2 shows typical integration of the Environmental and Social (E&S) considerations in the project cycle.

DISCUSSION QUESTIONS

- *Expand Table 3.1 to include the safeguard policies at EIB and FMO. Describe each of the policies in brief for their coverage and intent.*
- *Prepare a table that lists policies in rows and DFIs in columns as a checklist to understand the coverage of policies across the DFIs.*

World Bank categorizes projects based on the extent of their environmental impact. The Environmental Assessment (EA) requirements vary for each of these categories. The project categories are[10]:

- *Category A (High risk projects)*: A project is classified as a Category A project if it is likely to have significant adverse environmental impacts that are sensitive, diverse, or unprecedented. These impacts may affect an area broader than the sites or facilities subject to physical works. An EA is required for Category A projects.
- *Category B (Moderate risk projects)*: A project is classified as a Category B project if its potential adverse environmental impacts on human populations or environmentally important areas—including wetlands, forests, grasslands, and other natural habitats—are less adverse than those of Category A projects. These impacts are site-specific, few if any of them are irreversible, and in most cases mitigatory measures can be designed more readily than for Category A projects. The EA scope is narrower than the EIA scope for Category A.

TABLE 3.1
Safeguard Policies at DFIs

Organization	Overview of Safeguard Policies & Year of Introduction or Amendment
World Bank	**Operational Policies/Bank Procedures**[a] Environmental: 4.01 Environmental Assessment (1999) 4.04 Natural Habitats (2001) 4.09 Pest Management (1998) 4.11 Physical Cultural Resources (2006) 4.36 Forests (2002) 4.37 Safety of Dams (2001) Social: 4.10 Indigenous Peoples (2005) 4.12 Involuntary Resettlement (2001) Legal: 7.50 International Waterways (2001) 7.60 Disputed Areas (2001) **Environmental and Social Standards (ESS) (Introduced on August 4th, 2016)**[b] ESS 1: Assessment and Management of Environmental and Social Risks and Impacts ESS 2: Labor and Working Conditions ESS 3: Resource Efficiency and Pollution Prevention and Management ESS 4: Community Health and Safety ESS 5: Land Acquisition, Restrictions on Land Use and Involuntary Resettlement ESS 6: Biodiversity Conservation and Sustainable Management of Living Natural Resources ESS 7: Indigenous Peoples/Sub-Saharan African Historically Underserved Traditional Local Communities ESS 8: Cultural Heritage ESS 9: Financial Intermediaries ESS 10: Stakeholder Engagement and Information Disclosure
IFC/MGA	**Policy and Performance Standards on Social and Environmental Sustainability (2006)** Environmental and Social: PS 1: Social and Environmental Assessment and Management System Environmental: PS 3: Pollution Prevention and Abatement PS 6: Biodiversity Conservation and Sustainable Natural Resource Management PS 8: Cultural Heritage Social: PS 2: Labor and Working Conditions PS 4: Community Health, Safety and Security PS 5: Land Acquisition and Involuntary Resettlement PS 7: Indigenous Peoples
ADB	**Safeguard Requirements (2009)**[c, a] SR 1: Environment *(Environmental Assessment, Environmental Planning and Management, Information Disclosure, Consultation and Participation, Grievance Redress Mechanism, Monitoring and Reporting, Unanticipated Environmental Impacts, Biodiversity Conservation and Sustainable Natural Resource Management, Pollution Prevention and Abatement, Health and Safety, Physical Cultural Resources)* SR 2: Involuntary Resettlement *(Compensation, Assistance and Benefits for Displaced Persons, Social Impact Assessment, Resettlement Planning, Negotiated Land Acquisition, Information Disclosure, Consultation and Participation, Grievance Redress Mechanism, Monitoring and Reporting, Unanticipated Impacts, Special Considerations for Indigenous Peoples)* SR 3: Indigenous People *(Consultation and Participation, Social Impact Assessment, Indigenous Peoples Planning, Information Disclosure, Grievance Redress Mechanism, Monitoring and Reporting, Unanticipated Impacts)* SR 4: Special Requirements for Different Finance Modalities *(Multitranche Financing Facilities, Emergency Assistance Loans, Existing Facilities, Financial Intermediaries, General Corporate Finance)*

(Continued)

TABLE 3.1 (*Continued*)

Safeguard Policies at DFIs

Organization	Overview of Safeguard Policies & Year of Introduction or Amendment
IDB	**Operations Policies**[d] OP-703: Environment and Safeguards Compliance Policy (2006) OP-710: Involuntary Resettlement Operational Policy (1998) OP-765: Operational Policy on Indigenous Peoples (2006) OP-704: Disaster Risk Management Policy (2007) OP-761: Operational Policy on Gender Equality in Development (2010)

[a] Based on Safeguard Policy Statement, 2009, Asian Development Bank, online source: https://www.adb.org/sites/default/files/institutional-document/32056/safeguard-policy-statement-june2009.pdf

[b] Based on Environmental and Social Framework: Setting Environmental and Social Standards for Investment Project Financing, 2016, World Bank, online source: http://consultations.worldbank.org/Data/hub/files/consultation-template/review-and-update-world-bank-safeguard-policies/en/materials/the_esf_clean_final_for_public_disclosure_post_board_august_4.pdf;

[c] Based on Operations Manual Bank Policies (BP), 2013, Asian Development Bank, online source: https://www.adb.org/sites/default/files/institutional-document/31483/om-f1-20131001.pdf

[d] Based on OVE: Office of Evaluation and Oversight, Environmental and Social Safeguards including Gender Policy, 2013, Inter-American Development Bank, online source: https://publications.iadb.org/bitstream/handle/11319/5870/IDB-9%3a%20Environmental%20and%20Social%20Safeguards%2c%20Including%20Gender%20Policy.pdf?sequence=1

- *Category C (Low risk projects)*: Category C, if it is likely to have minimal or no adverse environmental impacts. The EA process required for Category C projects are limited to only screening and incorporation of best practices is deemed to be sufficient.
- *Category FI*: Category FI is applicable if made through a Financing Intermediary (FI) in subprojects, not known a priori, that may result in adverse environmental

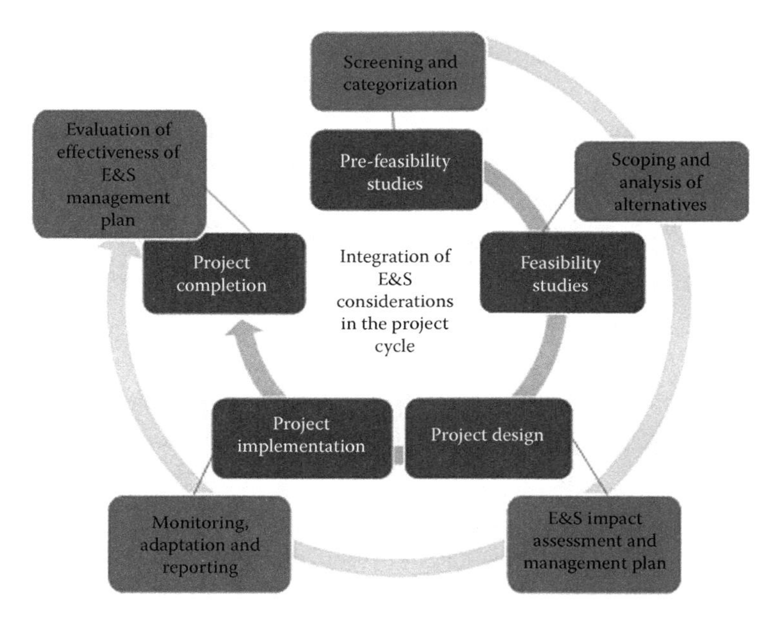

FIGURE 3.2
Integration of environmental and social perspectives in the project cycle. (Credit: Environmental Management Centre LLP, Mumbai.)

impacts. A FI is then expected to set up an Environmental and Social Management Framework (ESMF) with adequate institutional capacity and appropriate operating procedures following the steps outlined earlier in Chapter 2 (Section 2.12) for Project EIA. It is the category FI of the Bank that has significantly influenced the national level DFIs.

The above categorization is also followed by DFIs like the ADB.

When the project gets categorized as FI, the World Bank works with the national level DFIs and provides technical and financial assistance to set up ESMF reflecting the Bank's environmental and social safeguards. Figure 3.3 shows the structure of a typical ESMF at the FI.

We present in Box 3.1 a case study of Environmental and Social Safeguards Framework (ESSF) at Infrastructure Development Company Limited in Bangladesh.

3.4.1 Safeguards Approach of the Development Financing Institutions

As discussed in the previous section, safeguards were introduced to minimize adverse social and environmental risks associated with DFI funded projects and to introduce a

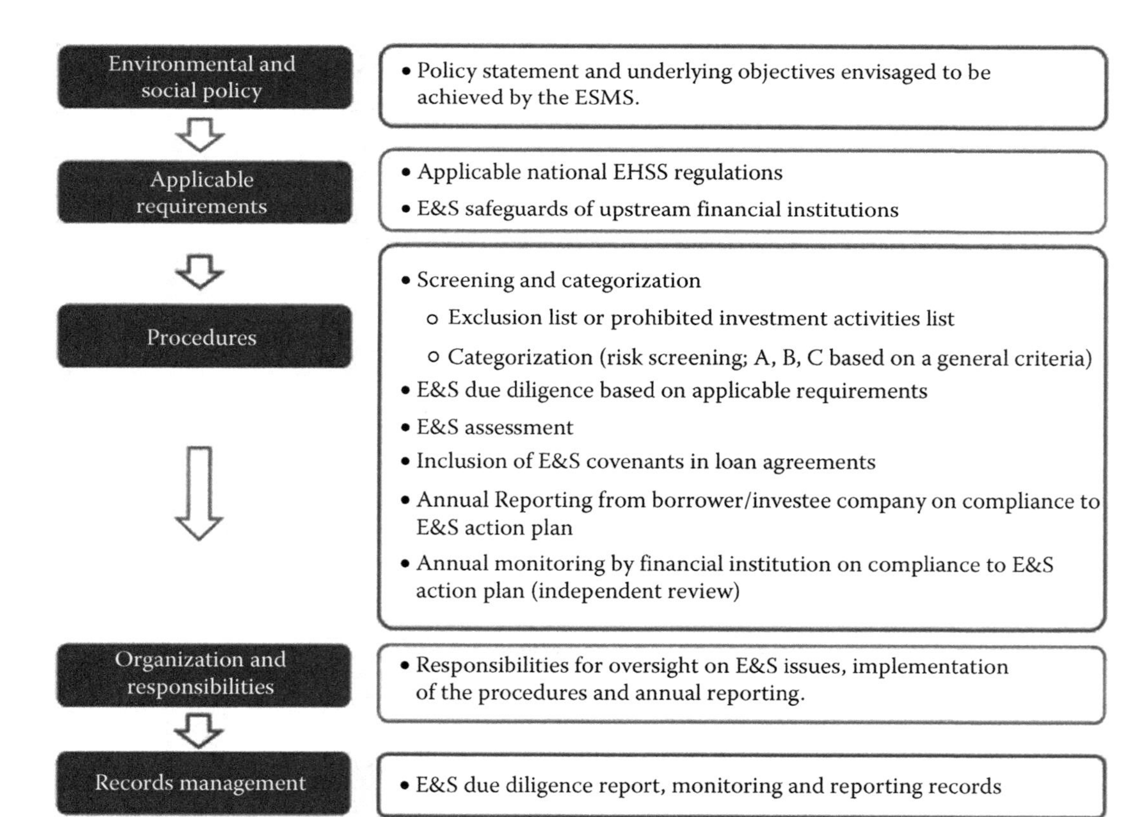

FIGURE 3.3
Typical structure of environmental & social management framework at a financing institution. (Credit: Environmental Management Centre LLP, Mumbai.)

BOX 3.1 ENVIRONMENTAL AND SOCIAL SAFEGUARDS FRAMEWORK (ESSF) AT INFRASTRUCTURE DEVELOPMENT COMPANY LIMITED, BANGLADESH

The Government of Bangladesh and the World Bank jointly established Infrastructure Development Company Limited (IDCOL) in 1997 to facilitate economic development in Bangladesh. IDCOL serves as a government owned Non-Banking Financial Institution that finances small-, medium-, and large-scale infrastructure projects like power generation, telecommunications, roads and bridges, gas related infrastructure, water supply and distribution, and renewable energy projects in Bangladesh and also encourages private sector investments. IDCOL offers a full range of financing solutions to private-sector owned infrastructure and energy projects including long-term local and foreign currency loans, working capital loans, debt and equity arrangement, debt restructuring, takeover financing, and financial advisory services. IDCOL also provides training courses on project financing and environmental and social safeguards.[11]

INFLUENCE OF MULTILATERAL BANKS ON IDCOL ESF

IDCOL sources funds from the Government of Bangladesh and DFIs like World Bank, ADB, Japan International Cooperation Agency (JICA), German International Cooperation (GIZ), Global Environment Facility (GEF), and others and has to comply with the mandates of its donors.

In 2000, IDCOL developed an Environmental and Social Appraisal Manual (ESAM) as mandated by the World Bank to process the Private Sector Infrastructure Fund (PSIF). Adoption of ESAM ensured adherence to World Banks's environmental and social safeguard policies.

Similarly, IDCOL developed an Environmental and Social Safeguards Framework (ESSF) in 2008 as mandated by one of its largest donors, ADB, who was financing $165 million through the Public-Private Infrastructure Development Facility (PPIDF). ESSF was built on ESAM that IDCOL had already developed.[12] For a second tranche of the loan in the year 2011, ADB required IDCOL to update the ESSF in line with ADB's SPS of 2009. To assimilate these multiple frameworks and policies, IDCOL developed a harmonized ESSF (Figure 3.4).

ENVIRONMENTAL AND SOCIAL SAFEGUARDS FRAMEWORK[11]

The ESSF included the following elements:

a. Integrated Environmental and Social Policy
b. Screening of Projects activities under the List of Prohibited activities and for the Environmental and Social Risk Rating (high, moderate or low risk) for the Project
c. Operational Procedures for Risk Mitigation and Control like EIA, site visits, due diligence, corrective action plans development and more
d. Institutional structure for operation of the ESSF, including allocation of roles and responsibilities

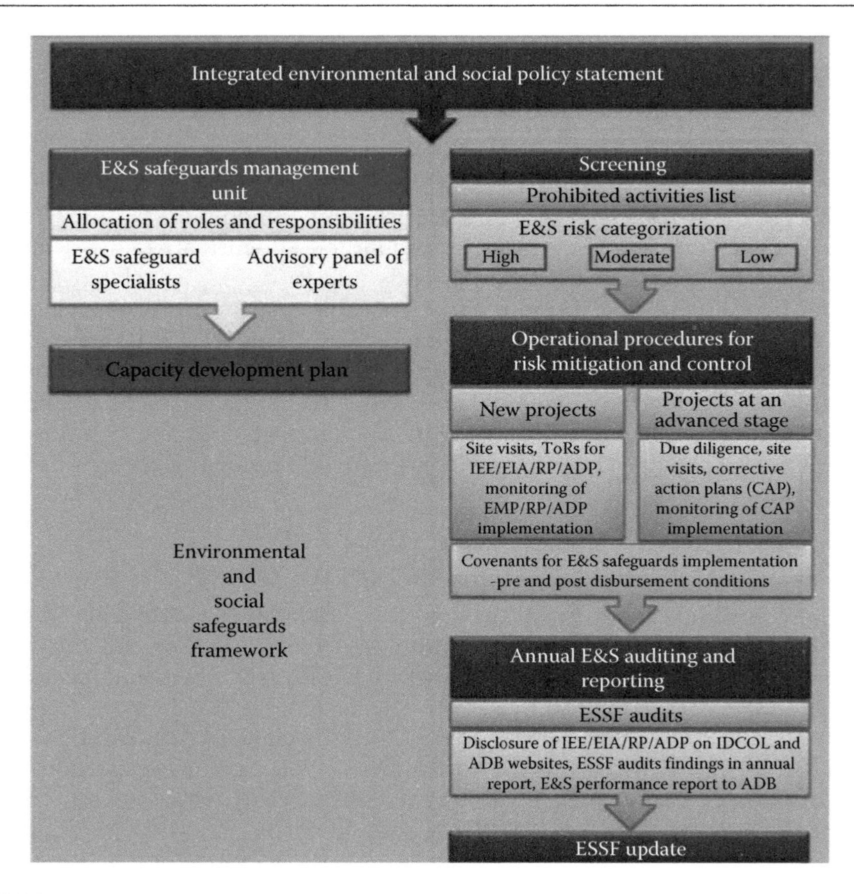

FIGURE 3.4
IDCOL's environmental & social safeguards framework structure. (From Infrastructure Development Company Limited (IDCOL), Environmental and Social Safeguards Framework (ESSF), Policy and Procedures, pp. 40, August 2011. With permission, online source: http://idcol.org/download/1d8514287c3e7cda76423b33a781f79c.pdf.)

e. Advisory Panel of Experts to guide the Environmental and Social Safeguards Management Unit
f. Capacity development plan at IDCOL for continued strengthening
g. Annual E&S auditing and reporting
h. Updating of the ESSF based on feedback and experience of its application to projects

DISCUSSION QUESTIONS

- *Prepare case studies like IDCOL on national level DFIs in your country that have developed ESSFs or equivalent due to requirements of category FI.*
- *To what extent are these ESSFs effective?*
- *Does operation of ESSFs add to a significant cost? Does investment in ESSF make business sense?*

set of regulations, monitoring, and reporting procedures to guide organizations that seek funds from DFIs. Introduction of safeguards can be viewed as establishing an accountability mechanism that protects the interests of DFIs, affected communities, and the environment.

DFIs often fund development projects in developing countries that lack stringent laws, governmental policies, and effective project planning, execution, and monitoring processes and systems. For these reasons, DFIs extended their support to countries over time following multiple approaches to implement the safeguard systems. The four principal approaches to safeguard systems that DFIs followed are described below:

- *Traditional approach*: DFI determines the regulations and procedures for the recipient government to follow without relying on national laws and institutions.
- *Country systems approach*: DFI approves use of the borrower government's laws and institutions as adequate and provides support as required.
- *Principles based approach*: DFI provides broad guiding principles that the borrower government can use to develop a safeguard system of its own and provide monitoring reports to the DFI.
- *No safeguards approach*: DFI relies completely on the local laws and institutions without any guiding principles.

In the early stage, most DFIs followed the traditional approach. The No Safeguards approach is not used anymore. More recently, attempts have been made to follow the Country Systems and the Principles-based approach. We discuss the Country Systems approach in the following section.

DISCUSSION QUESTIONS

- *Research one example of each of the above safeguard approaches (Visit http://www.bankinformationcenter.org/en/Document.102926.pdf).*
- *Discuss the merits and demerits of each approach.*

3.4.1.1 Use of Country Systems

With the objectives of scaling up the development impact (in terms of effectiveness and sustainability), increasing country ownership and reducing banks operational or supervision costs, the World Bank introduced *Use of Country Systems (UCS)*. The UCS was intended to respond to emerging needs and provide flexibility to borrowers whose systems for environmental management were well developed and functional.

In 2004, the World Bank piloted the UCS. For the UCS, the World Bank carefully assessed the borrower government's laws and institutions to address social and environmental risks and approved its use instead of World Bank's safeguards. This required establishing the policy equivalence and assessment of implementation capacities to arrive at the acceptability. In January 2008, the World Bank issued an evaluation report of the initial phase of the Pilot Program for UCS. By 2010, World Bank piloted this approach in 13 countries.[13] Boxes 3.2 and 3.3 show examples of UCS pilots conducted in India and Bhutan.

BOX 3.2 UCS PILOT IN INDIA[14,15]

Power Grid Corporation of India Ltd. (POWERGRID) is the largest electrical power transmission utilities in India. POWERGRID was the first Asian company to establish an Environmental and Social Policy and Procedures (ESPP) to manage the environmental and social impact of its activities in 1998. The company upgraded ESPP in 2005 and 2009 to ensure that it is synchronous with national policies and with the safeguards of the DFIs that were providing funds.

POWERGRID was chosen as a UCS pilot in India. The World Bank conducted an acceptability assessment of POWERGRID's ESPP considering the following:

1. Institutional Framework and Capacity
2. Processes and Procedures
3. Outputs: Environmental Assessments (EAs)/Environmental Management Plans (EMPs), Rehabilitation Action Plans, and, where applicable, Tribal Peoples Development Plans
4. Outcomes: Observed impacts of projects on the natural and social environment

Based on this assessment, World Bank proposed an Action plan to meet certain gaps in the ESPP to ensure that ESPP was fully in compliance with World Bank's safeguards. Use of UCS was subsequently approved.

 **DISCUSSION QUESTION**

- *What are the advantages of the UCS approach to the Development Financing Institution such as the World Bank and to the Borrower? What are the risks?*

3.4.2 Community Driven Development Approach by the World Bank

When the community members are provided with guidance, access to information, and technical assistance and financial support, they can effectively identify and address issues as well as opportunities and address them by working in partnership with local governments and financial institutions. World Bank developed an approach based on this premise called Community-Driven Development to enhance inclusivity in project planning and implementation.

Community-Driven Development[18] is an approach that gives control of development decisions and resources to community groups. Poor communities receive funds, decide on their use, plan and execute the chosen local projects, and monitor the provision of services that result. It improves not just incomes but also people's empowerment, the lack of which is a form of poverty as well.

Community-Driven Development (CDD) programs operate on the principles of transparency, participation, local empowerment, demand-responsiveness, greater downward accountability, and enhanced local capacity.

BOX 3.3 BHUTAN SECOND RURAL ACCESS PROJECT PILOT FOR COUNTRY SAFEGUARD SYSTEM

Bhutan was selected by the World Bank for the UCS as a two-year pilot in 2005. The World Bank team conducted their assessment and approved the use of Bhutan's environmental and social policies for the Second Rural Access Project (RAP II).

Under RAP II, Bhutan had in place the Environmental Friendly Road Construction (EFRC) techniques that were designed and developed to ensure that the road constructions in Bhutan were environmentally friendly. The EFRC factored in site conditions such as geologically sensitive areas, wet areas, rocks, paddy fields, and areas with high environmental and cultural value in order to prevent and mitigate environmental damage to these sensitive areas. The technique also adhered to Environmental & Social (E&S) principles like conducting EIA to assess the social and ecological risks and avoiding areas that are geologically and culturally sensitive. Good practices like minimum cutting of trees, controlled blasting to minimize damage to the environment, bioengineering to ensure structural stability of the sloped terrains, implementation of proper drainage systems to avoid flooding and soil erosion were included. Health and safety measures such as use of helmets, safety shoes, and eye and dust protection gear were also incorporated. The EFRC was expected to enhance the life span of the road giving a reasonable economic return while mitigating adverse environmental and social impacts.[16]

The EFRC was found to meet the requirements of the UCS. The World Bank helped the Department of Roads of Bhutan to develop their institutional capacity to implement EFRC across all rural access projects.[17]

 **DISCUSSION QUESTIONS**

- *Is a formal EIA process necessary for a EFRC based project?*
- *Can the EFRC approach be followed nationwide for all road projects irrespective of whether the projects were sponsored by development financial institution such as the World Bank?*

The World Bank recognized that CDD approaches and actions are important elements of an effective poverty reduction and sustainable development strategy. The Bank has supported CDD across a range of low- to middle-income, and conflict-affected, countries to support a variety of urgent needs, including water supply and sanitation, school and health post-conflict construction, nutrition programs for mothers and infants, rural access roads, and support for micro-enterprises.[19]

We illustrate in Box 3.4 some examples of the CDD approach.

3.4.3 Harmonization of Safeguards

Various DFIs followed their own set of safeguard policies and operating procedures. Many borrowing governments worked with multiple DFIs in the same program/project. DFIs soon realized the borrowing countries had difficulties to meet a wide range of donor requirements, many times duplicating the effort, sometimes leading to procedural conflicts for preparing, delivering, and monitoring the projects. There was a need to harmonize the

BOX 3.4 EXAMPLES OF COMMUNITY-DRIVEN DEVELOPMENT PROGRAMS

KECAMATAN DEVELOPMENT PROGRAM IN INDONESIA

The Kecamatan Development Program (KDP) in Indonesia, the world's largest CDD program, began in 1998. The program is currently in its third phase and has expanded to a nation-wide program. As of 2008, the program covered 32 of 33 Provinces, benefiting 38,000 villages. The program directly benefited an estimated 18 million people (as of December, 2006). The major factors responsible for the benefits were provision of basic infrastructure, including around 37,000 kilometers of local roads and 8,500 bridges, 9,200 clean water supply units, and 3,000 new or improved health posts. Through micro-financing, more than 1.3 million people obtained loans to start or complement local businesses. Furthermore, around 101,500 young people received scholarships to attend school and 5,100 schools have been built or rehabilitated.[20]

ANDHRA PRADESH IN INDIA

The Andhra Pradesh District Poverty Initiatives Program in India increased incomes for almost 90% of poor rural households, including 10 million women as of April, 2009. The number of households with access to credit rose from fewer than half a million in 2000 to more than 10 million in 2008. The annual credit flow to poor households increased from $42 million to $2.7 billion in 2008. Before the project less than 1,000 people had death and disability insurance coverage. Post project this number has risen to more than 9 million. Income sources shifted from subsistence wages to diversified self-employment and ownership of assets. Having seen the results of this CDD effort, the State of Andhra Pradesh is now using a similar approach to taking measures against climate change, especially in the area of water management in arid agricultural areas.[21]

NATIONAL SOLIDARITY PROGRAM IN AFGHANISTAN[22]

Since its launch in 2003, the National Solidarity Program (NSP), the government's principal community development program, has mobilized almost $2.05 billion in donor and government funding. The program worked through more than 35,000 community-elected Community Development Councils (CDCs) in all the 34 provinces of the country to finance over 88,000 community-level infrastructure schemes in the areas of water supply and sanitation, rural roads, irrigation, power, health, and education.

From 2003 to 2016, the NSP helped construct or rehabilitate almost 53,600 kilometers of roads; provide access to improved water sources to more than 11.7 million people by constructing approximately 86,300 improved community water points; generate 32 megawatts of power; irrigate more than 524,000 hectares of land; and build almost 2,000 classrooms.

In 2015, the NSP responded to the government's launch of the *Jobs for Peace* initiative by providing a *Maintenance Cash Grants scheme* for the maintenance of community infrastructure. Operational in 45 districts in 12 provinces, the scheme is meant to quickly disburse funds directly to rural communities to generate short-term employment. As of July 2016, a total of 1.9 million labor days have been generated. With an additional financing of $57 million, the grants scheme is expected to expand to all provinces and create over 11 million paid labor days that will directly benefit over 440,000 households.

NATIONAL INITIATIVE FOR HUMAN DEVELOPMENT IN MOROCCO

The National Initiative for Human Development (INDH) was launched in 2005 to improve the living conditions of poor and vulnerable groups through enhanced economic opportunities, better access to basic and social services, and improved governance. Phase 2 of INDH (2011–2015) expanded the target population and geographic scope, from 667 to 1,234 communities, and almost doubled resource allocation from $1.2 billion to $2.1 billion over five years. During the period 2005–2015, approximately 44,600 community-driven sub-projects were financed, providing more than 10 million beneficiaries with access to basic social and economic infrastructure services and training.

The impact evaluation of the first phase of INDH showed a 20% increase in revenues for participating rural households, and 10% increase for urban households, a reduction in child mortality from 6.2 to 0.9 deaths per 1,000 births, and a decrease in malnutrition from 14.4% of the population to 4.3%.

The World Bank supported the second phase of INDH through the first Program-for-Results operation; a third phase of INDH has been planned for the period 2016–2020, thanks to strong political support and a high-level of financial mobilization.

DISCUSSION QUESTIONS

- *How does one apply the environmental and social safeguards to the CDD projects? Can a UCS approach be used?*
- *Read more on the Program-for-Results approach introduced by the World Bank. In the context of CDD, what could be the risks as well as opportunities?*
- *Discuss the potential of the CDD model to improve climate resilience of vulnerable communities. Cite examples where such efforts have been carried out.*

safeguard policies and operating procedures amongst DFIs to benefit the borrowing countries and improve the effectiveness of the development assistance.[23]

3.4.3.1 Rome Declaration on Harmonization

In 2003, major DFIs gathered in Rome to reaffirm their commitment to developmental growth and to harmonizing their policies and processes to improve the effectiveness and successes of their individual and combined effort towards developmental activities in developing partner countries. In Rome, DFIs agreed to review their safeguard frameworks and identify ways to adapt them to borrowing countries' priorities, policies, and procedures. They agreed to deliver development assistance as per the partner countries' immediate requirements. They committed to providing assistance to national governments to enhance their leadership capacity and a sense of ownership towards developmental projects that were funded.[24]

3.4.3.2 Paris Declaration on Aid Effectiveness

In 2005, at the Paris Convention on Aid Effectiveness, multilateral organizations (donors) and borrower countries (or partner countries) gathered with the objective to co-ordinate and streamline development activities and processes between donors and the borrower countries. In addition to reforming and simplifying the policies and procedures mandated by donors to align with national priorities, another key objective was to enhance aid effectiveness during complex and emergency situations like tsunamis and other disasters.

The DFIs committed to UCS as much as possible to increase reliance on national policies and safeguards. Further, they committed to harmonize performance assessment frameworks for UCS to avoid conflicts that may arise for partner countries in complying with the UCS related procedures of different donors. In turn, the borrower countries committed to undertake any reforms required to ensure that the national systems, institutions and policies are effective, accountable, and transparent.[25]

One of the key differences between the impact of the Rome Declaration and Paris Convention was that the latter was more action-oriented. The Paris Declaration on Aid Effectiveness Convention recommended following five principles to make aid more effective:[82]

1. *Ownership*: Developing countries set their own strategies for poverty reduction, improve their institutions, and tackle corruption.
2. *Alignment*: Donor countries align behind these strategies and use local ESG systems.
3. *Harmonisation*: Donor countries coordinate, simplify procedures and share information to avoid duplication of efforts.
4. *Results*: Developing countries and donors shift focus to development results and measurement of the impact.
5. *Mutual accountability*: Donors and borrowers are accountable for development results.

Despite the fact that the international community addressed the effectiveness issue through the Paris Declaration and the subsequent Accra Agenda for Action, the implementation of the agreed actions has been difficult.

Governments and aid agencies have made commitments at the leadership level, but little has happened pursuing top-down, aggregating the targets. Decades of development have shown that if countries are to become less dependent on aid, they must follow a bottom-up approach, where they determine their own priorities and rely on their own systems to deliver that aid.[83]

It is not surprising therefore that the report of 2011 of the OECD-Development Assistance Committee's "Aid Effectiveness 2005-2010: Progress in Implementing the Paris Declaration" revealed that only one out of the 13 targets established for 2010 was met.

Participants formulated a mechanism to monitor progress of results and also create stronger mechanisms for transparency and accountability based on 12 indicators of progress with targets that were set to be achieved by 2010.[26]

DISCUSSION QUESTIONS

- *Visit http://www.oecd.org/dac/effectiveness/34428351.pdf for details on the 12 indicators and related targets developed under the Paris Declaration on Aid Effectiveness.*
- *Make a commentary on their relevance in the context of the 17 SDGs.*

3.5 Evolution of Environmental and Social Safeguards at Private Sector Financing Institutions

PSFIs followed the trend of DFIs in formulation of the environmental and social safeguards. In the initial phase of 1970s and 1980s, many PSFIs faced challenges due to the experience of land assets turning into liabilities because of site contamination. The PSFIs often used land parcels as collateral against loans. Many of these sites were found to be contaminated due to some previous instances of pollution or disposal of hazardous wastes. When the Comprehensive Environmental Response, Compensation, and Liability Act of 1980 (CERCLA) was introduced in the United States, banks were held liable for the environmental pollution created by their clients or borrowers and had to pay environmental fines or costs associated with remediation activities. Many banks went bankrupt due to such costs that were not anticipated. This liability prompted them to take into account site assessments and a comprehensive social and environmental appraisal. In the 1980s American banks were amongst the first to adopt such policies followed by European banks in the mid-1990s for better credit risk management.[27]

The environmental and social considerations at banking institutions thus emerged to mainly address the business risks. Later, public pressure and scrutiny, tightening of environmental regulations, and reputational risks were the principal drivers. The global financial economic crisis in 2007–2008 added an impetus. More recently, climate change related risks got added to the checklist of risks. PSFIs also realized that compliance to national laws and regulations was not enough to insulate from the environmental and social risks. There was need to set up environmental and social risk management systems beyond compliance. This led to establishment of Environmental, Social and Governance (ESG) frameworks in appraisal of investments.

Recognizing the importance of ESG, many initiatives emerged to integrate sustainability in the financial sector in the form of consortiums, working groups, formation of guiding principles or frameworks, performance reporting and products and services centered around sustainability.

Figure 3.5 shows the trail:

Some of the key financial sector initiatives that focus on sustainable development are:

- United Nations Environmental Program Financial Initiative UNEP FI (1992) (http://www.unepfi.org/)
- Global Reporting Initiative Financial Services Sector Supplement (2002) (https://www.globalreporting.org/information/g4/sector-guidance/sector-guidance/financial-services/Pages/default.aspx)
- Global Alliance for Banking on Values (2008) (http://www.gabv.org/)

Amongst the above, UNEP FI has played a catalytic role influencing the banking sector. We will discuss the role of UNEP FI in the next section.

3.5.1 United Nations Environment Program's Finance Initiative

Established in 1992 at the Earth Summit in Rio, Geneva-based United Nations Environment Program Finance Initiative (UNEP FI)'s objective is to integrate environmental and social sustainability into the global financial sector. UNEP FI recognizes the significant role that

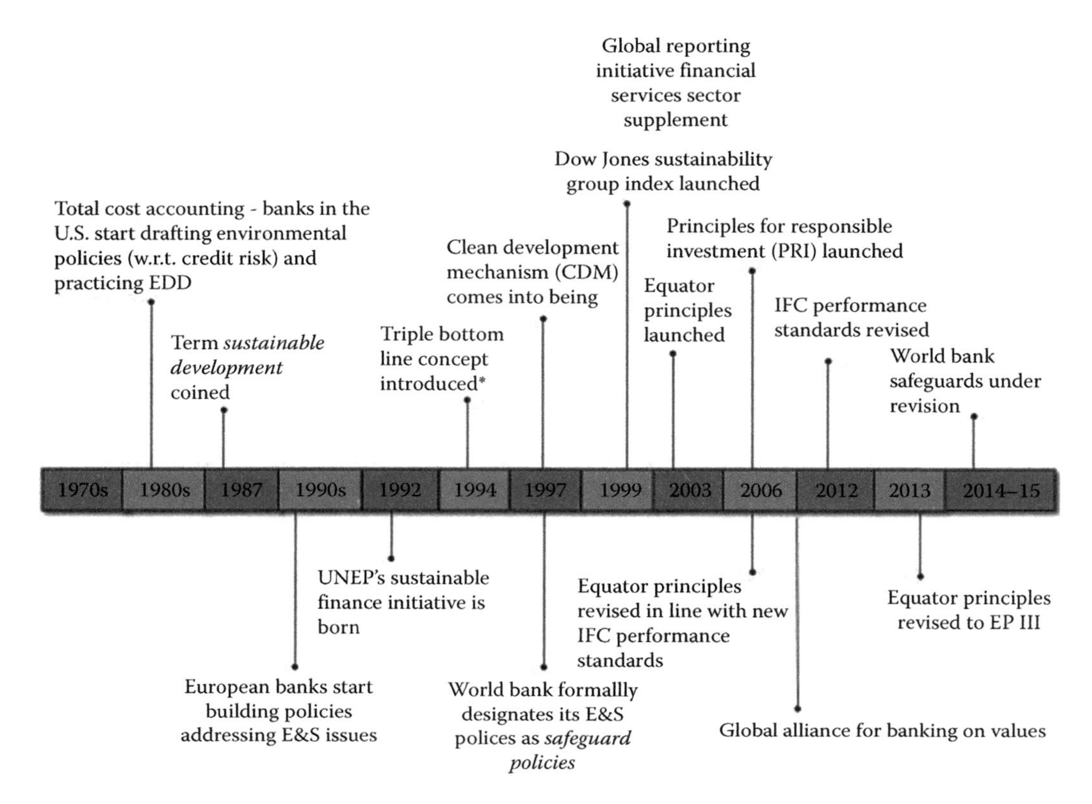

FIGURE 3.5
Key milestones in the evolution of environmental and social consideration at PSFIs.

the financial sector can play in addressing ESG challenges. UNEP FI focuses on policy and facilitates dialogues between stakeholders such as financial practitioners, policy makers, and regulators to meet its objectives.[28]

As of 2016, UNEP FI has 215 private and public organizations as members including banks, insurance companies, and investment funds from both developed and developing countries. Some of the members include JPMorgan Chase & Co., Bank of America, Barclays Group plc, Deutsche Bank A/S, Development Bank of Japan, the Development Bank of the Philippines, YES BANK Limited, and IL&FS in India and many others.

UNEP FI members sign and have to adhere to the UNEP FI Statement of Commitment, which is the backbone of UNEP FI. By signing this Statement, members commit to integrating environmental and social considerations into their business practices. The Statement provides guiding principles for the private sector to follow towards sustainable development, sustainability management practices, creating public awareness, and communication of performance and best practices and to recruit more financial companies to join this initiative.[29]

The three key industry areas that UNEP FI operates in are banking, insurance, and investment. We will discuss the insurance related initiatives in Section 3.5.7. Led by UNEP FI's Statement of Commitment, many global financial institutions have incorporated ESG standards into their decision-making. Many financing institutions are making investments in climate-risk related insurance instruments and in sectors such as renewable energy and biodiversity, etc. Some of the high impact initiatives projects and activities undertaken by UNEP FI are[30]:

- The incubation and launch of the UN-backed Principles for Responsible Investment (PRI) presented in Box 3.5.
- The Natural Capital Declaration (NCD) signed by CEOs of over 40 financial institutions to show their commitment towards factoring in natural capital considerations into the private sector, accounting, and decision-making by 2020 (2012).
- Launch of Sustainable Stock Exchange (SSE) Initiative that uses stock exchanges to enhance transparency and ESG performance of companies and encourage sustainable investment (2012). Section 3.8 discusses the SSE.

BOX 3.5 UN PRINCIPLES FOR RESPONSIBLE INVESTMENT (PRI)[31]

The UN PRI is a global platform that developed six voluntary and aspirational principles to encourage its signatories to factor in ESG issues into their investment decisions and to create a sustainable global financial system. Launched in 2006, the UN PRI has about 1,700 signatories collectively representing $62 trillion. The six UN PRI principles are:

- *Principle 1*: We will incorporate ESG issues into investment analysis and decision-making processes.
- *Principle 2*: We will be active owners and incorporate ESG issues into our ownership policies and practices.
- *Principle 3*: We will seek appropriate disclosure on ESG issues by the entities in which we invest.
- *Principle 4*: We will promote acceptance and implementation of the Principles within the investment industry.
- *Principle 5*: We will work together to enhance our effectiveness in implementing the Principles.
- *Principle 6*: We will each report on our activities and progress towards implementing the Principles.

UN PRI offers actionable guidance to its members to support compliance with all the principles. For example, to comply with Principle 1, UN PRI provides suggestions like development of financial tools, metrics, and analyses to incorporate ESG issues, to assess external and internal management capacity to incorporate ESG issues, and some other suggestions. UN PRI also developed a platform PRI Clearinghouse to encourage collaborative initiatives amongst its signatories such as pooling resources, knowledge share, and engaging with companies, stakeholders, and policy makers.

Owing to the complex nature and scale of FIs, UN PRI still faces many challenges including translation of principles to practice, improvement of member accountability, and clarity and consensus about its purpose and mission.[32]

DISCUSSION QUESTION

- *The PRI are often criticized because the Principles are essentially aspirational and not operational. Attempt creating Operational Guidelines for the six PRIs listed above. What are the benefits of such Operational Guidelines?*

- The Portfolio Decarbonisation Coalition (PDC) to encourage the finance sector investments to pull out of companies and projects that are carbon-intensive or to re-allocate their capital to companies with low-carbon activities, assets and technologies (2014).
- The Principles for Sustainable Insurance (PSI) is to serve as a framework for the insurance industry to adopt ESG principles (2012).
- The work stream on promoting energy efficiency dialogue within the G20 to promote best practices and capacity building for successful energy efficiency investment and financing initiatives from both the demand side (borrowers) and the supply side (banks and investors).

DISCUSSION QUESTIONS

- *Visit https://wedocs.unep.org/rest/bitstreams/10620/retrieve to read more on some of the noteworthy milestones of UNEP FI between 1991 and 2016.*
- *Prepare a time chart with notes on the drivers that were responsible for these initiatives.*
- *Research the impacts of these initiatives.*

A third-party evaluation of UNEP FI and its impact was carried out in 2016. While UNEP FI was recognized as one of the most innovative initiatives to integrate sustainability into the financial sector, the evaluation identified gaps such as insufficient monitoring and evaluation of the outcomes and impacts due to resource constraints and lack of co-ordination between UNEP FI and other UNEP groups.[41]

Complementing the catalytic role played by UNEP FI, the IFC played a key role in leading the ESG initiatives in the banking sector. IFC formulated Performance Standards (PS) and led banks to adopt the Equator Principles (EPs). We will discuss the PS of IFC and EPs in the next sections.

3.5.2 International Finance Corporation

IFC is one of the five organizations of the WBG founded specifically to support private sector development. Conceptualized in 1948–1949, IFC was established in 1956 with an initial $100 million capital from a dozen countries. IFC provides loans, equity, and offers advisory services to stimulate private sector investment in developing countries. IFC has invested in numerous private firms. A handful of examples include investments in the Brazilian pulp and paper company Champion Cellulose, Spanish auto parts manufacturer Fábrica Española Magnetos, Korea's LG Electronics, India's TATA Group, housing finance projects in Colombia, and financial advisory services to set-up Jakarta's securities markets, climate reform investment advice to China, a thermal power plant in Cote d'Ivoire and investments in micro-funding in Latin America, the Caribbean, and Bosnia. Between 1956 and 2016, IFC provided more than $245 billion in financing to businesses in developing markets.

In 1998, IFC adopted its first environmental and social review procedures and safeguard policies. IFC policies were the basis of the *Equator Principles* that were launched by commercial banks in 2003. Later in 2006, IFC upgraded its E&S procedures and safeguards into new PS discussed below[33]:

- Performance Standard 1: Assessment and Management of Environmental and Social Risks and Impacts.
- Performance Standard 2: Labor and Working Conditions.
- Performance Standard 3: Resource Efficiency and Pollution Prevention.
- Performance Standard 4: Community Health, Safety, and Security.
- Performance Standard 5: Land Acquisition and Involuntary Resettlement.
- Performance Standard 6: Biodiversity Conservation and Sustainable Management of Living Natural Resources.
- Performance Standard 7: Indigenous Peoples.
- Performance Standard 8: Cultural Heritage.

The objective of the PS was to help IFC's investment and advisory clients manage and improve their environmental and social performance through a risk and outcomes based approach. Though compliance with the PS is the clients' responsibility, IFC monitors and supervises the business activities it finances to ensure that they are implemented in accordance with the requirements of the PS.[34]

3.5.3 Equator Principles (EPs)

The EPs are based on IFC's Performance Standards on Social and Environmental Sustainability and on the WBG Environmental, Health, and Safety Guidelines (EHS Guidelines). They provide a risk management and decision-making framework that guide FIs to assess, determine, and manage environmental and social risk in projects. FIs that were keen to ensure that their projects do not have an adverse environmental and social risk adopted the EPs. As of January 2017, 89 financial institutions from 37 countries have adopted EPs. These are called the Equator Principles Financial Institutions (EPFIs).

The EPs are a global framework that applies to all industry sectors and to four particular financial products:

1. Project Finance Advisory Services, where the project capital cost is $10 million or more
2. Project Finance, where the capital cost of the project is $10 million or more
3. Project-Related Corporate Loans
4. Bridge Loans

On June 4, 2003, the first generation of the Equator Principles (EP-I) was launched and adopted by the first ten financial institutions. Among the founding members were ABN AMRO, Barclays, Citigroup, Credit Lyonnais, Credit Suisse, HVB, Rabobank, Royal Bank of Scotland, WestLB, and Westpac. Some of the other EPFIs include FMO, Bank of Montreal, Bank of America Corporation, Standard Chartered PLC, IDFC Bank, Standard Bank of South Africa Limited, Korea Development Bank, JPMorgan Chase Bank, N.A., HSBC Holdings plc, National Bank of Abu Dhabi PJSC, and many others.[35] Figure 3.6 lists the EPs.

The EPs were revised in 2006 (EP-II), again in 2012, and then again in 2013 (EP-III) following a major revision in IFC's PS. A comparison between the three versions of EPs is shown in Table 3.2.

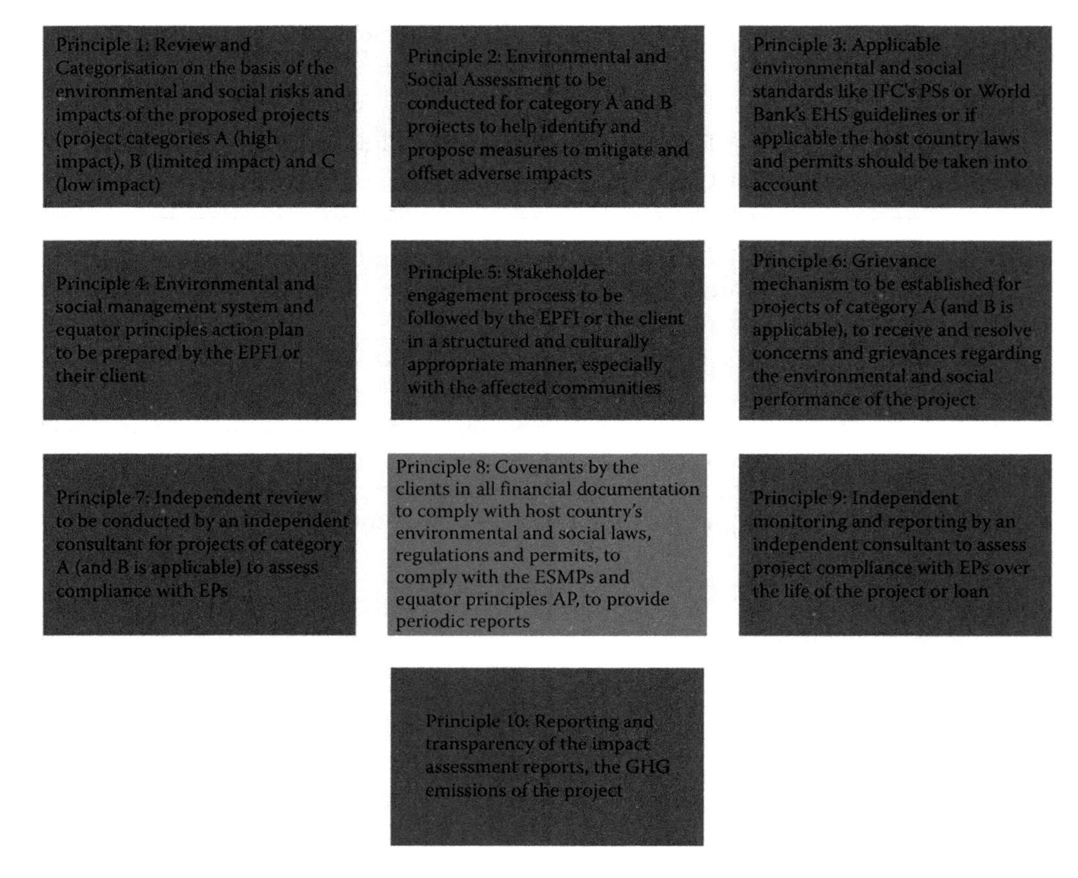

FIGURE 3.6
The equator principles. (From The Equator Principles, 2013. With permission, online source: http://www.equator-principles.com/resources/equator_principles_III.pdf.)

The EPs have received mixed responses from the members of the sustainability sector. While some view EPs as a visionary framework, others have critical opinions regarding the lack of implementation and enforcement. Some believe that EPs stayed more at the aspirational levels and less on operation and, hence, did not really change the way EPFIs assessed their projects. EPs thus failed to create a substantial impact. Lack of enforcement mechanisms has been identified and has been cited as one of the key drawbacks of EPs.[36]

3.5.4 Addressing Investment Portfolios and Supply-Chain

Boxes 3.6 and 3.7 includes an example of a European DFI that conducted a risk assessment of an industry it invested in to measure and enforce compliance with its environmental and social policies across the supply chain.

3.5.5 Investment Focus—Energy

Few financing institutions especially from the private sector target a specific sector for finance. We describe in Box 3.8 examples of three such institutions that finance renewable

TABLE 3.2

Changes in the Equator Principles Version I, II and III[a]

Issues	Version I	Version II	Version III
Project scope under EP	Project finance	Project finance, project finance advisory	Project finance, advisory, project-related corporate loans, bridge loans
Reporting and transparency	No specific reporting requirement	High level of reporting requirements—number of transactions screened and closed, project categorization and information on the EP implementation process	Reporting requirements included: Number of projects closed including information on categorization, sector, region and whether an independent review was conducted, project names for project finance deals (subject to client consent), information on EP implementation process including roles and responsibilities, staffing, policies and procedures and details of mandatory training during the first year of EP adoption.
Public Reporting	–	–	Online summary of Environmental and Social Impact Assessment and Greenhouse Gas (GHG) emission levels for projects emitting over 100,000 tons of CO_2 annually during the project operational phase.
Social impact assessment	–	Due diligence of social risks and prior consent of indigenous population required	Due diligence of social and relevant human rights and addressing identified issues, prior consent of indigenous population required and to follow UN principles on Business and Human Rights and the UN Protect, Respect and Remedy Framework.
Climate impact assessment	Environmental assessment of projects	EHS guidelines of the WBG to be followed and general due diligence	Alternative analysis for high carbon projects as per with IFC Performance Standard and Greenhouse Gas (GHG) emission levels reporting for projects emitting over 100,000 tons of CO_2 annually
Information sharing	–	Not mandatory but done informally	Formal approach introduced for members to share information related to environmental and social matters with other mandated financial institutions.

[a] Based on The Equator Principles III—2013, Equator Principles, online source: http://www.equator-principles.com/index.php/ep3

and clean energy and energy efficiency projects as the priority sectors. These include Triodos Bank from Netherlands, New York Green Bank, and the Swiss Reinsurance Company.

In the later sections of this chapter, we will discuss in detail bonds as a key financial instrument, for example, Green Bonds. It is worth, however, introducing here the Qualified Energy Conservation Bonds (QECB) introduced in the United States.

QECBs are financial instruments that are issued by the state or local government entities to raise funds for *qualified* energy conservation projects. QECBs are allocated to each US state by the US Treasury based on the population proportion. QECBs are debt instruments, and the investors make a profit via the interest against these debts. These are taxable bonds, which means that the investors will pay federal taxes on the interest earned. The projects that qualify for QECBs are improving energy efficiency in publicly owned buildings, implementing green community programs, electricity production in rural

BOX 3.6 ENVIRONMENTAL AND SOCIAL RED FLAG ANALYSIS OF INVESTMENT PORTFOLIOS

A European Development Finance Institution (EDFI) had invested in four companies through an Indian Private Equity Fund. The EDFI subscribed to a *Code of Responsible Investing* and wanted to identify any deviations or red flags from their environmental and social safeguards across the supply chain. The Code internally referred to IFC Environmental and Social Performance Standards.

The fund had invested in four companies which included:

- An engineering company manufacturing small to large gates for water treatment plants located in Central India
- A micro-finance institution located in North India
- An industrial foundry located in North India
- A healthcare services chain providing dialysis and ancillary services to End Stage Renal Disease patients located in North India

The red-flag analysis was aimed to:

- Assess compliance with the reference framework
- Identifying areas of non-compliance and suggesting options and actions to mitigate or remedy as well as potential financial cost associated with that option
- Identifying areas of potential lower order environmental and social risks, which require further assessment and additional information
- Recommending appropriate monitoring regimes and systems for each company
- Key areas of risk as identified by World Bank EHS guidelines (where applicable) were also assessed during the site visits

An assessment was conducted against:

- IFC Performance Standard 1 (E&S management systems) and current capacity of the company to manage risk and opportunities
- An analysis of labor and working conditions
- Occupational health and safety
- Resource efficiency and pollution prevention
- Client protection risk at the microfinance organisation

Key findings across the companies were lack of oversight by senior management on E&S compliance, lack of clarity on E&S governance, absence of dedicated human resources to manage EHS function, non-identification of EHS environmental impacts, and occupational hazards and risks linked to various operations. This assessment helped identify red-flags that were allocated high or medium priority for which action plans were undertaken to strengthen E&S legal compliance in the portfolio companies and establish an E&S management system to sustain effective E&S performance.

DISCUSSION QUESTION

- *Setting of red flags helps to understand risks/vulnerability, especially on the compliance and requirements of improvement. How should one move ahead to develop an action plan and evolve a system of periodic monitoring? Take an example of a garment supply chain such as at Marks and Spencer or retailers such as Walmart.*

BOX 3.7 SUPPLY CHAIN ASSESSMENT UNDERTAKEN BY A EUROPEAN DFI[37]

A European DFI invested in a company located in the State of Maharashtra in India manufacturing 100% recycled polyester filament yarn. The DFI recognized that there could be risks of violating children's rights through the operations.

As an initial step to insulate from such risks, the DFI conducted an internal risk assessment of its portfolio focusing on the risk of children being involved directly or indirectly in manufacturing operations. The risk assessment concluded the highest risks were found in companies using recycled materials as part of the raw materials. These risks primarily arose due to involvement of the informal sector.

The DFI, therefore, engaged a third-party consultant for conducting a supply chain audit of this with focus on child labor and children's rights. Suppliers were evaluated for environmental and social compliance and checked for any violation of children's rights. This evaluation was used to identify the top five high risk or critical suppliers. It was recommended that the company considers collaborating with non-governmental organizations or social enterprises.

DISCUSSION QUESTIONS

- *What makes a red flag analysis of supply chain relevant for PSFIs? Are social issues (e.g., related to labor and safety) more dominant than the environmental issues?*
- *Cite examples where red flag analyses have led to cancellation of supplier contracts. Is capacity development of the suppliers to meet the environmental and social requirements the right approach?*
- *How does one deal with differences between the E&S governance at the corporate level, investing PSFI, and the national laws? Who bears the cost of the incremental compliance?*

BOX 3.8 FINANCIAL INSTITUTIONS THAT INVESTED IN THE RENEWABLE ENERGY SECTOR

TRIODOS BANK

Triodos Bank is an independent financial institution based in the Netherlands. Established in 1980, Triodos Bank was launched by a group of individuals keen on developing a socially and environmentally sustainable society. Triodos Bank invested in the renewable energy sector, which was considered high-risk and was relatively unexplored at the time. Triodos has supported over 300 sustainable energy projects across Europe.[38]

Triodos Bank set-up two energy and climate funds—Triodos Green Funds or Groenfonds (TGF) in 1990 and Triodos Renewables Europe Fund (TREF) in 2006. TGF finances projects in areas like renewable energy, organic farming, sustainable real estate, and nature conservation projects. TREF provides investors an opportunity to invest in small-medium sized companies that produce green power through wind and solar energy power plants. Some examples of projects funded under TREF include a solar power plant in Belgium with a capacity of 2.7 MW producing 2.5 million KWh a year and onshore wind energy projects in the Netherlands.[39,40]

NEW YORK GREEN BANK

NY Green Bank (NYGB) is a state-sponsored investment fund which was set-up in 2014 as a specialized financial entity to foster increased private sector funding to support New York's clean energy projects. NYGB is a division of New York State Energy Research & Development Authority's Clean Energy Fund to develop a resilient and affordable energy system. Its investment portfolio includes residential efficiency, distributed or small wind, residential and commercial solar, and property assessed clean energy. NYGB works with private entities like banks, renewable companies, or project developers.

For example, NYGB collaborated with Bank of America Merrill Lynch to co-finance a Combined Heat and Power plant in a home for aged. It collaborated with Signature Public Funding Corp. (SPFC), a subsidiary of Signature Bank, towards a $12.9 million project to enhance the energy efficiency initiatives including lighting retrofits, building envelope improvements, energy management systems, water conservation units, and ventilator refurbishments. It worked with an energy software company on a $5 million project to offer homeowners a financial product to promote energy efficiency. Homeowners could make their payments for their energy efficiency measures based on the realized energy savings.[41]

SWISS REINSURANCE COMPANY

The Swiss Reinsurance Company (Swiss Re) has adopted several initiatives in the areas of energy efficiency, renewable systems, and business travel to reduce its own emissions. In 2007, Swiss Re incentivized its employees to *go green* by offering them 50% rebate on any of their personal investments to reduce their individual carbon emissions in relation to mobility, heating and electrical energy, including hybrid cars, use of public transport, and the installation of solar panels and heat pumps.

DISCUSSION QUESTIONS

- *It has been often observed that investments in renewable energy result in relatively low pay backs or investment returns. Against this observation, does investing solely in the renewable energy sector make business sense?*
- *Compared to renewable energy, financing energy efficiency projects do provide higher economic returns. Make a comparative analysis and list your observations.*
- *Does linking project outcomes to reduced emissions of GHGs add any value? Is this aspect rather trivial or could it be strategic over a long term?*
- *Research examples similar to the three banks in your own country.*

areas, mass commuting facilities to reduce energy consumption, fuel production from agricultural waste, and many others.

3.5.6 Investment Focus—Biodiversity

Biodiversity conservation has been a rather incomprehensible and abstract subject to be incorporated into the financial sector. Some of the barriers have been low awareness and information of sound biodiversity investment projects, lack of experience and expertise on conducting economic evaluation of biodiversity, difficulties in incorporating biodiversity into assessment procedures, and lack of mechanisms for addressing priorities of the biodiversity and finance sector.

However, today the financial sector is playing a significant role in alleviating biodiversity risks and in setting up innovative financial mechanisms to promote biodiversity conservation. Box 3.9 shows such examples.

3.5.7 Insurance Sector[46]

In 2006, 16 business leaders from the insurance industry came together to form the Insurance Working Group under UNEP FI. Their aim was to undertake and promote research, education and awareness, and product development and methodologies on sustainability in the insurance industry. The 16 insurance companies include Achmea (Netherlands), Allianz SE (Germany), American International Group (USA), AXA (France), Folksam (Sweden), HSBC Insurance Brokers Ltd. (UK), Insurance Australia Group (Australia), Interamerican Hellenic Life Insurance Company (Greece), Lloyd's (UK), MAPFRE (Spain), Munich Reinsurance Company (Germany), Norwich Union (Aviva) (UK), Storebrand (Norway), Swiss Reinsurance Company (Switzerland), Tokio Marine & Nichido Fire Insurance Co., Ltd (Japan), and XL Insurance (Bermuda).

Many of these companies have previously demonstrated their commitment to integration of sustainability into their business practices with focus on the areas of climate change, microinsurance, lifelong income, health, emerging manmade risks, environmental liability, natural resources, recycling and internal efficiency. Some examples of their

BOX 3.9 BIODIVERSITY RELATED INITIATIVES IN THE BANKING SECTOR

EUROPEAN BIOLOGICAL RESOURCE INITIATIVE[42,43]

The European Biological Resource Initiative (EBRI) was launched within the framework of the Pan-European Biological and Landscape Diversity Strategy by European Centre for Nature Conservation (ECNC), a non-governmental organization working for the conservation and sustainable use of Europe's nature, biodiversity, and landscapes. This was an innovative initiative that focused on promoting collaborations between the banking sector and the biodiversity conservation sector. The key objective was to increase investment in pro-biodiversity projects and by mainstreaming biodiversity objectives into bank policies.

A European Task Force was established to guide the initiative combining Banking, Business, and Biodiversity. Under this initiative, DFIs piloted the establishment of a Biodiversity Financing Facility (BFF) to support the EBRI objectives. EBRD, EIB, and the Hungarian Development Bank were involved in these pilots.

Encouraged by this initiative, DFIs like EIB, national banks including the Rabo Bank Nederland, the Hungarian Development Bank, and the Deutsche Bank started investing in biodiversity initiatives in Europe.

BIOCARBON FUND INITIATIVE FOR SUSTAINABLE FOREST LANDSCAPES

Initiated in 2004, the BioCarbon Fund (BioCF) was the world's first carbon fund that had an objective of sustainable land use and afforestation and reforestation as its prime focus. It was managed by the Carbon Finance Unit at the World Bank and provided financial assistance to projects that sequestered carbon through sustainable forestry in countries that are significant for their biological diversity. The financial assistance provided under BioCF is result-based, that is, it is based on the tons of carbon emissions reduced. This result-based financial structure incentivizes grantees to manage and execute the project successfully. Under this fund, nine carbon accounting methodologies were developed that allowed grantees to calculate and monitor their carbon reductions in a cost-effective manner.

BioCF's new initiative was launched in 2013 and was called the Initiative for Sustainable Forest Landscapes (ISFL). ISFL is a multilateral fund managed by the World Bank with donor governments and private companies as its partners for implementation. The initial funding of $360 million towards this initiative was provided by Germany, Norway, the UK, and the United States. One of the key differentiating features of ISFL from the previous BioCF avatars is the involvement of the private sector for innovation, expertise, knowledge and mobilizing capital which was not previously done.[44]

One of the projects funded under ISFL was in Ethiopia and Kenya to boost the coffee farming sector and farmers in the countries. IFC, Nespresso and an international NGO, TechnoServe, have collaborated for this project to plant shade trees and to help coffee farmers with financing, training, and technical assistance to improve farming practices that will boost productivity and improve climate resilience. IFC will lend Nespresso $3 million through the Nespresso Sustainability Innovation Fund, while

ISFL is providing $3 million as grant funds. TechnoServe will play the role of providing technical assistance and overseeing training and planting of trees.[45]

DISCUSSION QUESTIONS

- *Why should a financing institution invest in protecting or enhancing biodiversity?*
- *Are there any linkages between biodiversity and business?*
- *What is conservation finance? Should the community be involved in financing biodiversity related projects?*

sustainability efforts under the respective focus areas both before and after the formation of the IWG are included below:

- *Environmental Liability*: Achmea along with the Dutch Insurance Industry developed an insurance product—Environmental Damage Insurance policy in 1998. The policy covers a portion of the cost of remediation of environmental damage at the sites or land owned by the policyholder. Such policies were only given to companies that had environmental risk assessment and management programs in place, thus influencing other industry sectors to adopt sustainability.
- *Climate Change*: To research the impact of hurricanes on climate change, American Insurance Group (AIG) and Lloyd's collaborated with the Insurance Information Institute, USA, and the Harvard Medical School Center for Health and the Global Environment to develop a Catastrophe Modelling Forum that will provide new insights on the climate change risks associated with hurricanes.
- *Micro-finance*: AIG in partnership with a microfinance institute developed a low-cost insurance policy combined with micro-loans in Uganda in 1997. The policy provides coverage in the event of the borrower's death and also covers property damage due to natural disasters like storm, flood, tsunami, and earthquake.

3.6 Financial Intermediary Funds

Financial Intermediary Funds (FIFs) are multilateral financial arrangements that are utilized to foster global programs focused on preventing communicable diseases, climate change abatement, food security, disaster relief, and others. FIFs leverage both public and private resources to support these global programs. FIFs follow a flexible structure where it raises funds from a variety of sources and disseminates it to recipients both in the public and private sector through the FIF governing body.

DFIs like the World Bank and ADB have tapped into the FIF model to support global developmental programs. World Bank's role in the FIFs is that of a Trustee of the funds. The World Bank provide financial intermediary services like receiving, holding, investing, and transferring the funds as and when required by the FIF governing body. As a Trustee, World Bank does not oversee or monitor the activities or projects that are funded by the FIFs.[47]

Box 3.10 describes the Global Environment Facility that is one of the earliest and major FIFs.

BOX 3.10 GLOBAL ENVIRONMENT FACILITY[48]

Global Environment Facility (GEF) is an international cooperation that partners with United Nations agencies, multilateral development banks, national governments and international NGOs to address the world's most challenging environmental issues. Some of the agencies it works with include UNDP, UNEP, World Bank, ADB, UNIDO, EBRD, WWF and many others. These agencies fund projects in the private sector, NGOs and recipient countries through the GEF that acts like a financial mechanism and channels grants through its climate funds like Least Developed Countries Fund or the Special Climate Change Fund discussed under Section 3.7.4.

The GEF acts as a financial mechanism for important conventions like the Minamata Convention on Mercury, the Stockholm Convention on Persistent Organic Pollutants (POPs), the United Nations Convention on Biological Diversity (UNCBD), the United Nations Convention to Combat Desertification (UNCCD), and the United Nations Framework Convention on Climate Change (UNFCCC).

Since its establishment in 1992, the GEF has had a positive impact through:

- 790 climate change mitigation projects contributing to 2.7 billion tons of GHG emission reductions
- Sustainable management of 34 transboundary river basins in 73 countries
- Improved cooperation and governance of one-third of the world's large marine ecosystems
- Sound management and disposal of 200,000 tons of highly toxic Persistent Organic Pollutants
- Climate change adaptation to reduce the vulnerability of more than 15 million people in 130 countries

 **DISCUSSION QUESTIONS**

- *Study the project portfolio of GEF and analyse statistics on projects approved over time, funds released, thematic distribution of projects and region-wise project distribution.*
- *Compare funds supported by GEF and those contributed by the national governments to assess the leveraging factor.*

3.7 Climate Finance

Climate finance refers to the financing provided by public or private entities at a local, national or international level to significantly reduce emissions causing climate change. Climate finance includes investments needed for climate change mitigation and adaptation.

In 1992, many countries joined the United Nations Framework Convention on Climate Change (UNFCCC) and the Kyoto Protocol in 1995 as a part of their commitment to contribute to climate change mitigation and adaptation and to reduce greenhouse gases. Under both the Convention and the Protocol, the onus of this endeavor was more on the developed countries to reduce their own emissions and to provide financial support to developing countries. Article 11 of the UNFCCC had provided guidance on setting up of a financial mechanism to facilitate this financial assistance, which is monitored by the GEF. The operating fund for the Convention is called Green Climate Finance (GCF). GEF manages two funds called the Special Climate Change Fund (SCCF) and the Least Developed Countries Fund (LDCF). The fund managed under the Kyoto Protocol is called the Adaptation Fund (AF). The list of other Climate funds includes[49]:

- *Adaptation Funds*:
  - Adaptation for Smallholder Agriculture Program
  - Adaptation Fund
  - Least Developed Countries Fund (LDCF)
  - Pilot Program for Climate Resilience (PPCR)
  - Special Climate Change Fund (SCCF)

- *Mitigation Funds*:
  - Clean Technology Fund (CTF)
  - GEF Trust Fund 5th Replenishment
  - GEF Trust Fund 6th Replenishment
  - Scaling Up Renewable Energy Program (SREP)

Many other climate funds were established to either mitigate or adapt the impacts of climate change or to protect forestry and ecosystems under the UN REDD program.

- *REDD-plus Funds*:
  - Congo Basin Forest Fund (CBFF)
  - Forest Carbon Partnership Facility—Readiness (FCPF)
  - Forest Carbon Partnership Facility—Carbon Fund (FCPF)
  - Forest Investment Program (FIP)
  - UN REDD

The financial instruments used to disburse climate funds are grants, concessional loans, non-concessional loans, and equity. In addition to public or multilateral funds, private investments under climate finance are channeled largely towards the renewable energy, energy efficiency, and transportation sectors.

Per a biennial assessment conducted in 2016 by the UNFCCC Standing Committee, the total global finance in the year 2013–2014 invested by public and private entities together towards climate-related projects amounted to $714 billion. However, the funds required to meet the green infrastructure development needs is an average of $6 trillion per year for the next 15 years. The World Economic Forum in 2013 projected that around $5.7 trillion is needed annually to support the development of green infrastructure in the developing world.[50]

There is a huge investment gap that needs to be fulfilled not only with increased funding but also critical success factors like efficient technology and project management, monitoring and execution systems to increase the process efficiency while reducing costs. The current investment of $1.6 trillion into the fossil fuels sector can be assessed and channeled for green or renewable technology in a phased manner. Refer to Figure 3.7.

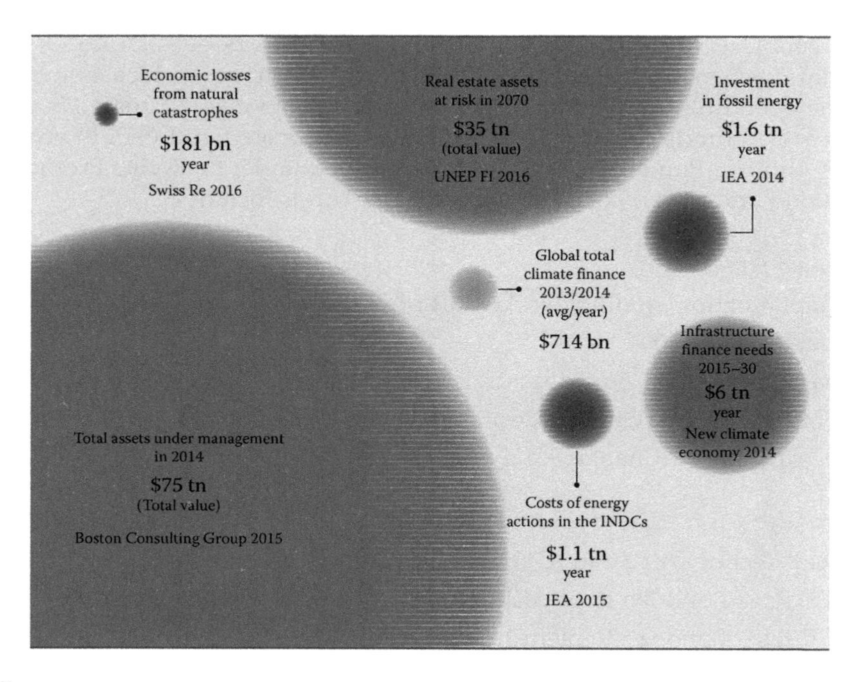

FIGURE 3.7
Overview of global climate finance flows. (From © UNFCCC Standing Committee on Finance, 2016 Biennial Assessment and Overview of Climate Finance Flows Report, 2016, United Nations Framework Convention on Climate Change, p.9. With permission, online source: http://unfccc.int/files/cooperation_and_support/financial_mechanism/standing_committee/application/pdf/2016_ba_technical_report.pdf.)

In the following sections, we will take a look at some of the Climate Finance funds and illustrate sample projects that are supported by these funds.

3.7.1 Green Climate Fund

The Green Climate Fund (GCF) was established by 194 countries in 2010 and is headquartered in Sorth Korea. It aims to provide equal support to climate change mitigation and adaptation projects and engages with the private sector. The financial instruments utilized under the GCF are grants, concessional loans, debt, equity and guarantees, and crowd funding. Box 3.11 shows examples of some of the projects funded under GCF.

3.7.2 Adaptation Funds

The Adaptation Fund (AF) was established in 2001 under the Kyoto Protocol with the objective of financing climate adaptation programs and projects in climate change vulnerable developing countries. Between 2010 and 2016, AF has funded projects worth $354.89 million to support 61 countries. AF is financed by governments, private donors, and through 2% of the proceeds from the sales of Certified Emission Reductions under the Clean Development Mechanism, which was also established under the Kyoto Protocol. The Adaptation Fund Board is the body that supervises and manages the AF. The actual project activities are carried out by Implementing Entities (IEs).[53]

BOX 3.11 EXAMPLES OF PROJECTS FUNDED BY GREEN CLIMATE FUND

CLIMATE RESILIENT INFRASTRUCTURE IN BANGLADESH

An amount of $80 million was allocated in 2015 to fund a 6-year long project to build 45 new cyclone shelters, renovate 20 existing centers and to safeguard critical roads (80 kilometers) in a rural coastal region in Bangladesh. In addition, the fund supports urban projects like improvements to drainage, flood protection, sanitation, water supply, and transport prioritizing the most vulnerable areas. GCF provided 50% of the finance and the rest will be provided by the German Ministry for Economic Cooperation and Bangladeshi Ministry of Local Government. The project will be executed and monitored by Local Government Engineering Department of Bangladesh.[51]

CLIMATE ACTION AND SOLAR ENERGY DEVELOPMENT PROGRAM IN THE TARAPACÁ REGION IN CHILE

GCF, with support of $50 million from the CAF Development Bank of Latin America, approved a $265 million project in Chile, to finance a private-sector led solar park of 143 MW in 2016. Located in the Atcama region, the project will be executed by the private company Atacama Solar S.A. The project, which is yet to be mobilized, is expected to reduce 3.7 million tons of CO_2.[52]

DISCUSSION QUESTIONS

- *Study the process to be followed and documentation to be submitted for accessing GCF in the context of your country.*
- *All proposals to GCF need Environmental and Social Impact Assessments (ESIA). Download some of the ESIAs submitted on the GCF website. How different are these ESIAs compared to the EIAs required to obtain Environmental Clearance?*

The AF runs another program called the Readiness Program for Climate Finance. The aim of this program is to provide assistance to national and regional IEs to manage the climate financing they receive. The program entails introductory seminars for climate finance, peer to peer learning forums, small grants to support project formulation, tools and guidance documents, and knowledge sharing. The three Readiness grants are[54]:

- South-South Cooperation grants to increase peer to peer support from the accredited National Implementing Entities (NIEs) and to the NIEs that are seeking accreditation.
- Project Formulation Assistance grants to help build the capacity of accredited NIEs in terms of project preparation and design.
- Technical Assistance grants to strengthen the capacity of NIEs in the areas of environmental, social and gender risk management of adaptation projects and programmes.

Examples of some of the projects funded under AF are shown in Box 3.12.

BOX 3.12 EXAMPLES OF PROJECTS FUNDED BY THE ADAPTATION FUND

BELIZE MARINE CONSERVATION AND CLIMATE ADAPTATION PROJECT

AF allocated $6 million for a marine conservation project to improve the climate resilience of the Belize Barrier Reef System. The project supports activities to improve the reef's eco-system through measures to protect and build its resilience to climate change. The Marine Protected Areas (MPAs), that is, areas that would have restricted human activities would be expanded and better enforced. The project aims to promote and create awareness of local communities residing near the reef regarding sustainable alternative livelihoods and how to maintain and enhance the health of the reef ecosystem. Essentially, the project focuses on creating a behavior change for improved management of the reefs. This five year project was approved in 2014, started in 2015 and as of February 2017, over 50% of the allocated funds had been transferred to the trust executing the project.[55]

FOOD PRODUCTION AND FOOD SECURITY IN NEPAL

To help the affected communities develop adaptive capacity to food insecurity created by climate change, AF allocated nearly $9.5 million towards a project to be executed by Ministry of Science, Technology and Environment; Ministry of Federal Affairs and Local Development; World Food Program. The project's aim is to improve local capacity to identify and mitigate climate risks, to diversify livelihoods and to improve food security and climate change resilience of natural systems like agriculture, forestry, livestock, and irrigation.[56]

 **DISCUSSION QUESTIONS**

- *What could be the global requirement of funds needed for adaptation?*
- *Who makes contributions to an Adaptation Fund? Why?*

3.7.3 Least Developed Countries Fund

Established in 2001, the Least Developed Countries Fund (LDCF) has the objective to help improve the adaptability of the least developed countries (LDCs) across the world that are most vulnerable to climate change. LDCs are identified based on three criteria—low income, weak human assets, and high economic vulnerability.

One of the specific tasks under the LDCF is to help the LDCs prepare and execute the National Adaptation Programs of Actions (NAPAs). NAPAs are country-specific strategies that identify the most critical and immediate interventions required by that country to adapt to climate change. Some of the focus sectors are water, agriculture and food security, health, disaster risk management and prevention, infrastructure, and fragile ecosystems. By 2012, LDCF has supported preparation of 49 NAPAs. Box 3.13 shows examples of LDCF funded projects.

BOX 3.13 EXAMPLES OF PROJECTS FUNDED BY LDCF FOR CLIMATE CHANGE ADAPTATION

IMPROVED WEATHER FORECASTS FOR MALAWIAN FARMERS

Farmers in Malawi faced many challenges like ecosystem degradation, inappropriate agricultural practices, and limited knowledge of climate change that affected sustainability of their livelihoods. The LDCF funded a project aimed at improving the weather forecasts provided to small-scale farmers to help them better plan their agricultural activities. This was done by strengthening Malawi's weather, climate, and hydrological monitoring capabilities, and ensuring delivery of available weather information to help farmers respond to extreme weather and warnings of floods, droughts, and strong winds. By delivering accurate information, the farmers could plant the right crops at the right time, hence, improving the climate change adaptability of the country's agricultural sector. In addition to radio and television channels, the project incorporated text messages to local farmers containing weather information. The 2013 project with a duration of 4 years is expected to benefit over 5 million people in 11 districts in Malawi and help improve food security of the country.[57,58]

CLIMATE CHANGE ADAPTATION PROJECT IN THE AREAS OF WATERSHED MANAGEMENT AND WATER RETENTION IN SENEGAL

Like in most African countries, climate change has affected Senegal's water resources, agriculture, and coastal areas. The LDCF funded project aims at improving the resilience of agricultural systems by focusing on improved water and knowledge management. The project objectives included improving the national awareness and knowledge about climate change impact on agricultural practices, and its capacity to tackle this challenge. The project aimed at improving the water systems by building 15 wells, ten retention basins, introduction of 60 apiculture hives per basin, treating 100 hectares of secondary rice plots through levelling and drainage, and producing 10,000 vitro plants. With proper monitoring and management methods, the project aims at improving the water scarcity issue plaguing the area.[59]

 **DISCUSSION QUESTIONS**

- *Study NAPA of your country and look for projects that may be funded by the LDCF.*
- *Check whether these projects have congruence with other national schemes.*
- *What is the value add of the LDCF or is it only limited to gap financing?*

3.7.4 Special Climate Change Fund (SCCF)

The Special Climate Change Fund (SCCF) was established under UNFCCC in 2001. Unlike the AF and LDCF, which funded only adaptation projects, SCCF financed programs under the areas of climate adaptation, technology transfer and capacity building, energy, transport, industry, agriculture, forestry, and waste management, and economic diversification. SCCF was also managed by GEF. SCCF projects are implemented through various GEF agencies like UNDP, World Bank, ADB, UNEP or EBRD. SCCF characterizes projects under four windows.

- Window A: Adaptation
- Window B: Transfer of technologies
- Window C: Energy, transport, industry, agriculture, forestry, and waste management
- Window D: Window A—Economic diversification for fossil fuel dependent countries

Of the four windows, Window A and B projects are most active and not much action was taken under C and D due to lack of funding.

3.8 Stock Markets

Stock market exchanges are increasingly playing a significant role in the sustainability sector by developing indices, ratings, and products that help investors who are looking to make investments in green organizations make better informed decisions. Stock exchanges are an interface between companies and investors, and they are well-positioned to drive industries to create a more transparent environmental performance reporting culture in global businesses.

Sustainable stock indices were developed by exchanges to recognize businesses that were addressing issues related to global warming and the environment. These indices serve as a benchmarking product for investors seeking detailed information about the sustainable performance of companies. Many stock market indices have sustainability metrics in place that the listed companies need to comply with.

Sustainability-related indices tend to fall into one of four categories[60]:

- *Broad-based*: benchmarks firms from all sectors (e.g., Dow Jones Sustainability Index)
- *Sector-based*: benchmarks firms from one sector (e.g., sustainable real estate or finance)
- *Sustainability Sector-based*: benchmarks firms based on a sustainable sector (e.g., *green* lean tech, renewable)
- *Sustainability Issue-based*: benchmarks firms based on a sustainability theme (e.g., water scarcity, diversity, good governance)

The World Federation of Exchanges (WFE) is a global industry association for stock market exchanges. In 2014, WFE established a Sustainability Working Group (SWG). The SWG convenes regularly to share best sustainability practices amongst the exchange markets. The WFE SWG listed key indicators under the ESG criteria to guide exchanges towards sustainability. Table 3.3 gives an overview of those indicators.

Companies that are listed on WFE's stock markets must comply with ESG norms and perform well against the above indicators in order to improve or have sustained access to capital. Compliance with the norms helps in corporate reputation, branding, and enhancing stakeholder relationships. Many WFE member stock exchanges had adopted sustainability metrics and reporting guidance for their listed companies prior to the WFE SWG's

TABLE 3.3

Key ESG Criteria for Stock Market Exchanges[a]

Direct & indirect energy consumption	Injury rate
Direct & indirect GHG emissions	Global health
Carbon intensity	Child & forced labor
Energy intensity	Human rights policy
Primary energy source	Human rights violation
Renewable energy intensity	Board—Diversity
Water management	Board—Separation of Powers
Waste management	Framework Disclosures
Environmental policy	Board—Confidential Voting
Environmental impact	Incentivized pay
CEO pay ratio	Fair labor practices
Gender pay ratio	Supplier code (SC) of conduct
Employee turnover ratio	Ethics code (EC) of conduct
Gender diversity	Bribery/anti-corruption code (BAC)
Temporary worker rate	Tax transparency
Non-discrimination	Sustainability report

[a] Based on WFE Sustainability Working Group Exchange Guidance & Recommendation—October 2015, online source: http://www.sseinitiative.org/wp-content/uploads/2015/11/WFE-ESG-Recommendation-Guidance-Oct-2015.pdf

issued guidance. Some of those exchanges are BM&FBOVESPA S.A. (Brazilian Exchange), Johannesburg Stock Exchange, Singapore Exchange, Shenzhen Stock Exchange, The Stock Exchange of Thailand, TMX Group Inc. (Toronto Stock Exchange), BSE India Ltd. (Bombay Stock Exchange), National Stock Exchange of India, Colombian Securities Exchange, Deutsche Börse AG (German Exchange), and others.[61]

In 2015, the UN agencies including UN Principles for Responsible Investment, the United Nations Conference on Trade and Development (UNCTAD), the United Nations Environment Program Finance Initiative (UNEP-FI) and the United Nations Global Compact (UNGC) initiated the Sustainable Stock Exchanges (SSE) Initiative to explore how stock exchanges can encourage corporate sustainable performance by strategies like mandating sustainability reporting. Over 60 global exchanges have committed to this Initiative. SSE provides a platform for exchanges to publicly commit to promote improved sustainability performance and disclosure among their listed companies.

One of the most important milestones in the stock exchange sector has been the establishment of the Dow Jones Sustainability Index (DJSI) in 1999. Indices of DJSI were developed to take into account the sustainability performance of the listed companies. DJSI was a collaborative initiative of the international investment company RobecoSam and S&P Dow Jones Indices.[62]

Box 3.14 gives an overview of DJSI, and how it assesses the sustainability performance of listed companies.

Since then many stock exchanges have developed sustainability indices to benchmark the sustainability performance of listed companies. Figure 3.8 shows a list of sustainability indices along with the year of establishment.

Box 3.15 includes examples of initiatives by stock markets to develop sustainable indices that have motivated hundreds of companies to adopt sustainable practices for economic, environmental, social and reputational reasons.

BOX 3.14 DOW JONES SUSTAINABILITY INDEX[63]

DJSI is a family of indices that assess the sustainability performance of global companies on an annual basis. The first index under DJSI, called the DJSI World, was published in 1999. Subsequently, many regional indices were established in Europe, Nordic areas, North America, and the Asian Pacific. It was established and is managed cooperatively by the S&P Dow Jones Indices and RobecoSAM.

DJSI assesses 2,500 publicly traded companies that belong to 60 different industry sectors annually based on financially relevant economic, environmental, and social factors. To facilitate this assessment, an annual questionnaire is issued to the invited companies who are required to fill out the questionnaire, which includes both general and industry specific criteria, as applicable to that industry.

The listed companies are monitored on a daily basis, and they stand at the risk of being taken off the list if they are involved in issues like economic crime or corruption, fraud, illegal commercial practices, human rights issues, labor disputes, workplace safety, catastrophic accidents, or environmental disasters. In 2010, the oil company BP was taken off DJSI due to the oil spill disaster in the Gulf of Mexico and the incident's long term repercussions on the environmental and local community.[64]

DISCUSSION QUESTION

- *May companies doing well on DJSI also do well on stock performance? Why?*

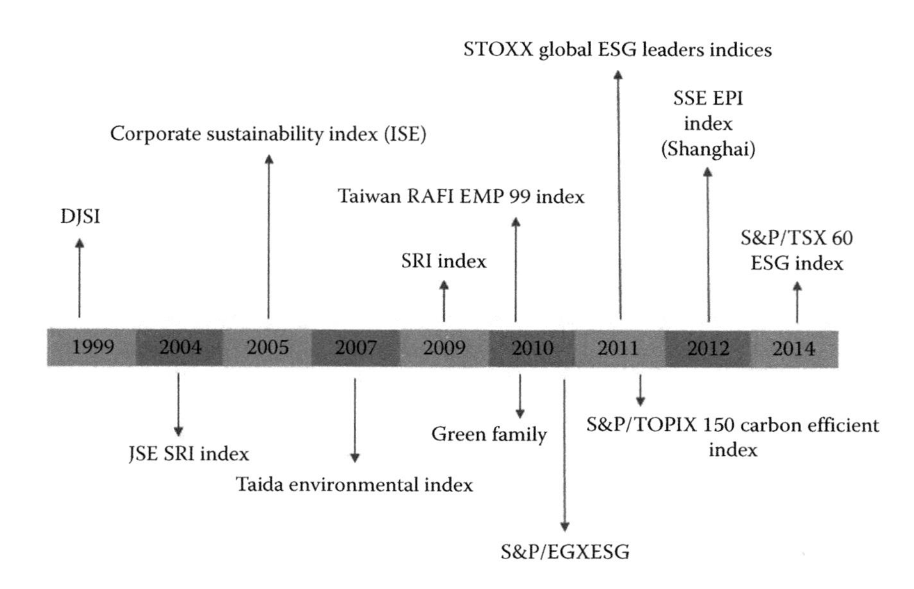

FIGURE 3.8
Global stock market exchanges and their sustainability indices. (Based on Exchanges and ESG Initiatives—SWG Report and Survey, 2015, World Federation of Exchanges, Sustainability Working Group (SWG).)

BOX 3.15 INITIATIVES TAKEN BY STOCK MARKET EXCHANGES TO INCORPORATE ESG FRAMEWORK INTO THEIR LISTED COMPANIES[65]

THE STOCK EXCHANGE OF THAILAND (SET)

The Stock Exchange of Thailand (SET) has been working on fostering sustainable development in Thailand since the late '90s. In the early 2000s, they introduced 15 Principles of Good Corporate Governance to support the right and fair treatment of shareholders, clarify the role of stakeholders including the Board of Directors, and promote disclosure and transparency. SET has a sustainability policy which is applicable to all its subsidiaries (includes other exchanges and trading companies). The Sustainability Framework focuses on five key areas, namely, market value, business operations, employees, society, and environment. SET is the first ASEAN stock exchange to partner with the UN Sustainable Stock Exchanges Initiative.

To incentivize listed companies to improve their ESG performance, SET has developed the Thailand Sustainable Investment List, which awards companies that perform exceptionally well on the ESG criteria. To qualify for this award, firms need to achieve at least 50% scores in the economic, social, and environmental dimensions. In 2016, firms from the oil, chemicals, cement, aviation, electronics, and many other sectors were on this list.[66]

To help Thai companies improve their organizational capacity and knowledge on sustainability reporting, SET has been training companies to create sustainable corporate strategies and process improvement, as well as producing sustainability reports. To increase international competitiveness, SET trains and encourages firms to get listed on DJSI. Since 2012, SET has trained 1,350 participants from 427 Thai listed companies.[67]

JOHANNESBURG STOCK EXCHANGE

The Johannesburg Stock Exchange was one of the early stock exchanges to form a Socially Responsible Investment Index (SRI Index) in 2004. The objective of the SRI Index was to integrate the principle of sustainable development (or triple bottom) into the regional business practices. Companies were measured against categories like environment, society, governance, and climate change and then listed on the exchange. SRI Index assessment increased the companies' awareness of sustainability issues and provided necessary guidelines on what measures can be introduced to improve their performance. It allowed companies to increase visibility and engage with more investors.

In 2015, JSE upgraded the SRI Index and launched the JSE/FTSE Responsible Investment Index series in partnership with the global index management firm, FTSE Russell. They adopted the FTSE ESG ratings methodology which is based on three pillars—environmental, society, and governance.[68]

BURSA MALAYSIA

Bursa Malaysia, Malaysia's stock exchange company located in Kuala Lumpur, introduced a Corporate Social Responsibility (CSR) Framework in 2006. The objective of the CSR Framework was to encourage companies to initiate or develop their CSR activities and subsequently report them to stakeholders via the exchange. The

Framework focused on four sustainability areas—environment, workplace, community, and the marketplace. All public listed companies on Bursa Malaysia had to report their CSR activities in their annual report.[69]

CARBONEX[70]

The S&P BSE CARBONEX (Carbon Index) was developed in India in 2012. The development of this index was funded by the UK's Prosperity Fund. Carbonex is designed to provide a cost-effective way for investors to assess a company's commitment to mitigating climate change impacts and associated risks in order to ensure their investments were not harming the environment or community.

Carbonex tracks the performance of the companies within the Bombay Stock Exchange (BSE) 100 index based on their carbon performance, which is measured through their carbon policies and GHG emissions. The BSE 100 listed companies' performance is modified based on this and benchmarked via Carbonex. The companies (also called constituents) are measured on the basis of:

- Data provided by the Carbon Disclosure Project (a data management company)
- Corporate Responsibility Reports
- Corporate Annual Reports
- Corporate Websites

The companies' climate management practices are assessed against four themes—Reporting & Disclosure, Strategy & Governance, Performance & Achievement, and Ecosystem action.

 **DISCUSSION QUESTIONS**

- *How do stock indices compare with ratings such as PROPER (discussed in Chapter 2)?*
- *Compare the structure of indices such as DJSI and Socially Responsible Index of Johannesburg Stock Exchange.*

3.9 Green Bonds

Green Bond is an innovative fixed-income financial instrument issued by governments, DFIs, or any company to raise capital investment for projects, build assets, or undertake business activities that will create a positive environmental impact. Some of the project categories that are funded under Green Bonds are renewable energy, energy efficiency, pollution prevention and control, sustainability waste management, sustainable land use (afforestation), biodiversity, clean water, and clean transportation.

The key stakeholders in the Green Bonds market are the issuers (entities with green projects that need funding or refunding), underwriters (financial institutions arranging the issuance of the green bonds), external reviewers (verifying the environmental impact or authenticity of the proposed projects) including rating agencies, intermediaries (such as stock exchanges), and investors (organizations with a mandate to invest in or purchase green projects for financial returns).

While issuing Green Bonds, the issuer clearly defines the objective and the kind of projects that will be funded with the proceeds. The projects are thoroughly screened, and the approved projects are invested in using the bond proceeds. The project execution and impact is monitored by the issuer.[71]

Green Bonds provide investors with an option to integrate environmental and social criteria into their investment plans. Green Bonds became an attractive option for investors who were reluctant to invest directly into individual green projects and were looking for options other than equities and funds. With organizations like DFIs and governments acting as principal issuers, the investors viewed Green Bonds as a low-risk investment opportunity.

EIB issued a Climate Awareness Bond in 2007 worth euros 600 million for funding energy efficiency and renewable energy projects. World Bank's first issue of Green Bond appeared in 2008 during the launch of the Strategic Framework for Development and Climate Change. IBRD of the WBG issued green bonds of value SEK 2.325 billion or $350 million for a 6-year period at a 3.5% interest.[72] The issuance of Green Bonds has increased from $3 billion in 2007 to $46 billion in 2015. Figure 3.9 shows the green bond issuance per year between 2007 and 2015.[73] Box 3.16 includes some examples of Green Bonds focusing on renewable energy.

The automotive sector is looking at Green Bonds as an attractive instrument to raise finance. Box 3.17 provides the details.

The Green Bonds market had its own challenges. Despite growing interest in the concept and booming market, the standards and safeguards were undefined and unclear. Validating the authenticity of the positive environmental impact claims by issuers was one of the key issues in the Green Bonds market. Typically, Green Bonds undergo a third-party certification to verify the environmental impact of the projects funded by Green Bonds.

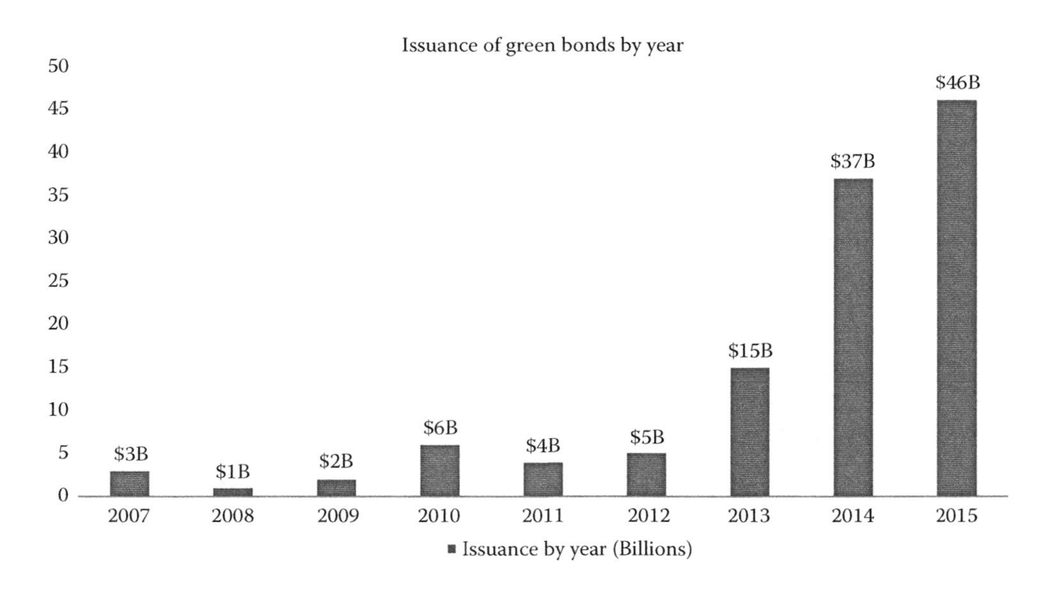

FIGURE 3.9
Green bonds issuance value from 2007 to 2015. (From Impact Report Update 2015, Bloomberg LP, p.13. With permission, online Source: https://data.bloomberglp.com/sustainability/sites/6/2016/04/16_0404_Impact_Report.pdf#page=10.)

BOX 3.16 EXAMPLES OF GREEN BONDS FOCUSING ON RENEWABLE ENERGY

Between 2008 and 2016, World Bank (IBRD) had issued 125 Green Bonds totaling \$9.1 billion. World Bank's investments in ten renewable energy projects are expected to generate ~2,400 MW of renewable energy capacity. Thirteen public transportation projects funded by World Bank Green Bonds are expected to increase public ridership by 2.3 million passengers per day.[74]

IFC launched a Green Bonds program in 2010 and between 2010 and 2015, IFC raised \$3.8 billion. The project categories eligible under this program are energy efficiency, renewable energy, resource efficiency, cleaner technology production, and sustainable forestry. All IFC funded Green Bond projects need to comply with IFC's PS for environmental and social issues and IFC's Corporate Governance Framework.

Mexico's state-owned development bank, Nacional Financiera S.N.C (Nafin), issued the country's first green bond in 2015. The bonds were solely focused on wind energy generation projects and were worth \$500 million with 5-year tenure and an interest rate of 3.41%. Investors were keen in investing in this bond to an extent that the demand for bond reached five times its value.[75]

Yes Bank pioneered the Green Bond issuance in India in 2015 when they issued Green Infrastructure Bonds for \$160 million. They also issued the world's first Green Masala Bonds of \$48.2 million to IFC to fund renewable projects like solar and wind power.[76] In 2016, Yes Bank also issued Green Bonds worth \$50 million to FMO with a tenure of 7-years for renewable projects.[77]

 **DISCUSSION QUESTION**

- *What is the process of launching a Green Bond? What documentation and approvals are needed in your country? What is the advantage of launching a Masala Bond?*

Unfamiliarity of investors in clean energy or green projects and lack of transparent information regarding the risk-return and the project's impact has resulted in reduced interest in making investments.

To address these issues, stakeholders have developed voluntary information-based instruments like Green Bond Principles and Green Bond Indices to help investors make informed decisions and to standardize the market.

Green Bond Principles (GBP) were developed in 2014 in consultation with issuers, investors and environmental groups as an attempt to standardize the Green Bond market and to serve as voluntary process guidelines for improved transparency and disclosure, and to promote sector integrity. The four key components of GBP are[79]:

- *Use of Proceeds*: To ensure that Green Bonds issued are appropriately invested in projects that fall under the approved environmental categories of alternative energy, energy efficiency, pollution prevention and control, sustainable water and green building
- *Evaluation and Selection*: To encourage the issuer to outline an appropriate project evaluation and selection process with well-defined eligibility criteria and environmental objectives

BOX 3.17 AUTOMAKERS REV UP GREEN BONDS FOR HYBRIDS AND ELECTRIC VEHICLES[78]

Green Bonds come as the markets for both electric vehicles and Green Bonds are growing. About $46 billion of Green Bonds were issued in 2015 and, according to Bloomberg New Energy Finance (BNEF), the issues could rise to $56 billion in 2017. Global electric vehicle sales are projected to reach 41 million by 2040, representing 35% of new light duty vehicle sales, according to BNEF.

The financial arm of Toyota Motor Corp., for instance, sold a $1.6 billion Green Bond to fund consumer purchases and leases of energy-efficient Toyota and Lexus vehicles. Zhejiang Geely Holding Group, which manufactures London taxis, raised $400 million in a Green Bond sale to finance development of zero-emission black cabs in the UK. The Chinese automaker's bond was close to six times oversubscribed. Hyundai Motor Co., which plans to launch 26 alternative fuel vehicles by 2020, also ventured into the market in March when its financial arm issued $500 million in Green Bonds.

 **DISCUSSION QUESTIONS**

- *Which sectors of investments dominate Green Bonds? Why do you see such a trend or distribution?*
- *If the returns for a subscriber to a conventional bond and the Green Bond are the same, then why do we see an increasing interest in investing in Green Bonds? Are they more secure or rather respected instruments?*
- *Analyse the performance of some Green Bonds in terms of economic returns as well as the impact that may not be monetized.*

- *Management of Proceeds*: To move the Green Bond proceeds to a sub-portfolio or otherwise attested by a formal internal process that is transparent
- *Reporting*: To encourage the issuer to annually report on the use of Green Bonds proceeds and their anticipated impact

3.10 Green Bond Indices[80]

In 2014, a consortium of banks, third-party evaluation agencies, and service providers launched Green Bond indices with the objective to develop and provide metrics for investors to track the performance of Green Bonds and to provide them with accurate information regarding the risk-return data of a Green Bond. Between 2014 and 2015, four Green Bond indices were launched:

- Bank of America Merrill Lynch Green Bond Index
- Barclays MSCI Green Bond Index
- S&P Green Bond Index and Green Project Bond Index
- Solactive Green Bond Index

These indices have different eligibility criteria and methodology to assess the performance of a Green Bond. Factors like the bond size, issuers' process transparency, coupon rate, and project category are some of the eligibility criteria. Refer to Special Report by Altis Investment Management on Green Bonds: a look at the unique aspects of an innovative asset class (web link: https://www.altis.ch/img/portfolio/ALTIScope-2016Q1-Special1-en.pdf) to read more about the Green Bond indices.

3.11 Conservation Finance

Conservation Finance is the concept of raising financial capital and deploying it for environmental conservation and restoration activities. Conservation finance aims at generating social and environmental benefits while making sound economic returns.

This for-profit approach is a shift from donor-based funding towards investor driven funding. Earlier, environmental conservation projects were dependent on grants and funds from public or private entities. These funding streams were not dependable because of changing priorities of investors and economy. There was a need for a dependable source of funding, which was also sustainable. Conservation finance mechanisms or business models like sale of sustainable agricultural products, textiles, ecotourism, or direct funding from development finance institutions and private investors or through conservation bonds or green bonds are on the rise. This model has been observed at smaller scales and have untapped potential.

Conservation bonds are issued by stakeholders who are looking to borrow money from investors at attractive rates to fund resource conservation projects like energy, land or water conservation. The investors range from governments, development finance institutions (DFI) or private investors who wish to make profitable investments for environmental conservation.

A study conducted in 2014 showed that DFI and private investment in conservation activities has increased since 2004. Between 2009 and 2013, $23.4 billion was invested in creating a positive conservation impact. Of this, DFI funding constitutes 92% of conservation finance, while private funding comprises the remaining 8% though the latter has an annual rate of increase of 26%. Private investors' focus has been on the sustainable food and fiber market.

Attractive investment options are helping the conservation finance market grow globally with investors investing in power, waste management, renewable and sustainable agricultural projects. However, it faces challenges like unclear risk and return rates of projects, lack of systems to conduct cost of conservation and valuation of natural ecosystems, smaller scales of investment, lack of standardized and proven systems to measure the project impact and lack of teams that have demonstrated a successful track record. Policy interventions will be required to facilitate lucrative options and market conditions to encourage impact investing.

The Nature Conservancy's Conservation Note (NCN) is one such impact investment product that allows individual and institutional investors to channel their capital towards high-priority conservation projects around the world at a pre-decided fixed rate of interest. This product was developed by Credit Suisse and Althelia Ecosphere, and the funds raised will be channeled through Althelia Climate Fund (ACF). Individual investors have committed a total of 15 million euros, or $17.5 million to purchase NCN through the

BOX 3.18 THE ALTHELIA CLIMATE FUND (ACF)

The Althelia Climate Fund (ACF) uses an innovative approach called Pay for Performance. Under this approach, funds allocated to a project are tied to predecided project milestones like project commencement, 50% project completion, upon 100% completion or post 1 year monitoring. The undrawn cash deposited with ACF is invested in green bonds elected from the Barclays/MSCI green bond index.

ACF has invested 5 million euros in a biodiversity and forest conservation project in Madre de Dios, the biodiversity capital of Peru. The project is committed to conserving 570,000 hectares of forest covering the National Reserve of Tambopata and the National Park of Bahuaja-Sonene. The project aims to restore agroforestry systems to generate deforestation free cocoa of nearly 3,200 tons annually and to prevent 4 million tons of carbon emissions over a period of 7 years.

 **DISCUSSION QUESTION**

- *ACF has invested 5 million euros in a biodiversity and forest conservation project in Madre de Dios. ACF uses an innovative approach of Pay for Performance. Develop a framework of indicators for the performance assessment.*

bank Credit Suisse, which is the first major bank that has offered individual clients a conservation investment option with market-rate returns. NCN will support conservation and agricultural activities and economic development of local communities in countries like Brazil, Peru, and Indonesia. Agricultural products like sustainably grown coffee, cocoa, and beef are sold at a premium price to generate financial returns for NCN's investors. NCN, when initially offered, was oversubscribed in just ten days indicating a huge demand by investors to make social and environmental impact investments along with generating revenue. Additionally, revenue is generated by selling carbon credits associated with these sustainable projects. We present in Box 3.18 more details on the ACF.

3.11.1 Environmental Impact Bond

Washington, D.C., has a combined sewer system that is used to collect and transport sewer and stormwater for treatment. During wet seasons, the increased sewer flow increases the load on the wastewater treatment plants and forces the overburdened system to discharge untreated sewer into the local rivers. The city has a grey infrastructure made of concrete-like holding tanks and non-permeable surfaces. The city explored green infrastructure for stormwater management like rain gardens and permeable pavement which will prevent stormwater from entering the sewer lines. However, this low cost green infrastructure has not been well-proven, and is not easily predictable and measurable. To test the performance of this green infrastructure, the District of Columbia Water and Sewer Authority (DC Water) got a pilot project approved to include this option into a stormwater control plan. This modification can result in cost savings but also bears performance risk.

DC Water issued an Environmental Impact Bond (EIB) which links financial pay-outs to the project's environmental performance. A $25 million tax-exempt EIB was sold to Goldman Sachs and the Calvert Foundation in September 2016 to fund a green infrastructure pilot project costing a total of $2.6 billion.

Unlike the grey concrete infrastructure, which is straightforward in terms of performance and measurement, green infrastructure is dependent on the site's climate, soils, and nature of vegetation that will determine the quantity of storm water absorbed. The nature of this process makes the measurement of water absorbed into the natural systems more challenging.

In the pilot, a system has developed a channel to direct all the stormwater from the site in a single flow pipe to enable measurement. Also, 12 months' runoff baseline data was taken before the pilot to compare it to 12 months of (ideally reduced) runoff after the green infrastructure installation. Using a software model, it is estimated that the green infrastructure will reduce the stormwater run-off by 30% with a 95% confidence interval. This can significantly reduce the load on the city's wastewater treatment systems.

The EIB will perform like a conventional municipal bond wherein if the project performs as expected, investors will profit from a 3.43% return on investment. The investors did not enter into this agreement with a philanthropic intent, but viewed it as a social impact bond, a debt instrument with a risk-adjusted market rate of return fully aware that they might even get zero profit. The pilot project can serve a dual purpose—to measure how green infrastructure performance and impacts runoff management at a large scale and also to measure the risk associated with green infrastructure projects to establish a baseline for future projects.

3.11.2 Rhino Impact Bond Project

The Zoological Society of London is working with other partners like Conservation International (CI), Fauna and Flora International (FFI), International Union for Conservation of Nature (IUCN), The Nature Conservancy (TNC), Wildlife Conservation Society (WCS), and the World Wildlife Fund (WWF) towards rhino conservation by reducing illegal poaching and wildlife trade. The project objectives include on-ground production of rhinos and reducing demand for products like rhino horns in countries where there is high demand.

To fund such rhino conservation initiatives, the project partners have developed a financial tool—Rhino Impact Bonds—to raise funds for wildlife protection in parts of Africa and Asia. Investors, mostly private, would invest upfront capital in rhino conservation projects for a period of 8–10 years. They would receive profits if the desired outcomes of the project, that is, increased rhino population and reduced poaching and illegal trade, are realized.

3.11.3 Forest Bonds

Forest Bonds are issued to generate funds to conserve, avoid deforestation and sustainably manage forests. Bonds are issued to investors and returns are promised based on revenue generated from forest-based activities.

IFC issued a forest bond in 2016, and it was oversubscribed and raised $152 million indicating that institutional investors can prove to be an important resource for funding forestry projects. One of the forests in Kenya, along the Kasigau Corridor, will be a recipient of these funds to protect it from cutting and agriculture destruction.

3.12 Impact Investing

Impact investments are investments that are intended to create a positive social and environmental impact in addition to economic returns. Impact investments are focusing on numerous sectors like sustainable agriculture, renewable energy, conservation, microfinance, and affordable and basic social services including housing, healthcare, and education. The investors receive returns on their investments at market rates and sometimes even below market rates as a lot of them focus more on the social and environmental aspect of their projects.

Like any other investments, measurement of the progress and performance is critical, and it involves defining social and environmental objectives to be met and metrics and reporting requirements. Impact investing is defying norms that investments in social and environmental issues are not profitable and are done for philanthropic reasons.

In 2015, $31 billion was committed towards impact investing as per a survey conducted by the Global Impact Investing Network (GIIN). These funds were committed to by 2,644 investors. Impact investors either target projects with a more programmatic objective, that is, their priority is to achieve the program objectives rather than financial gains. Financial mechanisms that are used for this approach are grant programs, equity, subordinated loans, senior loans, and cash. Other investors invest in social or environmental ventures but expect returns on par with market rates like any other investment they would have made. Guarantees, cash, fixed income, public equity, and private equity are the financial instrument more suitable for such investments.

In addition to ventures that create a positive impact on the environment and population, impact investments include investments in ventures that operate without any negative impact on the two ecosystems (Box 3.19).

BOX 3.19 IMPACT INVESTOR—ROOT CAPITAL

Root Capital is a not-for-profit impact investor that lends capital, provides financial training and marketing support to small and growing agricultural businesses like coffee and cocoa farmer cooperatives, mango exporters, and companies that sell drought-resistant hybrid seeds in Africa and Latin America. Small business like these lack access to steady capital as commercial banks do not lend in rural areas or require collaterals that such businesses cannot provide. In some areas, where microfinance options are available, many times the financial requirement of these businesses are beyond the lending limit, in some cases $25,000.

Since its inception in 1999, Root Capital has provided around $900 million to over 600 businesses in Africa and Latin America. The small businesses that have a network of farmers are able to better support farmers by paying them premium prices for their crops and providing sustained market access. The farmers in turn provide high quality crops to their customers and increased food security to local populations. By working with this sector, Root Capital is improving the social and environmental situation in that area.

Root Capital evaluates the social and environmental performance and practices of its borrowers using scorecards.

DISCUSSION QUESTION

- *Read about the Root Capital Impact Assessment Framework and the Scorecards. Can this framework be applied across various sectors? Discuss how you could improve the Scorecard approach and indicators.*

BOX 3.20 IMPACT INVESTOR—LOK CAPITAL

Lok Capital is an impact investment fund that was founded in 2001 in India. Over the years, the company has offered funds for various social ventures. Lok Capital invests in social enterprises to deliver basic social services that are scalable, affordable, and commercially viable. Lok Capital focuses on providing microfinance to low-income households in India though micro-finance institutions (MFIs). Lok I was floated for this purpose, wherein it made equity investments in MFIs. In 2006, Lok Capital invested in ten MFIs and also faced setbacks in some parts of the country due to loan defaults.

One of the MFIs that Lok Capital invested in 2014 is Utkarsh. The not-for-profit arm of Utkarsh worked with communities for vocational training, conducted fortnightly polyclinics to provide medical consultation and medication to its customers, educated its customers on matters of financial awareness like financial planning, budgeting, savings, debts, insurance, and banking. The impact of this financial awareness program was that individuals learnt to plan for their financial needs and started saving in a bank account.

Having invested only in MFIs, Lok Capital diversified and invested in sectors like healthcare, education, livelihood, and the affordable housing sectors. In 2016, Lok Capital invested $3.5 million in an affordable housing project in Delhi, the capital of India, which has high real estate rates.

DISCUSSION QUESTION

- *Look for examples where there are more than one impact investors involved in sponsoring a project. How do the investors assess their individual performance in such instances? Is this possible?*

We explain in Box 3.19 the case of Root Capital that is active in impact investing in Africa and Latin America. The case of Lok Capital in India is described in Box 3.20.

Endnotes

1. For further reading: Development finance institutions and private sector development, 2016, OECD, online source: http://www.oecd.org/dac/stats/development-finance-institutions-private-sector-development.htm
2. Based on: Members of the United Nations (193 nations), 2017, Infoplease, online source: http://www.infoplease.com/ipa/A0001295.html
3. For further reading: ADB Annual Report 2015: Members, Capital Stock, Voting Power, 2017, ADB, online source: https://www.adb.org/about/members
4. Based on: Who we are, FMO Entrepreneurial Development Bank, online source: https://www.fmo.nl/profile
5. Based on: About IFC—Overview, 2017, IFC: International Finance Corporation—World Bank Group, online source: http://www.ifc.org/wps/wcm/connect/corp_ext_content/ifc_external_corporate_site/about+ifc_new

6. Based on: Private Sector Operations Department, 2017, Asian Development Bank, online source: https://www.adb.org/site/private-sector-financing/private-sector-operations-department
7. Based on: History, 2017, The World Bank Group, online source: http://www.worldbank.org/en/about/archives/history
8. Based on: About the World Bank, 2017, The World Bank, online source: http://www.worldbank.org/en/about
9. For further reading: Herbertson, K., Will Safeguards Survive the Next Generation of Development Finance, International Rivers, 2012, online source: http://www.bankinformationcenter.org/en/Document.102926.pdf
10. For further reading: Higgins, P., Earth is our Business changing the rules of the game, Shepheard-Walwyn (Publishers) Ltd, 2012, online source: http://eradicatingecocide.com/wp-content/uploads/2012/06/New-World-Bank-Assessment-Rules.pdf
11. Based on: Products & Services, Infrastructure Development Company Limited (IDCOL), online source: http://idcol.org/home/infrastructure
12. Infrastructure Development Company Limited (IDCOL), Environmental and Social Safeguards Framework (ESSF), Policy and Procedures, pp. 10, 11, 15, 40, August 2011, online source: http://idcol.org/download/1d8514287c3e7cda76423b33a781f79c.pdf
13. For further reading: Herbertson, K., *Will Safeguards Survive the Next Generation of Development Finance,* International Rivers, pp. 7, 8, 28, 2012, online source: http://www.bankinformationcenter.org/en/Document.102926.pdf
14. Based on: Environment and Social Management, Power Grid Corporation of India Limited, online source: http://www.powergridindia.com/_layouts/PowerGrid/User/ContentPage.aspx?PId=361&LangID=english
15. Based on: Safeguards Diagnostic Review for INDIA: Piloting the Use of Country Systems to Address Environmental Safeguard Issues at the Power Grid Corporation of India, Ltd (POWERGRID), (2009), The World Bank, online source: http://siteresources.worldbank.org/PROJECTS/Resources/40940-1097257794915/SDR-PSDP-V-POWERGRID.pdf
16. For further reading: Visser, H. et al., Environmental Friendly Road Construction in Bhutan—Providing access to rural communities while protecting the environment, 2005, online source: http://www.piarc.org/ressources/documents/652,5_Visser-Augustijn-Rai.pdf
17. For further reading: Implementation Completion and Results Report (IDA—33090) on a credit in the amount of US $11.6 million (SDR 8.5 million) to Bhutan for Rural Access Project, 2006, The World Bank, online source: http://documents.worldbank.org/curated/en/590251468207264232/text/ICR27.txt
18. IDA At Work—Community-Driven Development: Delivering the Results People Need, online source: http://siteresources.worldbank.org/IDA/Resources/IDA-CDD.pdf
19. *Community-Driven Development—Context, 2016,* The World Bank, 2016, online source: http://www.worldbank.org/en/topic/communitydrivendevelopment/overview#1
20. Indonesia's Kecamatan Development Program: A Large-Scale Use of Community Development to Reduce Poverty, 2004, online source: http://web.worldbank.org/archive/website00819C/WEB/PDF/INDONE-4.PDF
21. Reddy, A.A., and Bantilan, M., Regional disparities in Andhra Pradesh, India, *Local Economy,* Vol. 28(1), 2013.
22. Community-Driven Development—Results, The World Bank, online source: http://www.worldbank.org/en/topic/communitydrivendevelopment/overview#3
23. Based on: Rome Declaration on Harmonization, 2003, online source: https://orangeprojects.lt/uploads/structure/docs/573_58deb52245f9992cece66d3310e35875.pdf
24. For further reading: Aid Effectiveness at a Glance, online source: http://www.dochas.ie/sites/default/files/AID_EFFECTIVENESS_AT_A_GLANCE.pdf
25. For further reading: The Paris Declaration on Aid Effectiveness and the Accra Agenda for Action, 2008, OECD, online source: http://www.oecd.org/dac/effectiveness/34428351.pdf

26. Based on: Aid Effectiveness: Three Good Reasons Why the Paris Declaration Will Make a Difference, 2005 Development Co-operation Report, OECD, Vol. 7(1), p. 52, 2006, online source: http://www.oecd.org/development/effectiveness/36364587.pdf;
27. For further reading: Jeucken, M. H. A., The Changing Environment of Banks, GMI Theme Issue: Sustainable Banking: The Greening of Finance, 1999, online source: http://www.sustainability-in-finance.com/gmi-jeucken-bouma.pdf
28. Based on: About United Nations Environment Programme Finance Initiative, UNEP FI, online source: http://www.unepfi.org/about/
29. Based on: UNEP Statement of Commitment by Financial Institutions (FI) on Sustainable Development, online source: http://www.unepfi.org/fileadmin/statements/UNEPFI_Statement.pdf
30. For further reading: Long, A.D., and Siddy, D., Evaluation of the UNEP Finance Initiative (UNEP FI), 2016, online source: https://wedocs.unep.org/rest/bitstreams/10620/retrieve
31. Based on: Principles for Responsible Investment, Brochure 2016, UNEP Finance Initiative and United National Global Compact.
32. For further reading: From Principles to Performance: An independent evaluation of the PRIs achievements and challenges at ten years, 2016, UN Principles of Responsible Investment, online source: https://10.unpri.org/wp-content/uploads/2016/04/PRI-final-report_-single-pages.pdf
33. Based on: Annual Report 2016, Experience Matters, IFC, online source: http://www.ifc.org/wps/wcm/connect/bf1bfb0b-216b-4cde-941b-dd55febe9d3a/IFC_AR16_Full_Volume_1.pdf?MOD=AJPERES
34. Based on: International Finance Corporation's Policy on Environmental and Social Sustainability, 2012, online source: http://www.ifc.org/wps/wcm/connect/7540778049a792dcb87efaa8c6a8312a/SP_English_2012.pdf?MOD=AJPERES
35. Based on: Equator Principles Association Members & Reporting, Equator Principles, online source: http://www.equator-principles.com/index.php/members-and-reporting
36. Based on: The Equator Principles: Do They Make Banks More Sustainable? online source: http://www.sustainablefinance.ch/upload/cms/user/201602_UNEP_The_Equator_Principles_Do_They_Make_Banks_More_Sustainable.pdf
37. Based on: Supply Chain Audit with focus on Child Labor and Children's Rights, Environmental Management Centre LLP, 2016.
38. Based on: Triodos Bank, Global Alliance for Banking on Values, online source: http://www.gabv.org/members/triodos-bank
39. For further reading: Energy and Climate: A pioneering track record in renewable energy investment, Triodos Bank, online source: https://www.triodos.com/en/investment-management/impact-investment/our-sectors/energy-and-climate/
40. For further reading: Renewable Energy: Funding and know-how from the bank that knows the sector best, Triodos Bank, online source: https://www.triodos.com/en/investment-management/impact-investment/looking-for-funding/renewable-energy/why-triodos/
41. Based on: Transaction Profiles, NY Green Bank, online source: https://greenbank.ny.gov/Investments/Transaction-Profiles
42. Based on: European Biological Resource Initiative, ECNC, online source: http://www.ecnc.org/projects/business-and-biodiversity/european-biodiversity-resource-initiative/
43. For further reading: Drucker, G. et al., Biodiversity and the Financial Sector: A relationship with mutual advantages, European Centre for Nature Conservation, Tilburg, The Netherlands, 2002, online source: http://www.ecnc.org/uploads/documents/2002-biodiversity-and-the-financial-sector.pdf
44. Based on: What Is the Initiative for Sustainable Forest Landscapes? BioCarbon Fund—Initiative for Sustainable Forest Landscapes, online source: http://www.biocarbonfund-isfl.org/isfl-design-principles

45. Based on: IFC, Nespresso, BioCarbon Fund Help Coffee Farmers Boost Productivity and Climate Resilience in Ethiopia and Kenya, International Finance Corporation—World Ban Group, 2016, online source: http://ifcextapps.ifc.org/ifcext/pressroom/ifcpressroom.nsf/0/4AF48BBA1A9DB49D85258044006F773A
46. Based on: Insuring for Sustainability: Why and how the leaders are doing it, 2007, The inaugural report of the Insurance Working Group of the United Nations Environment Programme Finance Initiative; online source: http://www.unepfi.org/fileadmin/documents/insuring_for_sustainability.pdf
47. Based on: Financial Intermediary Funds, The World Bank, online source: http://fiftrustee.worldbank.org/Pages/FIFSOverview.aspx
48. Based on: About Us, Global Environment Facility, online source: https://www.thegef.org/about-us
49. Taken from: UNFCCC Standing Committee on Finance, 2016 Biennial Assessment and Overview of Climate Finance Flows Report, 2016, United Nations Framework Convention on Climate Change, online source: http://unfccc.int/files/cooperation_and_support/financial_mechanism/standing_committee/application/pdf/2016_ba_technical_report.pdf
50. For further reading: The Green Investment Report: The ways and means to unlock private finance for green growth, World Economic Forum, online source: http://www3.weforum.org/docs/WEF_GreenInvestment_Report_2013.pdf
51. Based on: Project FP004: Climate Resilient Infrastructure Mainstreaming in Bangladesh, online source:http://www.greenclimate.fund/-/climate-resilient-infrastructure-mainstreaming-in-bangladesh?inheritRedirect=true&redirect=%2Fprojects%2Fbrowse-projects%3Fp_p_id%3D101_INSTANCE_Hreg2cAkDEHL%26p_p_lifecycle%3D0%26p_p_state%3Dnormal%26p_p_mode%3Dview%26p_p_col_id%3D_118_INSTANCE_4ZRnUzRWpEqO__column-2%26p_p_col_count%3D1%26_101_INSTANCE_Hreg2cAkDEHL_delta%3D30%26_101_INSTANCE_Hreg2cAkDEHL_keywords%3D%26_101_INSTANCE_Hreg2cAkDEHL_advancedSearch%3Dfalse%26_101_INSTANCE_Hreg2cAkDEHL_andOperator%3Dtrue%26p_r_p_564233524_resetCur%3Dfalse%26_101_INSTANCE_Hreg2cAkDEHL_cur%3D2;
52. Based on: Project FP017: Climate Action and Solar Energy Development Programme in the Tarapacá Region in Chile, online source: http://www.greenclimate.fund/-/climate-action-and-solar-energy-development-programme-in-the-tarapaca-region-in-chile?inheritRedirect=true&redirect=%2Fprojects%2Fbrowse-projects%3Fp_p_id%3D101_INSTANCE_Hreg2cAkDEHL%26p_p_lifecycle%3D0%26p_p_state%3Dnormal%26p_p_mode%3Dview%26p_p_col_id%3D_118_INSTANCE_4ZRnUzRWpEqO__column-2%26p_p_col_count%3D1%26_101_INSTANCE_Hreg2cAkDEHL_delta%3D30%26_101_INSTANCE_Hreg2cAkDEHL_keywords%3D%26_101_INSTANCE_Hreg2cAkDEHL_advancedSearch%3Dfalse%26_101_INSTANCE_Hreg2cAkDEHL_andOperator%3Dtrue%26p_r_p_564233524_resetCur%3Dfalse%26_101_INSTANCE_Hreg2cAkDEHL_cur%3D1;
53. Based on: About the Adaptation Fund, Adaptation Fund, online source: https://www.adaptation-fund.org/about/
54. Based on: Readiness Grants, Adaptation Fund, online source: https://www.adaptation-fund.org/readiness/readiness-grants/
55. Based on: Belize Marine Conservation and Climate Adaptation Project, Adaptation Fund, online source: https://www.adaptation-fund.org/project/belize-marine-conservation-and-climate-adaptation-project/
56. Adapting to climate induced threats to food production and food security in the Karnali Region of Nepal Adaptation Fund, online source: https://www.adaptation-fund.org/project/adapting-to-climate-induced-threats-to-food-production-and-food-security-in-the-karnali-region-of-nepal-3/
57. For further reading: Least Developed Countries Fund, Main Issue, online source: https://www.thegef.org/topics/least-developed-countries-fund-ldcf

58. Based on: Strengthening Climate Information and Early Warning Systems in Eastern and Southern Africa for Climate Resilient Development and Adaptation to Climate Change—Malawi, United Nations Development Programme, online source: http://adaptation-undp.org/sites/default/files/downloads/undp_malawi_ews_brief_c4es_10_october_2013.pdf
59. Least Developed Countries Fund: Financing Adaptation Action, Global Environment Facility, Page 49 online source: https://www.thegef.org/sites/default/files/publications/LDCF_Brochure_CRA_0.pdf
60. Based on: Corporate Responsibility: Best Practices and Guidance, Securities Commission Malaysia, online source: https://www.sc.com.my/corporate-responsibility/
61. Based on: ESG Ratings and data model: Integrating ESG into investments—Product Overview, FTSE Russell, online source: http://www.ftse.com/products/downloads/ESG-ratings-overview.pdf?66
62. Based on: DJSI Family Overview, The Sustainability Indices, online source: http://www.sustainability-indices.com/index-family-overview/djsi-family-overview/
63. Based on: CSA Guide—RobecoSAM's Corporate Sustainability Assessment Methodology, 2016, RobecoSAM, Version 4, online source: http://www.sustainability-indices.com/images/corporate-sustainability-assessment-methodology-guidebook.pdf
64. For further reading: BP removed from the Dow Jones Sustainability Indexes, 2010, SAM and Dow Jones Indexes, online source: http://www.sustainability-indices.com/images/100531-bp-statement.pdf
65. Based on: Sustainable Stock Exchanges Real Obstacles, Real Opportunities, online source: http://www.sustainalytics.com/sites/default/files/responsible_research__sustainable_stock_exchanges_2010.pdf
66. Based on: Thailand Sustainable Investment, The Stock Exchange of Thailand (SET), online source: http://www.set.or.th/sustainable_dev/en/sr/sri/tsi_p1.html
67. Based on: The Stock Exchange of Thailand Communication with Stakeholders (2012), Sustainable Stock Exchanges (SSE) Initiative, online source: http://www.sseinitiative.org/wp-content/uploads/2015/04/SET_Comm_Stake_Eng.pdf
68. For further reading: JSE launches new FTSE/JSE Responsible Investment Index Series, Johannesburg Stock Exchange, online source: https://www.jse.co.za/articles/jse-launches-new-ftse-jse-responsible-investment-index-series
69. For further reading: Corporate Responsibility: Best Practices and Guidance, Securities Commission Malaysia, online source: https://www.sc.com.my/corporate-responsibility/
70. Based on: S&P BSE CARBONEX, Bombay Stock Exchange India, online source: http://www.bseindia.com/indices/DispIndex.aspx?iname=CARBON&index_Code=77&page=130E3485-583D-4DED-98A2-F32F805D2324
71. Based on: What Is the Green Bond Process, The World Bank Treasury, online source: http://treasury.worldbank.org/cmd/htm/Chapter-2-Green-Bond-Process.html
72. Based on: World Bank and SEB partner with Scandinavian Institutional Investors to Finance "Green" Projects, online source: http://treasury.worldbank.org/cmd/htm/GreenBond.html
73. Based on: Impact Report Update 2015, Bloomberg, online source: https://data.bloomberglp.com/sustainability/sites/6/2016/04/16_0404_Impact_Report.pdf#page=10
74. For further reading: Green Bond Impact Report, 2016, The World Bank Treasury, online source: http://treasury.worldbank.org/cmd/pdf/WorldBankGreenBondImpactReport.pdf
75. Based on: Climate Bonds Welcomes First Mexican Green Bond, 2015, Climate Bonds Initiative, online source: https://www.climatebonds.net/files/files/Media%20Release-Climate%20Bonds%20Welcomes%20Nafin%20Mexico-Green%20Bond%205-11-15.pdf
76. Based on: YES BANK awarded the Green Bond Pioneer Award Held at London Stock Exchange (LSE), 2015, YES BANK, online source: https://www.yesbank.in/media/press-releases/fy-2015-16/yes-bank-wins-first-ever-green-bond-pioneer-award-in-london
77. Based on: YES BANK announced it will raise INR 330 crore to FMO, the Dutch Development Bank, 2016, YES BANK, online source: https://www.yesbank.in/media/press-releases/fy-2016-17/yes_bank_places_inr_330_crore_usd_50_million_equivalent_of_green_infrastructure_bonds_with_fmo_netherlands

78. The statistics is drawn from Morton, J., *Automakers Rev Up Green Bonds for Hybrids and Electric Vehicles*, Bloomberg, 2016, online source: https://www.bloomberg.com/news/articles/2016-05-27/automakers-rev-up-green-bonds-for-hybrids-and-electric-vehicles,
79. Based on: Green Bond Principles, 2016, International Capital Market Association
80. Based on: Green Bonds: a look at the unique aspects of an innovative asset class, 2016, Altis Investment Management, online source: https://www.altis.ch/img/portfolio/ALTIScope-2016Q1-Special1-en.pdf
81. Taken from: Development Co-operation Report 2014 Mobilising Resources for Sustainable Development, OECD, p. 64, online source: http://www.oecd-ilibrary.org/docserver/download/4314031e.pdf?expires=1488985252&id=id&accname=guest&checksum=97233A99AAEF6ED9CC97521065D745B0.)
82. Based on: Aid Effectiveness: Three Good Reasons Why the Paris Declaration Will Make a Difference, 2005 Development Co-operation Report, OECD, p. 52, Vol. 7(1), 2006, online source: http://www.oecd.org/development/effectiveness/36364587.pdf.
83. Deutscher, E., and Fyson, S., Improving the Effectiveness of Aid", *Finance and Development*, vol. 25, no. 3, The International Monetary Fund, September 2008.

4

Business Response

Business operations play an important role in the sustainability of the planet whether it is a manufacturing, product or service business at a large, medium, or small scale.

Businesses pollute the environment when the enforcement of environmental laws and standards set by authorities is non-existent or weak. They pollute to the point where the penalty for creating pollution becomes greater than the cost of pollution control. When the penalty is greater than abatement costs, businesses are forced to curb their emissions.

In the recent decades, large and small businesses have taken initiatives, voluntarily, or to comply with regulations, to minimize their impact on the environment. There are various reasons why most businesses today want to comply:

- Scarcity or non-availability of resources like fuels and industrial metals
- Increasing regulations and environmental penalties
- Inclination of employees, investors and consumers to be associated with sustainable brands
- Risks to brand or business reputation

4.1 Impact of Government Policies and Legislation

National governments strengthened policies and regulations to impose limits on resource extraction and use, emissions of pollutants, and use of hazardous substances in manufactured products. The regulatory authorities followed precautionary and polluter pays principles (discussed in Chapter 2) while dealing with environmental issues arising from business operations.

For instance, the European Parliament issued two Directives in 2003, Waste Electrical and Electronic Equipment (WEEE) and Restriction of Hazardous Substances (RoHS). WEEE Directive's objective was to regulate WEEE management by prevention, maximizing reuse, recycling, and other forms of recovery of wastes to reduce disposal. The objective of the RoHS Directive was to reduce the hazardous material content in consumer products.[1] Boxes 4.1 and 4.2 discuss the impact of WEEE and RoHS on business operations.

In June 2006, EU introduced the Registration, Evaluation, Authorization and Restriction of Chemicals (REACH) regulation to impose restriction and promote safe use of harmful chemicals that impact human health and the environment. The legislation required manufacturers and importers to register their products and a chemical safety report with European Chemicals Agency (ECHA). Box 4.3 presents the details.

Requirements such as WEEE, RoHS and REACH in the EU impacted not just the companies based in the EU, but across the supply chains, especially from Asia.

BOX 4.1 IMPACT OF WASTE ELECTRICAL AND ELECTRONIC EQUIPMENT (WEEE) EU DIRECTIVE

EU issued the WEEE Directive as a step towards sustainable production and consumption of electrical and electronic equipment (EEE). The objective of this Directive was to prioritize prevention of WEEE generation by re-use, recycling, or recovery of material to minimize disposal.

The Directive defined responsibilities of stakeholders in the EEE value chain like the producers, distributors, consumers, and individuals or businesses involved in the collection, treatment, recycling, and disposal of WEEE. The Directive included Articles that laid down requirements and procedures for sustainable product design to facilitate re-use, dismantling, and resource recovery of WEEE. Product labeling guidelines were included such as the *crossed-out wheeled bin symbol* to indicate that the product should not be disposed of in the bin. Systems were set up for segregated collection of WEEE with targets for collection, disposal, transportation, and treatment as well as resource recovery.[2]

This Directive required EEE manufacturing companies to re-look at their processes and re-design their products. A few examples of the impact of WEEE on business are described below (Figure 4.1):

- Global company Oracle started putting the WEEE label on its products to educate and inform customers that Oracle products should not be disposed into the dustbin.
- Canon operated take-back and recycling programs for many of its electronic equipment. Under the Canon Cartridge Recycling program, Canon recycles 100% of its cartridges, either into new cartridges or into other products.[3,4]
- Automobile companies like Volvo Truck, Nissan and Toyota collaborated to facilitate refurbishment and sales of second-hand parts. Volvo involved a

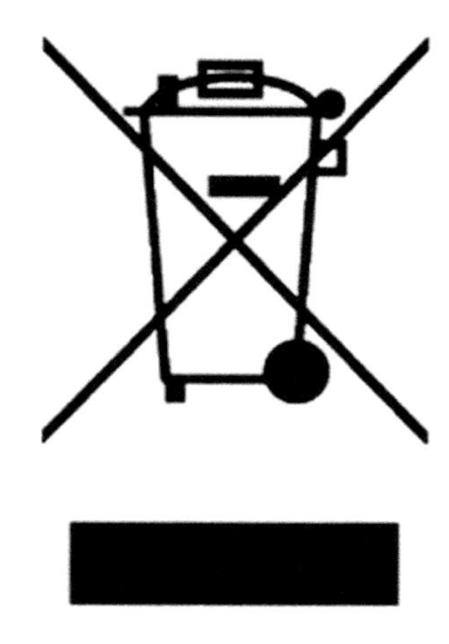

FIGURE 4.1
WEEE compliant product symbol (crossed out wheeled bin). (Adapted from The WEEE symbol, Waste Electrical and Electronic Equipment Directive, online source: https://en.wikipedia.org/wiki/Waste_Electrical_and_Electronic_Equipment_Directive#/media/File:WEEE_symbol_vectors.svg.)

dismantling company to outsource the task of dismantling end-of-life vehicles. This led to design changes such as reduction of number of screws and unification of materials used in a component to ease the dismantling process. Some Swedish companies established an in-house dismantling workshop, which not only saved it outsourcing costs but also gave the design team insight into how to design better vehicles for dismantling.[5]

DISCUSSION QUESTIONS

- *Discuss the supply chain challenges that manufacturers might have experienced to comply with the WEEE regulation.*
- *Cite an example of a successful and sustainable take-back program.*
- *Describe two market instruments (regulatory, economic or voluntary) used to incentivize consumers to participate in take-back programs.*
- *How did businesses internalize the costs of WEEE compliance and evaluate the benefits?*

BOX 4.2 IMPACT OF RESTRICTION OF HAZARDOUS SUBSTANCES (ROHS) DIRECTIVES IN EU

The RoHS Directive was issued by the EU in 2002–2003 to impose restriction on the use of certain hazardous substances in electric and electronic equipment. The EU Member States had to ensure that within 3 years of issuance of the directive, new electrical and electronic equipment put on the market would contain only allowable limits of lead, mercury, cadmium, hexavalent chromium, polybrominated biphenyls (PBB), or polybrominated diphenyl ethers (PBDE).[6]

Businesses had no choice but to comply with this Directive, if they wanted to continue operations in the EU. To comply with RoHS, electronics manufacturers had to track the amount of the restricted substances in their products, test them for compliance, keep records, and report to government agencies. This was a gargantuan task that many countries and companies like the United States, China, Korea, Norway, and Turkey undertook and continue to implement.[7] Some companies modified their products or withdrew them from the market, if they did not meet with RoHS Directives. For example:

- In 2006, Palm Inc., an American electronic manufacturer, withdrew their smart phone Treo 650 from the EU markets, as it did not comply with RoHS.[8]
- Apple's approach to use materials in electronics was precautionary and many Apple products were able to meet with RoHS. However, some products that did not meet the RoHS Directive like iSight, AirPort Base Station with Modem, iPod shuffle battery pack and others were withdrawn from the EU markets.[9]
- By 2010, Dell transitioned its Cold Cathode Fluorescent Lamp (CCFL) based laptop displays that contained mercury to mercury-free Light-Emitting

Diode (LED). This example demonstrated how environmental laws could lead to technological innovations.[10]

- The Thai RoHS alliance was set up to provide a common platform for EEE industry stakeholders, research institutions, testing laboratories to share experiences and pooling resources in order to comply with the RoHS Directive. To execute its objective, a Center of Excellence for Eco-materials and Capacity Building was set-up to facilitate training and capacity building of businesses in technology and production processes.[11]

Compliance with the two Directives came at a cost. The cost of document compliance, re-designing products, lost revenue due to withdrawal of products, or loss of time during the compliance phase globally cost the industry $32 billion.[12]

DISCUSSION QUESTIONS

- *Discuss the environmental and health impact of the hazardous substances mentioned in this Box.*
- *Did the RoHS Directive consider the economic and environmental impact of withdrawing millions of existing non-compliant products? What economic or technological support did the EU offer (or should have offered) to help manufacturers with this challenge?*

BOX 4.3 THE REACH LEGISLATION[13,14]

Under the REACH regulation of 2007, manufacturers that produce products containing Substances of Very High Concern (SVHC) are required to get authorization and replace these harmful substances. Non-compliance with REACH can disrupt business operations, leading to denial of product sales in the EU markets or attract litigation.

REACH applies to chemicals used in industrial processes or for domestic uses like cleaning products and paints. In Europe, EU manufacturers have to demonstrate the safety of their products or substances to ECHA and communicate safety procedures to the consumers.

Companies, especially of small and medium scales, experienced cost increases in order to comply with REACH, but they also improved their internal knowledge, risk management, and occupational health and safety policies.

The international clothing brand H&M created a Chemicals Restriction List that includes a list of chemicals banned or restricted for use or purchasing. H&M requires that its suppliers supply products that are free of harmful substances.

DISCUSSION QUESTION

- *Prepare a list of Substances of Very High Concern (SVHC) as per REACH regulation. See https://echa.europa.eu/candidate-list-table and check their Global Warming Potential and their potential for ozone depletion.*

4.1.1 Consequences of Non-Compliance

Regulatory non-compliance often led to protests from the public and subsequent intervention of the Judiciary. Box 4.4 illustrates an example of how regulatory bodies closed business operations that were violating the environmental laws and applicable standards in the textile industry. The case study also shows how innovations were made in achieving Zero Liquid Discharge (ZLD) with resource recovery to achieve compliance in a cost-effective way.

BOX 4.4 CLOSURE OF TEXTILE INDUSTRIES DUE TO VIOLATION OF ENVIRONMENTAL LAWS, INDIA LEADING SUBSEQUENTLY TO INNOVATION

The Indian Textile Industry contributes 4% to the country's Gross Domestic Product (GDP) and is the second largest provider of employment in the country. It is also a highly polluting, water and energy intensive industry that releases harmful effluents to water bodies and the atmosphere impacting the ecosystem and health of locals.

The Central Pollution Control Board (CPCB) of India issued Standards for discharge of textile effluents against parameters like pH, Bio-chemical Oxygen Demand (BOD), Chemical Oxygen Demand (COD), Total Suspended Solids (TSS), and Total Dissolved Solids in 1986. These standards were amended in 2015 and were applicable to textile industries that had effluent discharge of less than 25 kiloliters per day (KLD). Plants that discharged over 25 KLD and industrial clusters were required to set-up Zero Liquid Discharge (ZLD) units. They were required to re-use the treated effluent and ground water extraction was prohibited except for drinking and make-up water purposes.[15]

Tirupur is a major textile hub in India and contributes nearly 2% of the total foreign exchange earnings of the country. The local High Court and the State Pollution Control Board directed all the bleaching and dyeing units in Tirupur to set-up effluent treatment plants to meet the ZLD mandates in 2005. Though the industry set-up numerous treatment plants, due to technological challenges as well as high costs, it was unable to meet the ZLD norms. In 2011, the High Court ordered all the industries in Tirupur to shut down. They could only be re-opened after demonstrating successful trial runs that met the ZLD norms.

The local textile industry partnered with Tamil Nadu Water Investment Company Limited (TWIC) that demonstrated a ZLD compliant technology based on brine reuse. This technology was accepted by the authorities. Use of this technology even brought down plant operating costs due to recovery of salt and water. The Government announced funding for 20 Common Effluent Treatment Plants to modify existing treatment facilities and install systems to meet ZLD norms.[16,17] This is a good example to show how the judiciary drove businesses to innovate their operations and meet regulatory compliance.

 **DISCUSSION QUESTIONS**

- *Compare the allowable effluent discharge concentration range for the textile industry in a developing* (e.g., *Bangladesh) vs. a developed* (e.g., *the UK) country.*
- *Cite other examples of successful application of ZLD in textile industries.*

4.2 Impact of Economic Instruments

Governments have been using economic instruments (EIs) to solve environmental problems for a long time. EIs incentivize or penalize businesses to adhere to environmental standards, curb polluting activities and enhance natural resources like lands, forests, biodiversity, etc. EIs can complement environmental regulations by shifting the responsibility of environmental management on public and private polluters. It encourages prudent use of resources during purchase (like purchasing more energy efficient products that have subsidies or tax benefits) and disposal (like reduced quantities of waste disposal due to landfill tax or pay-as-you-throw schemes). Table 4.1 lists some of the economic instruments.

Box 4.5 illustrates a case study of the impact of pollution fee in reducing industrial pollution load into the Laguna Lake of Philippines and meeting environmental regulations.

TABLE 4.1

Illustrations of Economic Instruments

Economic Instruments	
Environmental Insurance	Environmental insurance covers costs associated with activities like clean-up, human or material damage compensation due to large-scale environmentally damaging accidents like the Exxon Valdez tanker accident, the Bhopal gas tragedy or the Gulf of Mexico oil spill.
Cap and trade systems	A system wherein a regulatory body sets a threshold (Cap) on the discharge of a particular pollutant on industries or businesses. Two businesses can trade emissions where one business can purchase the permit to pollute more from another business that is polluting lesser than the threshold. Carbon trading is the most common example of cap and trade system.
Environmental Funds	Environmental funds are allocated by governments or corporations to deploy programs involving renewable technologies, energy, water or resource efficiency solutions and biodiversity conservation at a town, city or national level.
Subsidies or Rebates	Subsidies can be allocated for issues like contaminated land remediation or installing solar PV systems. Rebates are offered on consumer products that consume less energy or water to encourage consumers to make green choices.
Performance Bonds	Environmental Performance bonds are funds issued by financing institutes to entities who commit to environmentally sound projects like landfill remediation, contaminated site remediation or green construction, and successfully complete them.
Fines or Penalties	Environmental fines and penalties are levied on entities that violate environmental laws or standards set by regulatory bodies.
Taxes and charges or pay-as-you-throw	These are taxes or charges imposed on resource use or pollutant discharges or on purchase of environmentally harmful products. Pay-as-you-throw is an instrument used for waste management, where entities are charged on the volume or weight of waste disposed by them.
Deposit-refund schemes	A deposit refund system involves charging consumers a surcharge or deposit fee on a product purchase that is refunded upon return of product for re-use or recycling or refurbishing.

BOX 4.5 PHILIPPINES ENVIRONMENTAL USER FEE SYSTEM AT LAGUNA LAKE

Laguna Lake Development Authority (LLDA), the manager of the Laguna de Bay in Philippines, imposed an Environmental User Fee System (EUFS) in 1997 to deal with the issue of industrial wastewater pollution. This system followed the *Polluters Pays* principle where industries were held accountable for their polluting activities and had to pay a water pollution fee for wastewater discharges into the Laguna Bay.

EUFS required all industries to obtain a *Wastewater Discharge Permit* from the LLDA that authorized them to discharge wastewater meeting the prescribed Effluent Standards (see DENR Administrative Order Mo. 35, Series of 1990) into any of the river tributaries of the Laguna Bay.

The EUFS included a fixed component and a variable component based on the annual volume discharged, the concentration and Biochemical Oxygen Demand (BOD) levels of the wastewater. The introduction of this economic instrument encouraged companies to install wastewater treatment and waste management systems, adopt cleaner production techniques, replace harmful chemicals used in their production process, and relocate non-compliant plants to other locations. These measures significantly reduced the wastewater volumes discharged by industries and the BOD levels discharged into the lake. In a survey conducted in 2004 across 89 firms in the area, it was found that the average wastewater volume discharged reduced from 621 cubic meters per day in 1999 to 349 cubic meters per day in 2004, and the average BOD concentration levels reduced from 48 grams per liter in 1999 to 11 grams per liter in 2004.[18,19,20]

DISCUSSION QUESTION

- *Out of the various economic instruments mentioned in Table 4.1, list the ones that you think will work best to promote waste reduction and recycling. Explain why.*

4.3 Impact of Information-Based Instruments

Some governments introduced ratings systems to recognize the environmental performance of businesses, both good and bad. These ratings were shared publicly that added pressure on the polluters and incentivized them to become compliant. The PROPER program of Indonesia discussed in Chapter 2 is one such example that forced businesses to improve their environmental performance and even go beyond the compliance requirements. The program had five ratings—Gold, Green, Blue, Red, or Black (Gold being the best performer and Black being the worst). In 1997, when the program was introduced, there were 4 plants in Blue, 39 in the Red and 9 in the Black category. The performance of the factories improved in 1998 with 26 in the Blue category, indicating that public disclosure is a strong tool for environmental management.[21] The EcoWatch program of the Philippines was similar to PROPER where the environmental performance of factories was publicly displayed.

4.4 From Meeting Compliance to Achieving Competitiveness

Businesses realized that they had to innovate both their upstream and downstream operations in order to ensure compliance with environmental laws. Upstream activities include raw material selection and extraction, product design and production, while downstream activities include energy, water and waste management, sales, transportation, pollutant discharge management, and others.

Factors like investor and public pressure, volatility in the prices and availability of essential resources like material, electricity, fossil fuels, and water required businesses to innovate their manufacturing processes and operations. Businesses had to innovate their marketing and branding strategy to maintain or even enhance their market competitiveness and brand image as a sustainable brand. As most large businesses included a network of suppliers, sustainability-based innovation had to penetrate to the supply chains network beyond the factory gates.

Companies like Dell, Toyota, and IKEA redesigned their products to make their upstream operations more resource efficient. Dell has designed laptops using recyclable and lesser material, IKEA has used renewable materials for their furniture, while Toyota has developed the energy efficient hybrid Prius. Toyota anticipated consumers concern for the environment early on and started marketing Prius, the world's most popular hybrid (gas and electric) vehicle, as a *green* product way back in 1997.[22]

On the downstream side, companies put in place energy, water and waste management systems focusing on Reduce, Reuse, and Recycling (3Rs). Companies strived to achieve similar or higher output with a lower input. Targets on Output to Input ratios were set such as Factor 4 and Factor 10. We describe the Factor concept in Box 4.6.

BOX 4.6 FACTOR 4 AND FACTOR 10: DO MORE WITH LESS[23]

The concepts of Factor 4, and later Factor 10, arose in order to promote resource efficiency.

The Factor 4 concept is to decrease the use of resources and energy and increase productivity by an overall factor of 4. For instance, manufacturing processes can be twice as productive with half the resources. Factor 4 strategies are developing more efficient products with higher longevity that can be reused or upgraded with the use of new technologies. For example, cars can be designed with ultra-strong materials with crashworthy materials, better tires, less weight, and good aerodynamics.

Factor 10 is a more aggressive concept and calls for a 10 times reduction in resource consumption to produce the same output. Factor 10 being more aggressive than Factor 4, requires technological, cultural, policy, and organizational changes. It needs changing of environmental policies, energy consumption policies and patterns, and consumer patterns, and reduction in population growth.

DISCUSSION QUESTION

- *Is the concept of Factor 10 realistic to achieve in practice? Give examples where a successful application of Factor 10 has been done with economic and environmental benefits.*

Due to the rising costs of resources and increasing resource insecurity, Micro-Small and Medium-Scale Enterprises (MSMEs) need to address this challenge by improving resource use efficiency and practicing cleaner production. There are pressures from the supply chain, where questions are raised on the compliance with *codes of conduct* in addition to the regulations. This requires a paradigm shift, often requiring a change in the way materials are sourced and handled, processes and technologies are selected and used, and products including product packaging are designed.

The MSMEs perceive such a transformation as costly, and feel that the return on investment may take some time. Apart from attitudinal or behavior change related barriers, there are reservations on technology, sometimes due to ignorance or lack of information. Technical assistance towards demonstration as well as adaptation is required. Further lack of financial literacy is also a major issue on the demand side that often leads to poor quality of proposals for investments, which hamper support from financial institutions and the co-investors.

4.4.1 Cleaner Production

The 1990s was an era when many leading business organizations across the world were interested in launching programs of interest to business and profitability could be integrated with the protection of the environment. It was also realized that unless such an integration was pitched, there was not going to be much interest or *buy in* by business.

UNEP's Cleaner Production was one such *smart* Program. The concept of Cleaner Production was established by UNEP's Division of Technology, Industry and Economics (DTIE) in Paris.

UNEP defined Cleaner Production as: "The continuous application of an integrated environmental strategy to processes, products and services to increase efficiency and reduce risks to humans and the environment."[24] This definition was pretty deep, yet expansive.

After ten years of successful promotion of CP, UNEP prepared the Global Status Report on Cleaner Production[25] and later a multimedia CDROM "Cleaner Production Companion"[26] that put a compilation of all the major CP resources across the world. UNEP and UNIDO launched a program of National Cleaner Production Centers (NCPCs) across developing nations and created Training and Guidance Manuals on how to set up and operate National Cleaner Production Centers,[27] amongst other publications.

4.4.2 Eco-Efficiency

In 1992, the World Business Council for Sustainable Development (WBSCD) devised the concept of Eco-Efficiency. The concept was based on promoting production and provision of goods and services that utilize fewer resources during production and create less waste and pollution post operations and use. Eco-efficiency is measured as the ratio between the value generated from the product or service, say GDP contribution to its environment impacts, for example, GHG or CO_2 emissions. Businesses have adopted this concept as a core operational philosophy as it positively impacts their bottom line and revenues.

There were clear intersections between CP and Eco-Efficiency. While both CP and Eco-Efficiency had origins primarily from the experience of countries in the European Union, in the United States, the term Pollution Prevention prevailed.

4.4.3 Green Productivity

In 1994, the Asian Productivity Organization (APO) came up with a definition of Green Productivity. Green Productivity was defined as a strategy for enhancing productivity and environmental performance for overall socio-economic development.[28] Green Productivity was considered as the application of appropriate productivity and environmental management policies, tools, techniques, and technologies in order to reduce the environmental impact of an organization's activities. In 2006, APO developed a Training Manual on Green Productivity, conducted training programs, and supported demonstration projects in industries and communities.

4.4.4 Resource Efficient Cleaner Production (RECP)

UNEP defined RECP as continuous application of preventive environmental strategies to processes, products, and services to increase efficiency and reduce risks to humans and the environment. RECP works specifically to advance production efficiency, management of environment, and human development. So, in some sense, RECP was an integrator of CP and GP. The intersections emphasized importance of the shared canvas of Productivity, Environment, and the Interest of Communities.

RECP was defined as the continuous application of preventive environmental strategies to processes, products, and services to increase efficiency and reduce risks to humans and the environment.[29]

Promoting RECP as a comprehensive environmental strategy requires

- Awareness raising among governments, businesses, financial institutions, academia and NGOs
- Dissemination of successful practices and technologies contextualized to national, regional or local needs
- Capacity building at various levels in all stakeholder groups

Internationally, UNEP and UNIDO spearheaded the RECP network[30] building on the NCPCs across the world and developed various RECP strategies, case studies, and tools.

4.4.5 PRE-SME Resource Kit

This generic electronic resource kit enables SMEs to achieve cleaner and resource efficient production. The kit is organized by resource category (water, energy, chemicals, wastes, and materials) with a clear methodological guidance, based on an in-depth survey and review of existing tools and techniques.[31]

The PRE-SME resource kit is targeted to (1) CEOs, (2) SME operations' managers, and (3) RECP consultants. This kit is available on the Web and can be downloaded.[32] There is also an Industrial Handbook that can be accessed.

Each of the above programs made dents in their own way. Some led to more outreach, acceptance and impact. The early definitions of these terms were tweaked during the course and re-interpreted especially to be reflected, in the Millennium Development Goals (and now the Sustainability Development Goals).

In the early phase of these programs, the business sector was asking for the evidence that would prove that it was profitable to integrate business with environmental and social considerations. UNEP DTIE created the International Cleaner Production Information Clearinghouse (ICPIC)[33] and came up with 400+ international case studies across more than 20 industrial sectors covering medium- and large-scale industries. These case studies did the job of convincing and were used in outreach and training programs. Today, we do not need any more convincing. We want to know more about *how to*.

Unfortunately, the concepts of CP, Eco-Efficiency, GP and RECP have not yet penetrated in the graduate level education programs, especially in the developing world. The ocean of resources created and the practice experience documented have not yet reached the student and community of young professionals. We need continuing education programs on these topics especially for the mid-level industry professionals. Those on the top levels of the hierarchy are generally aware of the benefits of the concepts, but we have a long way to go for mainstreaming sustainability in business.

4.4.6 Design for Environment

Businesses started exploring means to make their processes and products more sustainable. The first step was to innovate and alter the product design. Concepts such as Design for Environment (DfE) and Design for Sustainability (D4S) emerged. D4S involves designing a product, service, or process while considering its social and environmental impact across the product's lifecycle. For instance, the lifecycle of a product spans across stages like raw material procurement, manufacturing, packaging, transportation, sales, usage, re-use, and disposal. Use of renewable, biodegradable or recycled material, reducing the quantity of material required by altering design, designing for easy disassembly, re-use and recycling, minimizing quantity or increasing re-use of packaging, and enhancing resource efficiency of processes are some D4S strategies.

Use of materials other than virgin material for manufacturing or altering their products has been a prime focus of DfE and D4S strategies. In this approach, there were dual-objectives such as to improve the bottom line by procuring locally produced and easily available material and to lower the ecological impact by using sustainable or recycled material. Some of the material sources that are rapidly gaining popularity are from renewable and natural sources (like bamboo, hemp, cork, and agri-fibers) and use of pre-consumer or post-consumer recycled material (like steel, plastic from used products, or ocean waste).

There is no universally agreed definition of a sustainable product. Sustainable products are products that provide environmental, social, and economic benefits, while protecting public health and environment over their whole life cycle, from the extraction of raw materials until the final disposal. In designing sustainable products, the following principles are followed:

- Use of low-impact or low material intensity materials
- Minimum use of non-biodegradable or recalcitrant substances
- Use of local materials to the extent possible without adversely affecting local sustainability
- Design such that there is low energy demand/consumption and a higher energy efficiency
- Maximum possible use of renewable energy both in the making and use
- Design of the components for easy dismantling, reuse, and recycling
- Low carbon footprint through the products lifecycle

Examples of sustainable products are several.[34] The Bedol Water Clock is one case where a clock operates on water without batteries or electricity. The electrodes within the water reservoir of the clock convert ions into a current strong enough to power the clock for at least 3 months. A second example is the EcoKettle that was invented in the UK. It contains a special compartment in which the water is stored. You can then transfer a desired amount of water in the second compartment, which will be the only one to actually boil the water. This prevents energy that is wasted by boiling more water than you actually need.

Many designers are now looking to prolong the product's shelf life and make product and packaging refillable. An example is that of Bobble that is free of bisphenol A and complete with a carbon filter to remove contaminants from ordinary tap water. Bobble filters tap water and is 100% recyclable once it needs replacing and that is typically over 2 months. It is a smart solution to the current problem of disposable bottled water filling up our landfills.

Products are also designed in a modular style with flexibility to allow different configurations depending on the interest and use. AMAC's Rhombin Desktop Storage units are bins in the shape of equilateral triangles, and are designed to stack as well as cluster together in limitless configurations. These units are made in California using Cereplast, a plant-based bioplastic.

There is a lot of interest to make products using existing waste streams. In 2016, Adidas collaborated with Parley for the Oceans and innovated the world's first shoe using ocean plastic waste (see Box 4.11). Other remarkable examples in this category come from start-ups, Terracycle and Loopworks. One concern customers have is that there are no environmental standards prescribed for products made from waste to ensure health and safety of the consumer.

We present in Boxes 4.7 through 4.11 some interesting examples of DfE/D4S.

Hemp has been emerging as a super crop because of its multiple environmental benefits and product applications. Hemp can be used to produce food substances, textile, personal care products, textiles, bio-fuels, building material, paper, and medicines. It needs less water for cultivation, has a higher productivity in terms of crop produced per acre than traditional trees, is a natural substitute for cotton and wood pulp, is naturally pest resistant and so does not require pesticides and can grow in various climates and soil types. Companies in India, Australia, and UK manufacture a range of innovative products using various parts of the hemp plant like the seeds, oil, fiber from the bast, and wood. See Box 4.8

BOX 4.7 EXAMPLES OF DESIGN FOR ENVIRONMENT FOR RECYCLING, RE-USE OR DISASSEMBLY

- Xerox established a Recycling Design Guideline way back in 1995 that focused on strengthening design, easing product disassembly, standardizing parts to allow re-use in subsequent models, using sustainable materials, and improving product longevity and reusability. One example is redesigning of the caster (the wheel's part used at the bottom of printers or copiers) for disassembly. The wheel of the caster was made separable from the wheel to allow replacement and re-use of the caster instead of tossing the caster away in case of wheel rupture.[35]
- PUMA's Clever Little Bag was a packaging innovation that used 65% less cardboard required in shoeboxes. The shoebox structure was redesigned to use a recycled plastic bag that held the cardboard container that required no top. The new design also eliminated the need for plastic bags and tissue paper that is typically

stuffed in with shoes. The innovative design drastically reduced paper consumption and use of resources like energy, diesel, and water use by 60% (as estimated in 2010).[36]

- In 2008, Philips committed to designing sustainable products. Senseo Up, its first one-cup coffee machine, is an outcome of this commitment. Senseo Up designers successfully incorporated recycled plastics into the design, despite the aesthetic challenges associated with use of recycled plastic in a high-end product. The product uses 13% recycled plastic in its internal frame and baseplate. These are parts that do not come in contact with food or in the direct line of vision of users.[37]

DISCUSSION QUESTIONS

- *How can the concept of Design for Sustainability help manufacturers of electronics and electrical products comply with the WEEE Directive?*
- *Using an automobile as an example, explain how life-cycle-assessment can serve as a useful tool for design of a sustainable automobile. Research on the work done by Toyota and Ford and make a comparison.*

BOX 4.8 HEMPCRETE: BUILDING MATERIAL FOR LOW-IMPACT BUILDING

Hempcrete is a natural building material and is a *better-than-zero-carbon* material, because it sequesters carbon dioxide for the entire life span of the building. Hempcrete is the combination of the woody core of the hemp plant with water and a lime-based binding agent. Hempcrete is a good insulating material (reduces building energy demand) and prevents moisture and mold growth. Hempcrete weighs about 1/7th the weight of concrete.[38,39]

UK Hempcrete and Hempcrete Australia Pvt. Ltd. are some of the companies that create awareness about and supply Hempcrete. Companies in India and UK manufacture a range of innovative products using various parts of the hemp plant including textiles, food products, dyes, personal care products, and more.

DISCUSSION QUESTIONS

- *Discuss in quantitative terms how hemp is more sustainable than cotton with respect to water demand and productivity per acre of land.*
- *Look for two more product applications of hemp that have been adopted by businesses on a large scale.*

BOX 4.9 USE OF RECYCLED PRODUCTS TO MANUFACTURE CONSTRUCTION MATERIAL

Founded in 1953, Wausau Tiles is a Wisconsin-based manufacturer of construction products like steel-reinforced concrete, pre-cast concrete, coated metal furnishings, pavers, and terrazzo tiles. Over the years, the company explored use of recyclable material in its architectural products. The company attempted to reduce the use of virgin material and raw material costs and serve the increasing demand for *green construction products* in the building sector that helped clients with Leadership in Energy and Environmental Design (LEED) certification programs. They use recycled porcelain, glass, plastic, rubber, fly ash, and steel to manufacture products.

Glass recycling is a complex and expensive process. Wausau used broken glass as a concrete aggregate. Encouraged by the reduction in virgin material demand and the aesthetic appeal of this innovation, they re-designed a number of their products such as benches, tables, planters, pavers, and terrazzo tiles to include recycled glass. In some products, up to 56% of the product weight is recycled glass. Wausau used post-industrial porcelain in sinks, bathtubs, and toilet bowls. Their packaging material like pallets, cardboard cartons, and shrink-wrap are made of recycled material.[40,41]

DISCUSSION QUESTIONS

- *Should there be special incentives offered to companies that make products with recycled content? An example could be tax exemption.*
- *How can we ensure that the products made from recycled materials meet the basic quality requirements and are safe during use?*

BOX 4.10 BIOPLASTICS AND BIOCHEMICALS BY NOVAMONT

Novamont was established in 1989 with a mission to develop bioplastics and biochemicals with low environmental impact. The company developed biodegradable and compostable bioplastics and biochemicals that have found applications similar to conventional plastic material and lubricants. Bioplastics and biochemicals can play a huge role in abating issues like land and water contamination and waste management that are on a rapid rise on a global scale.

Novamont has developed three sustainable product lines of MaterBi, Matrol-Bi, and Celus-Bi. Materi-Bi is a biodegradable granular form of bioplastics made from renewable and/or non-renewable materials like starch blends, cellulose, vegetable oils or specific synthetic polymers. It can be manufactured into products like plastic packaging, cling films, toys, accessories, and biofilters. The material is certified under European (UNI EN 13432) and International standards. Its renewable material content is over 50% and its greenhouse gas emissions per kilogram of product is at least 54% lower than conventional plastics.[42]

Matrol-Bi[42] is their range of bio-lubricants and bio-greases that degrades in a few days and is a great product for use in mechanical equipment used in ecologically sensitive systems like motor boats used in lakes or rivers, motors used in tractors or for other agricultural uses, etc. In case there is accidental spillage, these products will have no adverse impact on soil or the aquatic life.

Novamont developed Celus-Bi that finds applications in the cosmetics sector. The product is made from renewable food crops grown in Europe. Its bio-based ingredients can be used in moisturizers, shampoos, and lipsticks. Celus-Bi microbeads are a biodegradable alternate to the plastic microbeads used in face washes and body scrubs. As discussed later in Chapter 5, plastic micro-beads are in the process of being phased out in many countries and by leading cosmetic businesses due to their impact on aquatic life.[43,44]

DISCUSSION QUESTIONS

- *How is biodegradability of a product measured? Is there any threshold to define biodegradability?*
- *Compare the extent of biodegradability (in terms of time) of a conventional plastic shopping bag and one made of bioplastics.*
- *Are materials that biodegrade always safe? Read Many Faces of Biodegradability at https://prasadmodakblog.wordpress.com/2014/09/11/many-faces-of-biodegradability/*

BOX 4.11 PRODUCT INNOVATION USING OCEAN PLASTIC

Plastic pollution in oceans is a menace. Plastic bags, bottles, cans, bottle caps, and lighters are accumulating in oceans year on year. An inventory study estimated that in 2014 the quantity of micro plastics in oceans was in the range of 93,000–236,000 metric tons, which is merely 1% of the waste entering the oceans in 2010.[45] Large and small businesses are attempting to clean up the oceans of plastic and to recycle the captured plastic into useful products.

In 2015, Adidas partnered with Parley for the Oceans, an organization dedicated to protecting oceans, to create running shoes out of ocean waste.[46]

The upper of the shoe (part that covers top part of the foot) is made of yarn and filaments that are manufactured from plastic collected from the coastline of Maldives. The shoes were sold commercially in 2016.[47]

Adidas created another revolutionary product using the material Biosteel. Biosteel is a high-strength, light-weight fiber-based material, which is 100% biodegradable and has high elasticity. It was developed by a German Biotech company, AMSilk. Adidas and AMSilk partnered to create a new range of running shoes made of Biosteel. This innovation was inspired from spider silk.[48]

A Miami-based company, the Osom brand, makes high quality upcycled clothing products from discarded garments that are landfilled. They converted discarded garments into recycled thread and then into new products like socks.[49]

DISCUSSION QUESTIONS

- *Discuss the impact of ocean plastic on marine life by taking sea turtles as an example.*
- *Read Technological Innovation to Ease Ocean Clean Up https://www.theoceancleanup.com/technology/*

4.5 Life Cycle Thinking

It was realized that environmental impacts of establishments, for example, factory, household, hotel, etc., are not limited to aggregation of direct input-output inventories of their unit operations within their boundaries, but expand to upstream acquisition of resources as well as a downstream fate upon the end of life. On this account, environmental impacts of service, products, or product-service systems having a life cycle of their own beyond the factory gate must be accounted over their life cycle, and this responsibility shall lie with all involved over the life cycle.

Response from the industrial sector came in the form of lean production, 3R, and industrial symbiosis. Industries benefited from such measures as that led to reduction in production costs due to reduction in material (or energy) inputs per service unit (MIPS)[50]. Environmental benefits were communicated in terms of reduction in ecological footprints[51] or CO_2 equivalents.

Post the 1990s, it became evident that decision-making at the consumption phase involves the consumer or a procurement agency that has the power. Sustainable lifestyle, consumption, and associated ethical and social responsibility gained prominence. Life cycle of the product or service from its cradle-to-grave formed the basis of application of systems thinking by sustainability practitioners in industries, and other stages of the supply chains. Policy makers in Europe, Japan, and the United States were the early adapters of the concept and its practical implementations. Government's participation came in the form of eco-industrial parks, procurement policies, and product and process certifications.

Today, the understanding of this concept has moved beyond the industrial ecology domain. Life cycle thinking (LCT) is an essential component of sustainable production and consumption, circular economy, and sustainable regional development. See Figure 4.2.

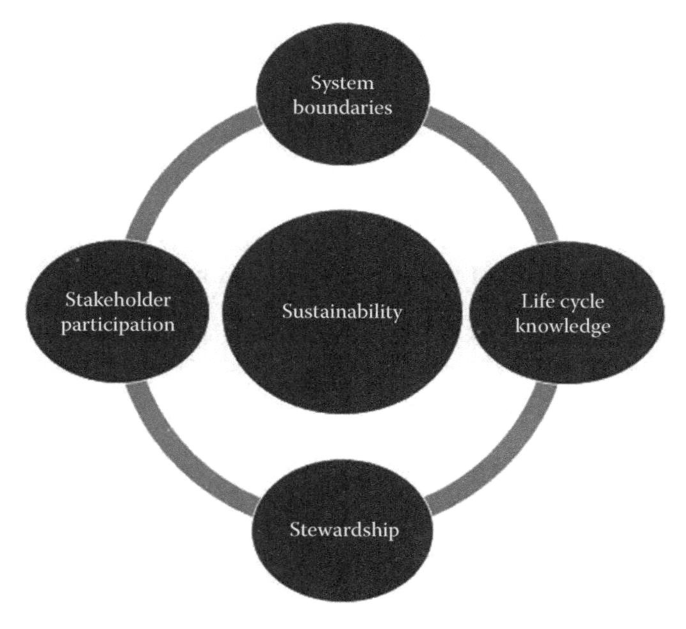

FIGURE 4.2
Conceptual components of life cycle thinking.

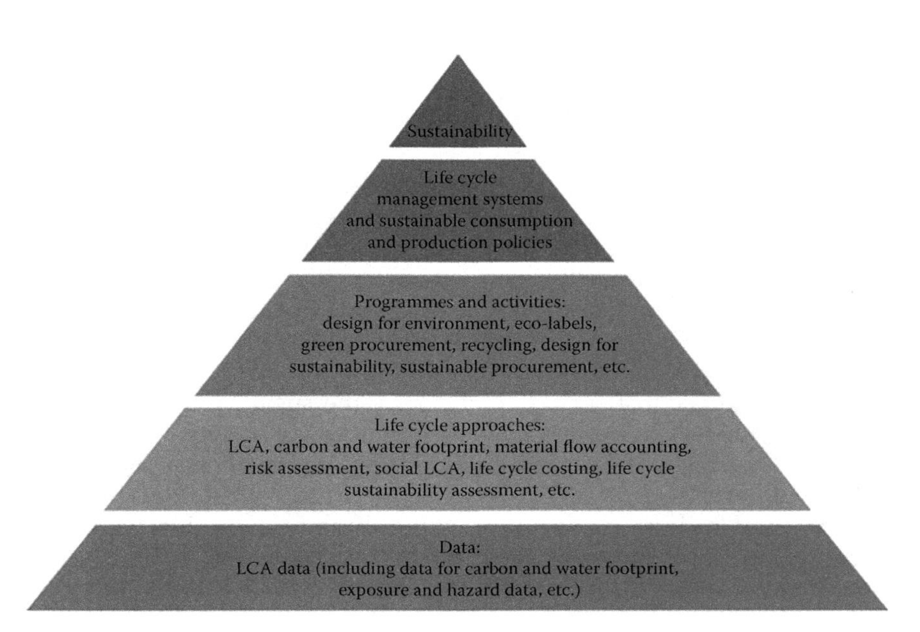

FIGURE 4.3
Sustainability framework supported by Life Cycle Thinking.

LCT is made operational through life cycle management (LCM).[52] Businesses, governments, and organizations can adopt the LCM approach in order to integrate LCT in their decision-making. Figure 4.3 presents the sustainability framework supported by LCT and associated approaches. Life cycle data forms the base of the pyramid, and is essential to generate life cycle inventories and to translate inventories into mid- or end-point impacts.

In 1993, the Society of Environmental Toxicology and Chemistry (SETAC) published a set of guidelines and in 1996 the International Organization for Standardization (ISO) standardized the life cycle assessment (LCA) methodology.[53] In 2002, the United Nations Environmental Program (UNEP) and SETAC launched an international partnership known as Life Cycle Initiative,[54] to promote application of life cycle thinking. In 2006, ISO published the ISO 14040 and the revised ISO 14044 standards.[55] Various national LCA networks like American Center for LCA, the Australian LCA network, and the Thai LCA network were established. In recent years, it has been realized that inclusion of social indicators is necessary to address a holistic decision-making. In 2006, the European Commission (EC) commissioned the Coordination Action for the Innovation in Life Cycle Analysis for Sustainability (CALCAS) project to structure the varying field of LCA approaches and to define research lines and programs to further LCA studies.[56] The project has produced a Life Cycle Sustainability Analysis (LCSA) framework. In 2009, UNEP also published a guideline document to conduct social life cycle assessment of products.[57]

4.6 Ecolabels

According to the Global Ecolabeling Network,[58] *Eco-labeling* is a voluntary method of environmental performance certification and labeling that is practiced around the world. An ecolabel identifies products or services that are proven to be environmentally preferable based on the LCA within a specific product or service category.

Most of the environmental regulations focused on processes and resultant wastes/ emissions. Stipulations through eco-labels became relevant as impacts beyond the factory gate were understood across the life cycle of the product.

Eco-labels instill trust and transparency in the market by differentiating between sustainable and unsustainable products. They increase consumer awareness and help them make informed and environmentally conscious purchasing decisions. They stimulate market competitiveness by incentivizing manufacturers and businesses to choose the greener path. More often than not, businesses reduce their operating and material costs when they opt for Eco-labels or certifications as they focus on resource efficiency.

Eco-labels are voluntary instruments that certify the sustainability or environmental impact of products, processes and services based on life cycle considerations. In contrast to a self-proclaimed environmental symbol developed by a manufacturer or service provider, an impartial third party awards the Eco-label if the product (or service) meets the established environmental leadership criteria that typically assess a product's sustainability.

Environmental labeling programs originated in 1978, with the label *Blue Angel* launched by Germany. Energy Star Label for electrical appliances, LEED or BREEAM for green buildings, Carpet and Rug Institute Green Label for carpets and cushions, EarthCheck for travel and tourism, Fairtrade for ethical trading and fair labor wages, and Forest Stewardship Council for sustainably managed forests are just a few examples of Eco-labels that businesses in various sectors are adopting globally.

The Organization for Economic Cooperation and Development (OECD) conducted an analysis in 2013 and found that the number of labels increased roughly fivefold between 1988 and 2009.[59] As of 2011, the Eco-label Index, a global directory of eco-labels, includes 377 schemes in 211 countries and 25 industry sectors. Both the consumers and retail stores supported the move towards eco-labels. Several national governments in the European countries supported eco-labels and countries in Asia such as Japan, Korea, and Thailand took the lead. The increasing number of eco-labels is a result of pressure from governments and consumers for greater environmental sustainability of manufacturing and product consumption systems.

Some of the common criteria on which Eco-labels are typically assessed are resource conservation and efficiency, use of sustainable materials, biodiversity conservation, product declarations (about the product ingredients), compliance with health and safety standards, biodegradability, adoption of an environmental management system through the product's (or service's) life cycle, and social inclusion. Boxes 4.12 through 4.14 include descriptions of a few Eco-labels.

The eco-labels have evolved through four *waves*. See Figure 4.5. The first wave focused on greenness of the product addressing resources (inputs) and wastes/emissions (outputs) showing preference to products that had low resource intensity[67] or low "ecological rucksack."[68] The next wave addressed the health, safety, and biodegradability related considerations, which led to phasing out and/or substitution of harmful and non-biodegradable substances. The third wave looked at social issues such as management of labor (working hours, fair wages and child labor) and came up with requirements to meet the codes of conduct[69] and ethical practices across the supply chains. The fourth wave came up with a need to reduce carbon footprints in product making, packaging, and transportation, showing preference for Low Carbon Goods and Services. Today, despite the

BOX 4.12 BLUE FLAG CERTIFIED BEACHES/MARINAS

Initiated in France in 1985, Blue Flag is a certification program for the coastal areas that focuses on sustainable development of beaches, marinas, and boating operators. The program's prime focus was in Europe but it expanded to include South Africa, Morocco, New Zealand, South America, Canada, and the Caribbean. It is operated by a non-profit organization, Foundation for Environmental Education (FEE).

The program has four focus areas to promote sustainable development: environmental education and information, environmental management, safety and services, and water quality. The program encourages cleanliness, sustainable transportation, sanitation, and safety in the coastal areas. Under the water quality standards, a beach can be certified as a Blue Flag beach, if there is no discharge of untreated domestic or industrial water. Water sampling is done every bathing season to ensure it meets the national standards for bathing water/microbiological parameters like faecal colibacteria (E. coli), total colibacteria and physico-chemical parameters like pH, oils, waste residues, color, and such others are monitored to ensure they are within the acceptable range.[60]

DISCUSSION QUESTIONS

- *Review all the criteria listed under the Blue Flag program. See—http://ec.europa.eu/ourcoast/download.cfm?fileID=1018*
- *Form a group of 4–5 students. Pick your favorite beach or one near you that you think can benefit from this program. As the beach manager, what would be the top five sustainable initiatives you would implement and why?*

BOX 4.13 EU ECO-LABEL OR EU FLOWER

EU Eco-label or EU Flower is a voluntary eco-label scheme established in 1992 by the European Commission for everyday products like cleaning detergents, all-purpose cleaners, personal care products, clothing, paints and varnishes, electronics, flooring, furniture, lubricants, and many more (Figure 4.4).

To qualify for the EU Eco-label, products have to comply with a tough set of criteria like the products should be energy efficient, water efficient, free from hazardous materials, should be bio-based, should use less packaging material, have reduced impact on aquatic life, should employ production processes that are less polluting, and should be easy to re-use or recycle.

These environmental criteria were set by numerous stakeholders including consumer organizations and industry and take into account the whole product life cycle—from the extraction of raw materials, to production, packaging and transport, right through to your use, and then your recycling bin.

FIGURE 4.4
EU Ecolabel logo. (From © European Union, 1995-2017, EU Ecolabel Products and Services, European Commission (EC): Environment, 2017. With permission, Online Source: http://ec.europa.eu/environment/ecolabel/eu-ecolabel-products-and-services.html.)

In 2015, 44,051 products were certified under EU Eco-label. The countries with high number of EU Eco-label certified products licenses include France (28%), Italy (17%), and Germany (11%).[61]

DISCUSSION QUESTION

- *List three Eco-labels used in the food and restaurant industry and compare the criteria.*

BOX 4.14 BLUESIGN CERTIFICATION FOR TEXTILES

Bluesign is a European Standard for sustainable textiles developed by the firm Bluesign Technologies (BT) based in Switzerland. Since 2000, the company has been helping businesses and brands in the textile industry improve their environmental, health and safety standards.

Bluesign's strategic approach is *Input Stream Management*. The Bluesign System is designed to keep harmful substances out of the lifecycle of textiles by not allowing it to enter the production line. When a company applies for Bluesign certification, BT partners with the member company to audit its textile mills, manufacturing process, raw material, energy and water inputs, chemical usage, pollutant discharge, and recommends ways to improve them.

For example, Bluesign has three ratings of Chemicals: Blue—safe to use, Gray—special handling required, and Black—forbidden. The system requires its partners to phase-out the black chemicals used and sets pre-conditions for use of the gray chemicals. As of 2014, the system has a Bluesign System Substances List (BSSL) that has over 900 approved chemicals and has banned the use of about 600 chemicals.[62,63]

Bluesign System partners make improvement in five key areas of resource productivity, consumer safety, water emissions, air emissions, and occupational health and safety. Once all these criteria are met with, the brand's product get a Bluesign Label.[64]

Many global textile brands are Bluesign System Partners. Patagonia is the first US based textile manufacturer to become Bluesign certified in 2007. By 2013, 15% of the brands' products contained Bluesign certified fabrics.[65]

North Face is another brand that has benefited itself and the environment by partnering with Bluesign. Between 2010 and 2015, they have saved 1,173 million liters (equivalent to 470 Olympic sized swimming pools), 69 million kWh (equivalent to 6,403 cars off the road for one year), and 4 million kgs of chemicals not used (equivalent to 212 tankers full of chemicals).[66] Other brands like Puma, Salomon, Nike, Omega, Wilson, Columbia, and Adidas have also partnered with Bluesign System.

DISCUSSION QUESTIONS

- *Discuss the three types of certifications—first, second, and third party certifications*
- *List five substances that are banned per BSSL? And why?*
- *Do you consider Bluesign as more of a lobby for price negotiations with the suppliers or an elite club?*

challenging economic environment, consumer demand for lower-carbon products and services is on the rise.

In terms of downstream of the supply chains, especially the Small and Medium Enterprises (SMEs), adherence to criteria set for the eco-labels became part of the requirements in vendor registrations. Eco-labels became like a benchmark that companies had to meet if they wished to continue to be suppliers to international retailers or organizations. Eco-labels also influenced the development of sustainability frameworks and standards in certain sectors like the Roundtable on Sustainable Palm Oil, or Bonsucro, the Better Sugar Cane Initiative. This sector wide impact influenced a large number of the SMEs asking for the change.

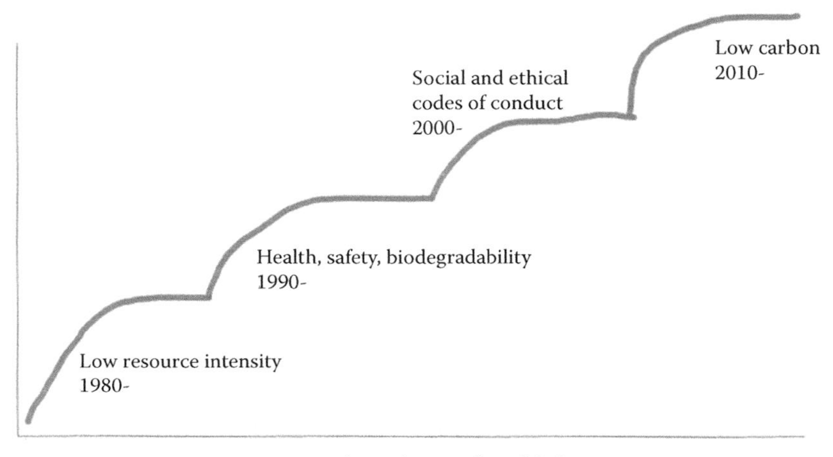

FIGURE 4.5
Evolution of eco-labels.

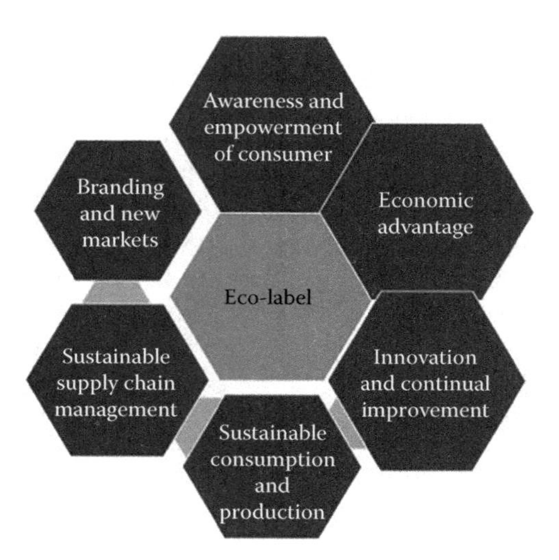

FIGURE 4.6
Benefits of eco-labelling.

Eco-labels offer many benefits. Fulfilling the eco-label criteria can be beneficial for businesses. This will help them adopt environmentally sound management practices and business models and also help them to improve productivity and efficiency. Eco-labels can also help in better branding and widening of the market—especially sectors like organic food. Figure 4.6 illustrates the benefits of eco-labeling. Amongst all, product innovation is considered to be the principal opportunity. We present in Box 4.15 a case study of eco-labelling in a textile industry in Egypt.

BOX 4.15 ECONOMIC AND ENVIRONMENTAL BENEFITS OF ECOLABELLING AT MISR MAHALLA IN EGYPT[70]

Misr Mahalla is the largest textile company in Egypt. The company had to meet the requirement to comply with Oeko Tex 100 standard as imposed by the German market. Misr Mahalla, under the support of the SEAM (Support for Environmental Assessment and Management) Program, achieved this requirement with significant economic and environmental benefits. These benefits are elucidated below.

CHEMICAL AND DYE SUBSTITUTION

Meeting the Oeko-Tex standard required a chemical audit followed by substitution of hazardous chemicals and certain objectionable dyes. The chemical substitution and process modifications resulted in annual savings of LE30,456. Savings that resulted from modifications to the bleaching process were LE89,820 on an annual basis. Against these gains, purchasing eco-friendly chemicals and dyes resulted in a yearly increase of LE59,364.

REDUCED SHIPMENT ANALYTICAL COSTS

Previously a number of German clients required that all shipments without an eco-label had to be tested, resulting in analytical costs of LE1,000 per shipment. This requirement was not required for articles with an eco-label. The annual savings to the company by reducing the analytical costs was around LE20,000.

POTENTIAL EXPORT MARKET GAINS

The annual value of the factory export market is around LE383 million. Of this, almost 15%, or LE57.5 million in value, was exported to Germany. Because of the grant of an eco-label, the market was expected to increase by 5%, equivalent to gain of LE2.9 million per annum.

IMPROVED PRODUCTION EFFICIENCY

Financial benefits also resulted from the following:

- 5% improvement in the Right First Time in the dyeing process;
- 20% reduction in the processing time;
- 14% reduction in steam consumption.

IMPROVED PRODUCT QUALITY

There was a noticeable improvement in the product quality. When sodium hypochlorite was used for bleaching, there were occasional incidences of low tensile strength, which at times was 20% lower than the required standard. Since the elimination of hypochlorite with hydrogen peroxide, such incidences have not occurred.

ENVIRONMENTAL IMPROVEMENTS

- Complete phase-out of sodium. This resulted in the elimination of AOX and a reduction of Total Dissolved Solids (TDS) in the effluent.
- Improvement in working conditions through the elimination of hazardous chlorine compounds.
- Reduction in water consumption leading to reduction in volume of generated effluent.
- Reduction in energy requirements.

Green products help to minimize risks, improve competitiveness, and widen the market for a brand. Greening of the products and associated supply chains leads towards Smart Sustainability (see Figure 4.7).

Many eco-labels have helped in raising the awareness and creating trust and creating a common framework around sustainability.

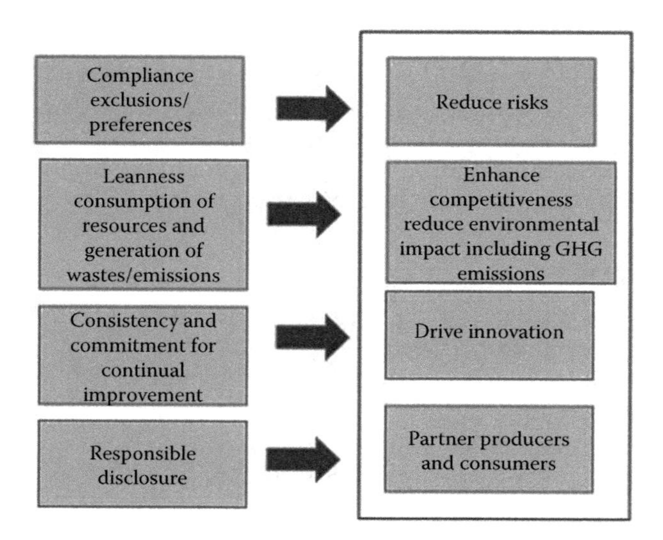

FIGURE 4.7
Leveraging on green products towards smart sustainability.

4.7 Influencing Supply Chains

Most large businesses include a network of small and medium-scale enterprises (SMEs) that form a part of the company's supply chain. Supply Chain Management in business-as-usual operations and in maintaining environmental compliance is a gargantuan task for any business. The first steps for the parent organization are to set sustainable goals and develop policy and guidelines to meet these goals. The next step is to disseminate the goals, policy, and technical know-how of the methods and technologies to achieve these goals within their supply chain network and use their influence and clout as a client to enforce it. Many businesses have developed Codes of Conduct and audit procedures to whet their suppliers prior to entering the corporate network and to monitor their performance. Boxes 4.16 through 4.19 illustrate some interesting examples of managing environmental compliance across supply chains.

BOX 4.16 WHY SONY REVAMPED THE SUPPLY CHAIN AFTER THE CADMIUM CRISIS

The events of December 2001 culminated in the introduction of a new supplier management system at Sony. Right before Christmas, the Dutch government seized 1.3 million PlayStations and 800,000 accessory packs of Sony valued at around $160 million. They were seized because the government found that the consoles and accessory cables contained the toxic element cadmium in excess of EU's allowable limit. Cadmium can cause kidney damage and when released into the atmosphere upon disposal, can contaminate the soil, groundwater, and food chain. The entire shipment was sitting in warehouses.[71]

Sony tried to salvage the issue by replacing the accessory with cadmium free parts and tried to trace the root cause of the problem in the supplier network that produced these products for Sony. However, the lack of supply chain transparency made the root cause analysis a complex operation and cost Sony $180 million and 18 months. Realizing the importance of a transparent and sustainable supply chain management process, Sony revamped its supplier network. They put in place supplier selection criteria that included Green Partner Standards, Green Partner audit, and adherence to Sony's Supplier Code of Conduct, which was based on the Electronic Industry Code of Conduct.[72]

 **DISCUSSION QUESTIONS**

- *Compare the Green Partner Standard, Sony's Supplier Code of Conduct, and the Electronic Industry Code of Conduct.*
- *Discuss another example where a business got impacted due to the RoHS Directive.*

BOX 4.17 IKEA'S ENVIRONMENTAL RESPONSIBILITY AND IWAY, ITS CODE OF CONDUCT

Founded in 1943, IKEA is a Swedish lifestyle furniture store, known for selling disassembled furniture with 328 stores in 43 countries across the world (as of 2015).[73]

IKEA developed their first Environment Policy in 1990 to ensure that the company factors in the environment in its operations. In that decade, IKEA launched an Environmental Action Plan and initiatives like[74,75]:

- Using Totally Chlorine Free (TCF) paper for their catalogues.
- Replacing polyvinylchloride (PVC) in textiles, shower curtains, and furniture.
- Designing a model of chair (OGLA) made from 100% pre-consumer plastic waste.
- Introducing a series of air-inflatable furniture products to reduce raw materials use and transportation weight and volume.
- Using wood from responsibly managed forests that replant and maintain biological diversity; IKEA became a founding member of Forest Stewardship Council in 1993.
- A Waste Management manual was introduced in 1999.
- Launching IWAY, a Code of Conduct for its suppliers and a special Code of Conduct on Child Labor in 2000.

The IWAY Standards specify minimum requirements and standards for the environment, social aspects, and working conditions to be followed by IKEA suppliers of products, components, raw material, and services. IWAY has led to improvements in its supply chain and has helped IKEA and its suppliers become more competitive. An overview of the IWAY Standards is included below[76]:

- Prevention of Child Labor; Forced or Bonded Labor.
- Safe and healthy working conditions; prevent exposure to hazardous conditions.
- Prevent harm in terms of air, noise, ground, and water and make efforts to reduce energy consumption.
- Standards for purchase, storage, handling, and transportation of chemicals.
- Standards for handling, storing, transportation, and disposal of waste (both hazardous and non-hazardous) to prevent harmful emissions to air, ground and water, risks of ignition/explosion, and ensure workers' health and safety.
- At least minimum wages and overtime compensation.

By the end of FY15, IKEA had 978 home furnishing suppliers in 50 countries with 11 years as the average partnership period. IKEA visits and audits their suppliers regularly to ensure compliance with IWAY. In 2012, more than 1,000 audits were carried out in supplier factories.

IKEA's 2015 report boasts of significant improvements in the supply chain. IKEA suppliers that are FSC certified increased from 34.9% in 2011 to 72% in 2015. Their home-furnishing suppliers increased their energy efficiency by 17.6% in 2015 when compared to 2012; and the water efficiency in liters consumed per cubic meters of product produced increased by 37.3%. Their suppliers have experienced that investing in working conditions and the environment has led to more orders, better productivity, and improved profitability, thereby improving competitiveness.[77]

 **DISCUSSION QUESTIONS**

- *Read IKEA FY 15 Group Sustainability Report—http://www.ikea.com/ms/en_US/img/ad_content/2015_IKEA_sustainability_report.pdf*
- *Can disassembled furniture reduce the overall environmental impact and, specifically, that of product transportation? Are there any social benefits? Corroborate your responses (yes or no) with reasons.*

BOX 4.18 MARKS AND SPENCER'S (M&S) GREEN SUPPLY CHAIN NETWORK

M&S's green strategy called Plan A is one of the best examples of businesses' commitment to environmental and social causes. Launched in 2007, Plan A consisted of 100 goals to be achieved over a 5-year period. Some of the goals included combating the pressure on finite natural resources, cutting down waste, saving energy, trading fairly, and animal welfare.

M&S sources products from more than 70 countries. To increase transparency of their supply chain network, they developed an interactive supply chain that lists details of all their suppliers across the globe.[78]

All M&S suppliers need to comply with their Global Sourcing Principles, which amongst the many clauses include providing healthy and safe working conditions, following M&S environmental standards, and compliance with local and national laws. All their suppliers go through performance audits.[79]

To achieve the Plan A goals, M&S needed the support of its wide supplier network. Some of Plan A's initiatives related to its supply chain network are discussed below:

- M&S is working with World Wide Fund for Nature (WWF) and Better Cotton Initiative to help their farmers produce cotton with less water and pesticide. Their goal is to get 70% of their cotton from sustainable sources by 2020.[80] They use organic cotton or recycled cotton in their textile-based products and follow the FairTrade policy by paying farmers a Fairtrade price and a premium to incentivize them to produce organic cotton and protect them from volatile market prices.
- M&S sources palm oil from only those suppliers that have adopted the Roundtable on Sustainable Palm Oil (RSPO) growing standard, which has an objective of *promoting the growth and use of sustainable oil palm products through credible global standards and engagement of stakeholders*. All their palm oil suppliers undergo audits to seek certification and authorization to supply to M&S.[81]
- M&S stores donate surplus products like fruits, vegetables, baked goods, and groceries like pasta, cereal, and cooking sauces to the community in need of food. They have a network of food banks, community centers, and community cafes across the UK to help them combat waste management via this social channel.[82]

DISCUSSION QUESTIONS

- *Trace the history of Better Cotton Initiatives and discuss the technological innovations that contributed to sustainable cotton farming.*
- *Read interesting stories of M&S's suppliers here— https://corporate.marksandspencer.com/plan-a/our-stories/about-our-farmers and use for discussions.*

BOX 4.19 CORPORATE SYNERGY SYSTEMS: A TAIWANESE INITIATIVE TOWARDS GREENING OF SUPPLIERS

Like in many Asian countries, the Taiwanese manufacturing sector consists of more small and medium enterprises (SMEs) than large industries. Even if the SMEs wish to adopt sustainable practices, they face technical and financial obstacles and need to be supported by large corporations and industry associations to accelerate adoption of sustainability.

One of the approaches heavily promoted by the Taiwanese government is formation of Corporate Synergy Systems (CSS), an association led by business leaders to help SMEs in their supplier network adopt sustainable practices. They help SMEs and suppliers through training, education, and demonstration programs, developing in-house expertise, introduction to technologies, financial incentives, and information exchange forums. They incentivize SMEs by giving business preference to suppliers that comply with CSS guidelines over the non-compliant suppliers.

Between 1995 and 1998, large electronics and electrical devices, paper products, and textile and automobile manufacturers organized CSSs to promote Integrated Waste Management systems in their supplier network. The attractive economic benefits, compliant supplier preference by large corporations, and increase in global competitiveness incentivized SMEs to adopt integrated waste management (IWM) practices if they wanted to stay in business.[83]

 **DISCUSSION QUESTIONS**

- *If you were a small business that manufactures coconut cookies, then what challenges would your company face to set-up a sustainable waste management system? Discuss the challenges in the operational, economic, and technological perspectives.*
- *What kind of government or corporate support would you seek to overcome these challenges?*

Transparency for Sustainable Economies (Trase) is software that was established as an interactive online platform to enable government, businesses, investors to understand the problem, and manage the environmental and social impacts of their supply chains. The objective of Trase was to create transparency in supply chains and provide data to relevant stakeholders regarding the flow of commodities like palm oil, soy, beef, and timber. The platform traces the movement of commodities from its origin to exporters and importers and finally its consumption markets. This data can help stakeholders identify risks, measure impact, and monitor progress of their sustainability supply chain and deforestation-free trading initiatives.[84]

4.8 Sustainability Assessment of Technologies

The need for promotion of Environmentally Sound Technologies (ESTs) in the context of sustainability was recognized in the early 1990s. The United Nations Conference on Environment and Development (UNCED) in 1992, stressed the need to promote ESTs and highlighted so in Agenda 21. Chapter 34 of Agenda 21 defines ESTs as those technologies that protect the environment, are less polluting, use all resources in a more sustainable manner, recycle more of their waste and products, and handle residual waste in a more sustainable manner than the technologies for which they are substitutes. ESTs include a variety of cleaner production process and pollution prevention measures, as well as end-of-pipe and monitoring technologies. There was, however, no explicit focus to embed sustainability considerations in the selection of technologies.

The International Environmental Technology Centre (IETC) of the United Nations Environment Program initiated the development of a methodology to address the above limitations. A new methodology known as Sustainable Assessment of Technologies (SAT) was subsequently developed and was tested on pilot projects across the world.[85] The focus of SAT was to emphasize both process as well as outcome, with an interest towards informed and participatory decision-making.

Unlike ESTs, SAT does not look at technologies in isolation but considers them in the form of a *system* where a combination of ESTs are considered. SAT follows a progressive assessment procedure, through tiers like screening, scoping, and detailed assessment. Such an approach allows entry points for diverse stakeholders and for optimizing the information requirements.

SAT could be applied by government as well as by business at both policy and operational levels. At the policy/government level, SAT can be applied for strategic decision-making. These strategic level decisions are often made by planners, civic body officials, and mayors/elected representatives.

Once decisions at the strategic level are taken, SAT could be applied at the operational level, especially by business and primarily by the technical/engineering staff, designers, and consultants to assess alternate technology systems. Target users could also include developmental as well as commercial financing institutions that often play a key role in funding projects and programs that make use of various technologies.

Last but not least, individual hamlets/villages and enterprises can also use the SAT methodology for comparing a number of available options for sanitation, water supply, and treatment or manufacturing technologies. See Figure 4.8 that describes possible applications of the SAT methodology.

The steps followed for application of SAT methodology are shown in Figure 4.9. The methodology follows the typical Plan-Do-Check-Act cycle of continuous improvement as recommended by systems like Quality/Environmental Management Systems (ISO[86] 9000/14000).

SAT methodology could be useful to businesses in making a choice on the technologies. Often the technology choices are made only on economic grounds and the environmental and social factors are not considered in the assessment framework. Experience has shown that choices made only on cost considerations may result only in short term gains and, in some cases, even pose unanticipated risks.

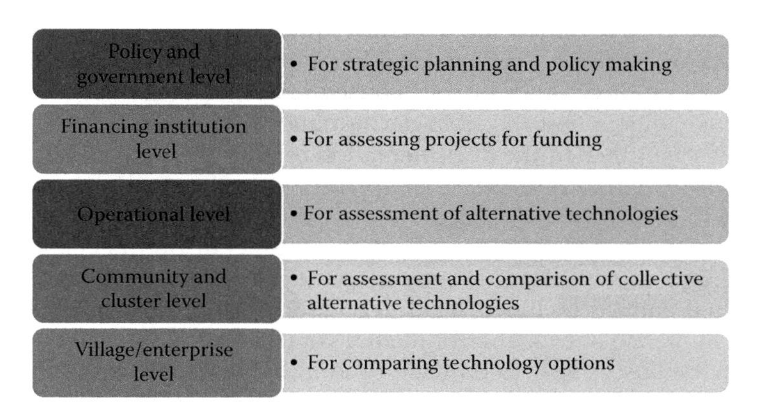

FIGURE 4.8
Possible applications of the SAT methodology.

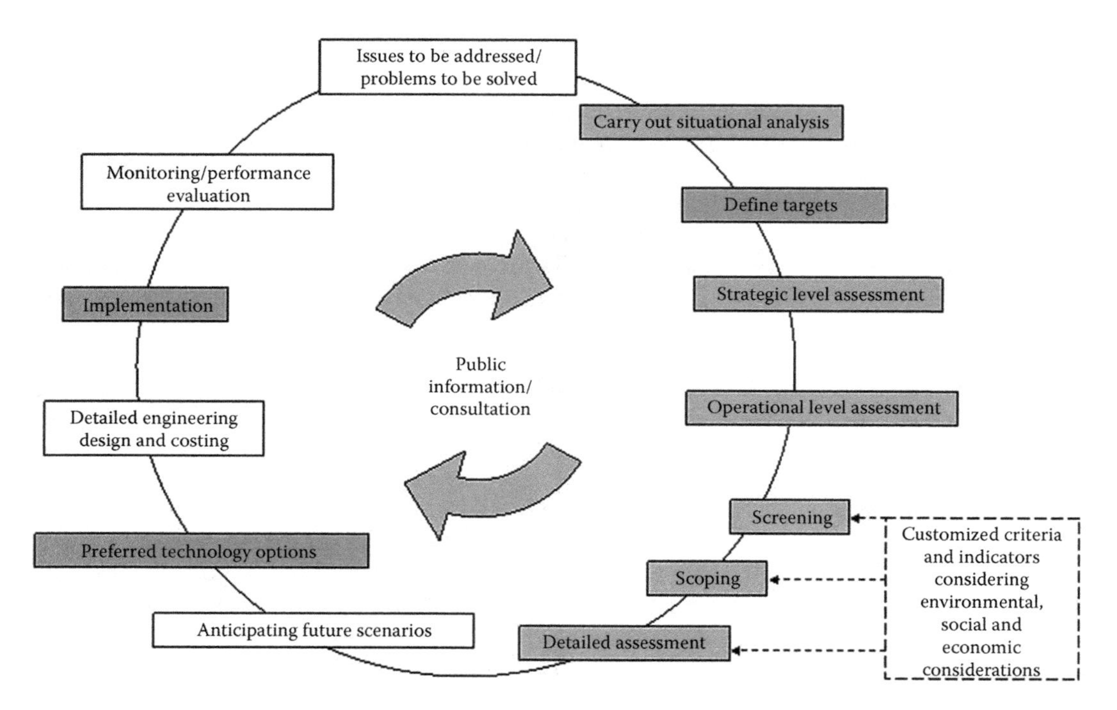

FIGURE 4.9
The SAT methodology.

4.9 Environmental Management System ISO 14001

4.9.1 Evolution of ISO 14001

The ISO 14001 Environmental Management System (EMS) was evolved on the basis of BS 7750 (1992) that was published by the British Standards Institute (BSI). In 1996, ISO 14001 was released that specified the general requirements for EMS. ISO 14001 considered what aspects and actions could be considered to lessen the organization's impact on the environment, both direct and indirect consequences, the legal requirements, and all effects on the stakeholders that the EMS performance will have. In 2004, ISO 14001:2004 was published. It strengthened the documentation requirements and established requirements for periodic evaluations to assess compliance with the standard and to investigate nonconformities.

In the internal context, ISO 14001 looked at any actions or products and services, strategic directions, and capabilities that may affect the organization's environmental performance. External context included legal, regulatory, economic, social, and cultural or political issues. In all the above, environmental context covered all other environmental aspects that may be susceptible to damage by the organization's environmental performance, including climate, air and water quality, land use, existing contamination, natural resource availability, and biodiversity.

Under Risks and Opportunities, elements such as compliance obligations to EMS and the business were considered. All risk assessments were to be recorded, actioned, and measured. Actions on improvement were driven by leadership and employee involvement.

EMS was developed with the following objectives:

- Enhancement of environmental performance
- Fulfilment of compliance obligations
- Achievement of environmental objectives

It was not surprising, therefore, that ISO 14001 EMS was sought after by the businesses as a strategy for continuous improvement as well as addressing the risks. ISO 14001 EMS also became a requirement by the investors and the markets asking the businesses to obtain such a certification. The business houses in turn demanded ISO 14001 EMS certification across the key companies of the supply chains.

In 2015, there was a major revision in the ISO 14001 standard. The following were the major changes:

- The system was designed for generating value for the organization.
- New concepts were considered, especially the risk-based approach.
- Requirements were stated around the product life cycle.
- Achieving intended outcomes was emphasized as a reason for establishing the EMS.
- A generic framework of standards and terminology was set out in Annex SL.
- A significant re-ordering was done of the key clauses.
- Expectation on organisations was expanded to commit to proactive initiatives to protect the environment from harm and degradation.
- Initiatives to address specific issues were encouraged such as: Sustainable resource use, Water and air quality and recycling, Climate change mitigation, Preservation of ecosystems and biodiversity, and Eco restoration.

As described above, ISO 14001 (2015) adopted a Lifecycle Approach where identification of environmental aspects/impacts was recommended using a life-cycle perspective. A road construction company, for example, was expected to come up with ideas such as recycling/reuse (where top layers of the road surface can be reused while resurfacing) and use recycled material (e.g., Use of plastics as binder material and use of fly ash). Table 4.2 shows a comparison between ISO 14001 (2004) and ISO 14001 (2015) EMS.

4.10 Sustainability and Corporate Social Responsibility

Corporate Social Responsibility is a corporate management philosophy that guides organizations to allocate their resources towards social and environmental projects. The drivers of CSR in the developed world are sustainability and corporate governance and good citizenship, while in the developing world, they are environmental protection, philanthropy, and market reputation.

TABLE 4.2

A Comparison between ISO 14001 (2004) and ISO 14001 (2015) EMS

ISO 14001:2015 CLAUSES		ISO 14001:2004 CLAUSES	Salient Points
Context of the organisation			
4.1	Understanding the organisation and its context	New requirement!	• Determine external and internal factors relevant to the EMS • Including possible environmental conditions
4.2	Understanding the needs and expectations of interested parties	New requirement!	• Determine interested parties relevant to the EMS • Determine their needs and expectations • Which ones are compliance obligations
Environmental conditions—state or characteristic of the environment as determined at a certain point in time			
4.3	Determining the scope of the environmental management system	4.1 General requirements	• Establish scope by determining boundaries and applicability of the EMS. • Consider external/internal issues, compliance obligations, function and physical boundaries, activities, products and services, and ability to control. • Scope must be documented and available to interested parties.
4.4	Environmental management system	4.1 General requirements	• Similar as previous version but introduces interactions between processes. • EMS must include external/internal issues and compliance obligations. • Achieving intended outcomes is now emphasized as a reason for establishing the EMS.
5. Leadership			
5.1	Leadership and commitment	New requirement!	• Leadership demonstrates commitment and is accountable for the EMS's effectiveness. • Ensures policies, objectives, continual improvement, resource availability, etc.
5.2	Environmental policy	5.2 Environmental Policy	• Similar requirements • The new standard stresses the importance of context and interested parties.
5.3	Organizational roles, responsibilities and authorities	4.4.1 Resources, responsibility and authority	• No requirement for a specific management representative. • However, those duties must be assigned. • Resources are now divided as a separate clause.
6. Planning			
6.1	**Actions to address risks and opportunities**		
6.1.1	General	New requirement!	• Consideration of identified internal and external issues, and the needs and expectations of interested parties. • Determine the risks and opportunities related to environmental aspects, compliance obligations and other issues. • Must document risks and opportunities and processes considered important

(*Continued*)

TABLE 4.2 (*Continued*)

A Comparison between ISO 14001 (2004) and ISO 14001 (2015) EMS

ISO 14001:2015 CLAUSES		ISO 14001:2004 CLAUSES	Salient Points
Risk is defined as the effect of uncertainty on objectives			
6.1.2	Environmental aspects	4.3.1 Environmental aspects	• Enhances the previous requirements • Needs to consider a life cycle perspective • Environmental aspects and associated environmental impacts must be documented.
6.1.3	Compliance obligations *Compliance obligations* is the new term for *legal and other* requirements	4.3.2 Legal and other requirements	• The requirements remain the same. • Gives equal weight to non-legislative mandatory obligations and voluntary obligations as legal requirements • Compliance obligations must be documented.
6.1.4	Planning action	New requirement!	• Requires the identification of risk and opportunities • Organization will have to plan actions to address the environmental aspects and compliance obligations • Evaluate the effectiveness of actions
6.2	**Environmental objectives and planning to achieve them**		
6.2.1	Environmental objectives	4.3.3 Objectives, targets and program(s)	• Requirements remain the same, but are further elaborated in the new version. • When setting objectives, consideration now needs to be given to the risk associated. • Specific requirements for the objectives to be documented, monitored, communicated and updated as appropriate
6.2.2	Planning actions to achieve environmental objectives	4.3.3 Objectives, targets and program(s)	• The term *targets* and *program(s)* is no longer used, however the requirements for these are included here • Requires planning to determine how the objectives will be achieved and how will the results be evaluated
7. Support			
7.1	Resources	4.4.1 Resources, roles, responsibilities and authorities	• Similar requirements. The new version emphasizes resource provision by dividing it into a separate clause.
7.2	Competence	4.4.2 Competence, training and awareness	• Enhances the previous requirements • Need for training has been expanded into a wider need for taking actions to acquire necessary competences. • Must now evaluate the effectiveness of action taken to address competence. • Documented information (records) must be retained for competence.
7.3	Awareness	4.4.2 Competence, training and awareness	• Specific clause but not otherwise different

(*Continued*)

TABLE 4.2 (*Continued*)

A Comparison between ISO 14001 (2004) and ISO 14001 (2015) EMS

ISO 14001:2015 CLAUSES		ISO 14001:2004 CLAUSES	Salient Points
7.4	**Communication**		
7.4.1	General	4.4.3 Communication	• Enhances the previous requirements • Requirements are now more prescriptive • Communication must be documented
7.4.2	Internal communication		• Specific clause but not otherwise different • Mechanisms for persons working under the organisation's control to make suggestions on improving the EMS
7.4.3	External Communication		• Specific clause, but not otherwise different
7.5	**Documented Information**		
7.5.1	General	4.4.4 Documentation	• EMS needs documented information required by the Standard and as determined by the organization.
7.5.2	Creating and updating	4.4.5 Control of documents	• Specific reference to the need for ensuring appropriate format and media
7.5.3	Control of documented information	4.4.5 Control of documents 4.5.4 Control of records	• Specific clause, but not otherwise different • Document control activities to be addressed by the system are specified.
Terms documents and records have been replaced by the term documented information			
8. Operation			
8.1	Operational planning and control	4.4.6 Operational control	• Similar intent but expanded clause • Addresses outsourced processes are controlled or influenced • Requirements for determining procurement activities and considering requirements in design activities, taking into account a life cycle perspective • Documented information is to be maintained as considered necessary.
8.1	Emergency preparedness and response	4.4.7 Emergency preparedness and response	• Similar requirements • Documented information to be maintained as considered necessary by the organisation.
9. Performance evaluation			
9.1	**Monitoring, measurement, analysis and evaluation**		
9.1.1	General	4.5.1 Monitoring and measurement	• Greater detail on requirements for monitoring and measurement activities is specified. • Specific requirement for the evaluation of performance and the use of indicators • Environmental performance is to be communicated internally and externally. • Records of monitoring and evaluation

(*Continued*)

TABLE 4.2 (*Continued*)

A Comparison between ISO 14001 (2004) and ISO 14001 (2015) EMS

ISO 14001:2015 CLAUSES		ISO 14001:2004 CLAUSES	Salient Points
9.1.2	Evaluation of compliance	4.5.2 Evaluation of compliance	• Similar requirements • Defined process with frequency of compliance evaluation determined • Evidence of compliance evaluation is to be documented (Records).
9.2	**Internal Audit**		
9.2.1	General	4.5.5 Internal audit	• Similar requirements • To be conducted at planned intervals
9.2.2	Internal audit program		• Well defined internal audit programme • Evidence of internal audits and results must be documented (Records).
9.3	Management review	4.6 Management review	• Similar intent but expanded clause • Changes in risk and opportunities to be considered during the review process. • Needs to include trends in nonconformities and corrective actions, monitoring and measurement results, conformity with compliance obligations, audit results, and communications. • Review resource adequacy • Management review results must be documented (Records).
10. Improvement			
10.1	General	4.5.3 Corrective action	• Opportunities for improvement must be determined • Action must be taken to achieve intended outcomes of EMS
10.2	Nonconformity and corrective action	4.5.3 Corrective action	• Similar intent but expanded clause • Nonconformities must be reacted to, applicable action taken, and deal with any mitigation. • Effectiveness of corrective action must be reviewed. • Change EMS, if required • Evidence of action taken from nonconformities must be documented. • Evidence of results of corrective action must be documented.
10.3	Continual improvement	New requirement!	• EMS needs to be continually improved in order to enhance environmental performance.

As per the UK's Department of Trade and Industry, CSR represents *the integrity with which a company governs itself, fulfils its mission, lives by its value, engages with its stakeholders, measures its impact and reports on its activities.* In India, as per the Company's Act of 2013, CSR must include projects that ensure environmental sustainability.

Businesses realize that development of the economy, society, and the environment is dependent on each of these factors. Therefore, businesses need to be socially and environmentally responsible.

4.10.1 CSR and Profits

A 2013 study by MIT business school and Boston Consulting group showed that 37% of companies that they surveyed found their sustainability efforts to be profitable. Encouraged by the positive results, 48% of companies modified their business models to incorporate sustainability principles into them.[87]

CSR also improves sales, employee interest, and loyalty, and attracts better personnel to the firm and enhances brand reputation. CSR activities are also a factor considered on sustainability indices like FTSE4Good or Dow Jones Sustainability Indices. This can in a way enhance the company's stock price, making executives' stock and stock options more profitable and shareholders happier and proud to be associated with that organization.

The benefits of CSR have been debatable with some believing that CSR are more about branding and public relations, and they take away from the main services that the company offers. Despite this, no large public company shies away from CSR activities.

CSR appeals to consumers. For example, Starbucks's C.A.F.E. Practices guidelines ensure that their resources are sustainably grown. Starbucks markets its coffee as beneficial to the growers who produce it. Such information encourages consumers to be loyal to Starbucks even though they charge a premium price for a cup of coffee. Tom's Shoes is another noteworthy example of a company that has CSR at its core. For every pair of shoes that a customer purchases, Tom's Shoes donates one pair of shoes to a child in need.[88] We discuss these examples in Chapter 6.

United Nations studies on trade and investment (2009)[89] identified that many businesses realize that local environmental degradation, global climate change, poor labor standards, inadequate health and education systems, and many other social ills can add directly to the costs and risks of doing business domestically or globally. Slippages in these areas can increase operating costs, raw material costs, hiring, training, and other personnel costs, security costs, insurance costs, and the cost of capital. They can create both short-term and long-term financial risks, market risks, litigation risks, and reputation risks. Companies that understand and address these challenges can improve their risk and reputation management, reduce their costs, improve their resource efficiency, and enhance their productivity, which can make the highest impact to society and a business's future. These can be done by integrating CSR into companies' core business. Strategic CSR accomplishes strategic business goals, as well as social goals; it benefits both the business and society.

4.10.2 CSR Promotes Social Entrepreneurship and Innovation

CSR and innovation are the foundation of business competencies. CSR, if strategized, can lead to *social innovation*. Companies of the future will be the ones that see CSR as an opportunity for innovation, rather than risks to be alleviated. Corporate Social Innovation, for instance, refers to a product innovation with a social purpose. The Corporate Social Innovation should focus on the low-income market, the Base of the Pyramid (BOP). The CSR should promote Social Entrepreneurship and, thus, catalyze innovation as the social entrepreneurs act as the change agents for society, inventing new approaches, and creating solutions to change society for the better. Eco–Innovation should form one of the drivers of the CSR to influence the business.

Salesforce has been a leader in social responsibility, creating the Salesforce.com Foundation in its growth years just after going public. The concept revolves around *giving back 1% product, 1% time, and 1% in equity*. The company in 2011 put over $24 million into community and global impact projects, ranging from non-profit philanthropy to for-profit social businesses. eBay invested significantly in CSR with the acquisition of WorldOfGood in 2010. The Founder/Chairman of eBay, Pierre Omidyar, has used his personal wealth to fund the Omidyar Network, which has given billions to social responsibility and sustainability programs. Since 2006, Google China has sponsored the Social Innovation Cup, which is a national competition aimed at empowering China's youth to address pressing social issues through grassroots, innovative solutions. Google's social venture and philanthropy organization was funded with $1 billion and has engaged employees in hundreds of engineering projects aimed at social innovation since 2004.

Many start-ups practice social responsibility as a core component of their competitive strategy. For example, Twitter and Facebook have all had early efforts involving their employees involved in community-based responsibility programs. Three years after its founding, Twitter partnered with the non-profit Room to Read, as an example of social responsibility in a program called The Fledgling Initiative. The trend is likely to continue in this direction. In the future, expect to see an emerging class of Social and/or Impact investors who will be looking to invest in companies that view social responsibility as a building block for their success.[90] When businesses tackle social problems as a central part of their competitive strategy, they achieve large-scale and fundamentally sustainable changes in society.[91]

4.10.3 Is Strategic CSR the Solution?

Earlier CSR was considered as corporate philanthropy, which means donations given by the organization to the public for their betterment, but now the scenario has changed, and CSR has become the hottest and most talked about issue all over the world. It first emerged in the United States in the arena of 1950 as philanthropy donation but now the concept of CSR has changed; its focus has shifted from obligation or responsibility to a strategy.

Traditional CSR (charity or philanthropic) is where the firm's objective is to produce a desired level of CSR with no regard for maximizing its social profits. Strategic CSR is where the firm identifies social activities that consumers, employees, or investors value and integrates those activities into its profit-maximizing objectives.[92]

Traditional CSR activities that encompass community development and philanthropy are usually seen as distinct and unrelated to core business operations. A business could have a CSR program of education and healthcare, while polluting the environment and treating workers poorly. Strategic CSR is meant to address this problem by addressing any negative value-chain impacts, while supporting the business strategy and the needs of the community. Thus, traditional CSR is differentiated in motivation, implementation, and impact from Strategic CSR.[93,94]

The success of a business depends on its relationships with the external world, the regulators, potential customers and staff, activists, and legislators. Decisions made at all levels of the business, from the boardroom to the shop floor, affect that relationship. For the business to be successful, decision-making in every division and at every level must take account of those effects. External engagement cannot be separated from everyday business; it must be part and parcel of everyday business. Digital communication has enabled individuals and nongovernmental organizations (NGOs) to observe almost every activity of a business, to rally support behind it, and to launch powerful global campaigns very quickly at almost zero cost. High expectations and scrutiny are here to stay. Successful companies must be equipped to deal with them.

CSR as defined and understood today, has not yet evolved to reap economic benefits to the company. Strategic thinking and the goal to increase the CSV of the company has still remained in the back seat. This is restricting innovation and opportunities, as well as making the companies treat CSR in a silo. It can be termed neither as *Traditional* nor *Strategic*, but rather termed as *egoistic*.[95]

Strategic CSR, if integrated into core business operations, can certainly add a value. If properly designed and implemented to fit the needs of the community and the business, CSR can become a source of opportunity, innovation, and lead to competitive advantage. Those that have acted already on these lines are now reaping the rewards. We, however, need more clarity and guidance in this direction, especially on the positioning of the CSV. Based on the experience of implementation of the rules, CSR Version 2.0 may emerge to address these issues.

Endnotes

1. Based on: Directive 2002/96/EC of the European Parliament and of the Council of 27 January 2003 on waste electrical and electronic equipment (WEEE)—Joint declaration of the European Parliament, the Council and the Commission relating to Article 9, EUR-Lex, 2003, online source: http://eur-lex.europa.eu/legal-content/EN/TXT/?uri=CELEX:32002L0096
2. Based on: Directive 2012/19/EU of the European Parliament and of the Council of 4 July 2012 on waste electrical and electronic equipment (WEEE), EUR-Lex, 2012, online source: http://eur-lex.europa.eu/legal-content/EN/TXT/?uri=CELEX:32012L0019
3. Taken from: Oracle, WEEE Symbol, WEEE and Environmental Compliance, p. 1, Online Source: http://www.oracle.com/us/products/applications/green/weeeandenvironmental-compliance-1973534.pdf
4. Based on: Recycle—Contributing to the Circular Economy, Canon, 2016, online source: http://www.canon.ie/about_us/sustainability/environment/recycle/
5. For further reading: Rossem, C., Tojo, N. and Lindhqvist, T., Extended producer responsibility: An examination of its impact on Innovation and Greening Products, GreenPeace International, Friends of the Earth Europe and the Environmental Bureau (EEB), 2006, online source: http://www.greenpeace.org/international/PageFiles/24472/epr.pdf
6. Based on: Directive 2002/95/EC of the European Parliament and of the Council of 27 January 2003 on the restriction of the use of certain hazardous substances in electrical and electronic equipment, EUR-Lex, 2003, online source: http://eur-lex.europa.eu/LexUriServ/LexUriServ.do?uri=CELEX:32002L0095:EN:HTML
7. Based on: Other RoHS Green Initiatives Worldwide, *RoHS Guide Compliance*, 2017, online source: http://www.rohsguide.com/rohs-future.htm
8. For further reading: Gohring, N., Palm stops shipment as Treo fails European pollution law, *IT World Canada*, 2006, online source: http://www.itworldcanada.com/article/palm-stops-shipment-as-treo-fails-european-pollution-law/6201
9. Based on: Marsal, K., Apple to halt sale of some products in Europe, appleinsider, 2006, online source: http://appleinsider.com/articles/06/06/21/apple_to_halt_sale_of_some_products_in_europe
10. Based on: Dell's Chemical Use Policy, DELL, 2013, online source: http://www.dell.com/downloads/global/corporate/environ/Chemical_Use_Policy.pdf
11. Based on: Center of Excellence in Eco-materials and Capacity Building for Thai RoHS Alliance, Thai Industrial Standards Institute, Ministry of Industry, National Metal and Materials Technology Center (MTEC)—a member of NSTDA, 2014, online source: https://www.mtec.or.th/en/2014-05-08-04-21-36/research-projects/683-center-of-excellence-in-eco-materials.html
12. Based on: Understanding RoHS and REACH: Impacts on Your Global Markets, Silicon Expert Technologies, 2009, online source: http://siliconexpert.com/sites/default/files/assets/pdf/SiliconExpert%20White%20Paper%20-%20REACH-RoHS.pdf

13. Based on: The Consequences of REACH for SMEs, Policy Department A: Economic and Scientific Policy European Parliament, 2013, online source: http://www.eesc.europa.eu/resources/docs/the-consequences-of-reach-for-smes.pdf
14. Based on: Surviving REACH A Guide for Companies that Use Chemicals, Global Development and Environment Institute, Tufts University, 2005, online source: http://www.ase.tufts.edu/gdae/Pubs/rp/SurvivingReach.pdf
15. Based on: Ministry of Environment, Forest and Climate Change, Notification of Amendment Rules 2015, 2015, online source: http://www.moef.nic.in/sites/default/files/Effluents%20from%20textile%20Industry.PDF
16. For further reading: Hussain, S., *Indian Case Study on ZLD—The Tirupur Textile Cluster Experience,* Tamil Nadu Water Investment Company Limited, 2014, online source: http://gpcb.gov.in/images/pdf/ZLD_PRESENTATION_11.PDF
17. Based on: Tirupur was the first industrial cluster to implement ZLD in India, *The Indian Textile Journal*, 2016, online source: http://www.indiantextilejournal.com/News.aspx?nId=AAW3M4OsICNBhWOHdJRr9A==&NewsType=test-India-Sector
18. For further reading: Santos-Borja, A. C., Environmental User Fee System for Laguna de Bay, Philippines, *Integrated Lake Basin Management: Training Materials*, Japan International Cooperation Agency (JICA) and ILEC: International Lake Environment Committee, 2009, online source: http://wldb.ilec.or.jp/ILBMTrainingMaterials/resources/Environmental_User_Fee_presentation.pdf
19. Based on: WEPA: Water Environment Partnership in Asia, Environmental User Fee System (EUFS), Ministry of the Environment of Japan, 1977, online source: http://www.wepa-db.net/policies/law/philippines/eufs.htm
20. For further reading: Catelo M. A. O., Impact Evaluation of the Environmental User Fee System: A Stakeholder Perspective, *Economy and Environment Program for Southeast Asia*, p. 21, 2007, online source: https://ideas.repec.org/p/eep/tpaper/tp200707t2.html#author
21. Based on: Folmer H. and Tietenberg T. H., *The International Yearbook of Environmental and Resource Economics 2006/2007,* Edward Elgar Publishing, Glos, UK and Masachusetts, USA 2006.
22. For further reading: Manget, J., Roche, C. and Munnich, F., *Capturing the Green Advantage for Consumer Companies*, Boston Consulting Group, 2009, online source: http://www.bcg.com/documents/file15407.pdf
23. SD Features Sustainability Concepts Factor 10, http://www.gdrc.org/sustdev/concepts/11-f10.html
24. See: Resource Efficient and Cleaner Production. UNEP DTIE, online source: http://www.unep.fr/scp/cp/
25. For further reading: Sustainable Consumption and Cleaner Production—Global Status 2002, UNEP DTIE, online source: http://www.unep.fr/shared/publications/pdf/3211-GlobalStatus02.pdf
26. Based on: Cleaner Production Companion, UNEP DTIE, 2004, online source: http://www.unep.fr/scp/publications/details.asp?id=DTI/0579/PA
27. *Guidance Manual: How to Establish and Operate Cleaner Production*, UNEP UNIDO, online source: http://www.unep.fr/shared/publications/pdf/WEBx0072xPA-CPcentre.pdf
28. Based on: Green Productivity GP: An Approach to Sustainable Development, Asian Productivity Organization, p. 9, 2002, online source: http://www.apo-tokyo.org/publications/wp-content/uploads/sites/5/ind_gp_aasd-2002.pdf
29. Resource Efficient and Cleaner Production (RECP) in Developing and Transition Countries, UNIDO, online source: http://www.gcpcenvis.nic.in/PDF/UNIDO_UNEP_RECP_Programme.pdf
30. Based on: RECPnet—The Network for Resource Efficient and Cleaner Production, online source: http://www.recpnet.org/
31. Download the kit from http://www.stenum.at/media/documents/UNEP%20PRE%20SME%20Industrial%20training%20handbook.pdf

32. For further reading: *PRE-SME—Promoting Resource Efficiency in Small & Medium Sized Enterprises: Industrial Training Handbook*, UNEP-UNIDO, 2010, online source: http://www.stenum.at/media/documents/UNEP%20PRE%20SME%20Industrial%20training%20handbook.pdf
33. Based on: *International Cleaner Production Information Clearinghouse ICPIC User Guide Version 1.0*, UN, 1997–1999, online source: http://www.un.org/Pubs/environ/923d4.htm
34. These examples have been drawn from article by Dimitar Vlahov http://www.sustainablebrands.com/news_and_views/blog/dimitar_vlahov/5_types_sustainable_products_follow_2015_beyond and by Rinkesh at http://www.conserve-energy-future.com/25-green-eco-friendly-products.php
35. Based on: Reuse/Recycling Design, Fuji Xerox, online source: https://www.fujixerox.com/eng/company/ecology/cycle/newstyle/design/
36. Based on: Puma's 'Clever Little Bag' Slashes Sneaker Packaging, *GreenBiz*, 2010, online source: https://www.greenbiz.com/news/2010/04/13/puma-clever-little-bag-slashes-packaging
37. Based on: *Senseo Up Coffee Maker: The Small Coffee Maker That's Big on Sustainability*, Philips, 2015, online source: https://www.90yearsofdesign.philips.com/article/6
38. Based on: Ippolito, A., Hemp "Vegan High Quality Protein", *nativebar*, 2015, online source: http://nativebars.com/hemp-vegan-high-quality-protein/
39. Based on: What is Hempcrete? *American Lime Technology*, 2012, online source: http://www.americanlimetechnology.com/what-is-hempcrete/
40. Based on: *OECD Sustainable Manufacturing Toolkit—Start-up Guide: Seven Steps to Environmental Excellence*, OECD, 2011, online source: http://www.oecd.org/innovation/green/toolkit/48661768.pdf
41. Based on: Green Initiative, WAUSAU MADE, online source: http://www.wausaumade.com/why-wausau-made/green-initiative
42. Based on: Mater-Bi—Carrier Bags, Novamont, 2017, online source: http://materbi.com/en/solutions/carrier-bags/
43. Based on: Mater-Bi, Novamont, 2017, online source: http://www.novamont.com/eng/mater-bi
44. Based on: Carrier Bags and Bags for Fruits/Vegetables in New Grades for High Environmental and Technical Performance with Saldoflex at K 2016, Novamont: Press Release, 2016, online source: http://www.novamont.com/public/Comunicati/k/PR_Novamont_K%202016_Saldoflex_eng.pdf
45. For further reading: Sebille E. et al., A global inventory of small floating plastic debris, *Environmental Research Letters*, 2015, online source: http://iopscience.iop.org/article/10.1088/1748-9326/10/12/124006/pdf
46. Based on: Mcalone, N.—Business Insider, Adidas is selling 7,000 of these amazing shoes made from ocean waste, *Business Insider, Science Alert*, 2016, online source: http://www.sciencealert.com/adidas-is-selling-7-000-of-these-amazing-shoes-made-from-ocean-waste
47. Based on: Parley for the Oceans, Adidas X Parley: Stopping the waiting game, Parley Ocean School, online source: http://www.parley.tv/updates/2015/12/21/adidas-x-parley-stop-waiting-start-acting
48. Based on: The Biosteel Story, *Biosteel-fiber*, online source: http://www.biosteel-fiber.com/home/#c221
49. Rudee, T., Trending: BioSteel, Self-Healing Protective Fabrics among Latest Innovations Revolutionizing Textiles, *Sustainable Brands*, 2016, online source: http://www.sustainablebrands.com/news_and_views/chemistry_materials/talia_rudee/trending_biosteel_self-healing_protective_fabrics_amo
50. Early Case Examples: Xerox, 3M, Veolia, etc.
51. Expressed in global hectares.
52. Based on: Greening the Economy Through Life Cycle Thinking, UNEP, p. 15, 2012, online source: http://www.unep.fr/shared/publications/pdf/DTIx1536xPA-GreeningEconomythroughLifeCycleThinking.pdf

53. Based on: Schepelmann P., Life Cycle Assessment, online source: http://www.ivm.vu.nl/en/Images/AT9_tcm234-161581.pdf
54. Based on: Life Cycle Initiative, online source: http://www.lifecycleinitiative.org/
55. Based on: ISO 14044:2006—Environmental management—Life cycle assessment—Requirements and guidelines, online source: https://www.iso.org/standard/38498.html
56. Based on: Reimann K. et al., Evaluation of environmental life cycle approaches for policy and decision making support, online source: http://publications.jrc.ec.europa.eu/repository/bitstream/111111111/15195/1/lbna24562enc.pdf
57. Guidelines for Social Life Cycle Assessment of Products, UNEP, 2009
58. Visit the website of Global Ecolabelling Network to get a comprehensive overview—https://www.globalecolabelling.net/
59. Based on: Ecolabels: A booming Business, EC Environment—Eco-innovation Action Plan, 2015, online source: http://ec.europa.eu/environment/ecoap/about-eco-innovation/policies-matters/eu/ecolabels_en
60. Blue Flag Beach Criteria and explanatory notes 2008–2009, European Commission, 2009, online source: http://ec.europa.eu/ourcoast/download.cfm?fileID=1018
61. Facts and Figures, EC Environment, 2017, online source: http://ec.europa.eu/environment/ecolabel/facts-and-figures.html
62. Based on: *Bluesign System, Version 1.0*, 2014, Bluesign Technologies ag
63. Based on: Input Stream Management, *bluesign technologies*, 2013, online source: http://www.bluesign.com/industry/bluesign-system/input-stream-management#.WDaPQSN94zY
64. Based on: Bluesign product, Consumer: How to find it, 2013, online source: https://www.bluesign.com/consumer/label
65. Based on: Bluesign Standard, Patagonia, online source: http://eu.patagonia.com/enGB/patagonia.go?assetid=70730
66. Based on: Manufacturing Recycled Materials, The North Face, online source: https://www.thenorthface.com/about-us/responsibility/product/manufacturing.html
67. Resource intensity is a measure of the resources (e.g., water, energy, materials) needed for the production, processing and disposal of a unit of good or service, or for the completion of a process or activity; it is therefore a measure of the efficiency of resource use.
68. An Ecological Rucksack is the total quantity (in kg) of materials moved from nature to create a product or service, minus the actual weight of the product. That is, ecological rucksacks look at hidden material flows. Ecological rucksacks take a life cycle approach and signify the environmental strain or resource efficiency of the product or service.
69. A code of conduct is a set of rules outlining the social norms and rules and responsibilities of, or proper practices for, an individual, party or organization. Related concepts include ethical, honor, moral codes and religious laws.
70. Visit http://www.eeaa.gov.eg/seam/Manuals/ecolabell/CONTENT-1.html for accessing the full case study
71. Based on: Smith, T., Dutch officials seize cadmium-packed PlayStation kit—Sony wonders what fuss is about, *The Register*, 2001, online source: http://www.theregister.co.uk/2001/12/05/dutch_officials_seize_cadmiumpacked_playstation/
72. Based on: Shoemaker, M. and Christensen, M., How Sony strengthened its supply chain and added value, *GreenBiz*, 2013, online source: https://www.greenbiz.com/blog/2013/03/24/how-sony-strengthened-its-supply-chain-and-added-value
73. For further reading: IKEA Group FY15: Sustainability Report, IKEA, 2015, online source: http://www.ikea.com/ms/en_US/img/ad_content/2015_IKEA_sustainability_report.pdf
74. Based on: Staffordshire University, International Supply Chain Management—BSB20123-7, Written Examination Stimulus Material Case Study—IKEA, online source: http://www.staffs.ac.uk/schools/business/resits/postgrad/InternationalSupplyChainMgmtIKEACaseStudy.pdf
75. For further reading: IKEA Sustainability Report 2010, IKEA Group, 2010, online source: http://www.ikea.com/ms/no_NO/pdf/sustainability_report/Sustainability_Report_fy10.pdf

76. Based on: *IWAY Standard: Minimum Requirements for Environment and Social & Working Conditions when Purchasing Products, Materials and Services*, IKEA Supply AG, Edition 4, 2008, online source: http://www.ikea.com/ms/en_JP/about_ikea/pdf/SCGlobal_IWAYSTDVers4.pdf
77. Based on: Isler, L., IWAY or how we support our suppliers and their suppliers, *Green Lifestyle*, IKEA, 2014, online source: http://lifeathome.ch/en/2014/02/people-and-communities-iway-or-how-we-support-our-suppliers-and-sub-suppliers/
78. Based on: M&S Supplier Map, Marks & Spencer, 2016, online source: https://interactivemap.marksandspencer.com/?parentFolderPiD=56fa5de072529d3e644d25d1
79. Based on: Global Sourcing Principles, M&S, 2016, online source: https://corporate.marksandspencer.com/file.axd?pointerid=fdc4fb2ad28c4db7bac11e8bb64416a4
80. Based on: BCI Cotton, Marks and Spencers, 2015, online source: https://corporate.marksandspencer.com/plan-a/our-stories/about-our-sourcing/cotton
81. For further reading: Wheatley, F., *Marks & Spencer: Improving Palm Oil Sourcing, SPOTT: Sustainable Palm Oil Transparency Toolkit*, 2012, online source: http://www.sustainablepalmoil.org/case-studies/marks-spencer/
82. For further reading: Doing Good with Food Surplus, M&S: Plan A, 2016, online source: https://corporate.marksandspencer.com/plan-a/our-stories/about-our-initiatives/food-surplus
83. For further reading: Chiu, S., The use of Corporate Synergy Systems in promoting industrial waste minimization in small and medium enterprises in Taiwan, Taiwan Environmental Management Association, Taipei International Green Productivity Association, Tapei, online source: http://www.parasnis.net/update181004/e_publi/gsc/0309RES_PAPERS.pdf
84. Based on: Furlong, H., New Platform Helps Companies 'Trase' Deforestation in their supply chains, *Sustainable Brands: The Bridge to Better Brands*, 2016, online source: http://www.sustainablebrands.com/news_and_views/ict_big_data/hannah_furlong/new_platform_helps_companies_trase_deforestation_supply
85. Based on: Chandak, S. P. Sustainable Assessment of Technologies: Making the Right Choices. IETC-UNEP. Presented at the *1st Stakeholder Consultative Workshop/Training Program of the Project on Converting Waste Agricultural Biomass to a Fuel/Resources in Moneragala District, Sri Lanka* funded by UNEP and coordinated by the National Cleaner Production Centre, August 21, 2009.
86. ISO stands for International Organization for Standardization.
87. Based on: Fellow, A., One-Third of Companies Report Profits from Sustainability: Survey, Bloomberg, 2013, online source: http://www.bloomberg.com/news/2013-02-06/one-third-of-companies-report-profits-from-sustainability-survey.html
88. Based on: Caramela, S., What Is Corporate Social Responsibility? *Business News Daily*, 2014, online source: http://www.businessnewsdaily.com/4679-corporate-social-responsibility.html
89. For further reading: United Nations ESCAP, Creating Business and Social Value: The Asian way to Integrate CSR into Business Strategies, 2009, online source: http://www.unescap.org/publications/detail.asp?id=1379
90. Based on: Rhyakin, *Innovation, Corporate Social Responsibility and Sustainability*, blog posted on May 17, 2011, online source: http://haykin.net/entrepreneursparks/blog/
91. Based on: Borgonovi et al., Creating Shared Value in India: How Indian Corporations Are Contributing to Inclusive Growth While Strengthening Their Competitive Advantage, October 2011, FSG, online source: https://sharedvalue.org/sites/default/files/resource-files/India_CSV.pdf
92. Definitions sourced from: https://www.stlouisfed.org/publications/re/articles/?id=1258
93. Based on: Samina, A., Traditional v/s strategic corporate social responsibility: In pursuit of supporting sustainable development, *Journal of Economics and Sustainable Development*, Vol. 4, no. 20, 2013, ISSN 2222-1700 (Paper) ISSN 2222-2855.
94. Based on: Alpana, Traditional CSR v/s creating shared value: A process of stakeholder engagement, *IOSR Journal of Economics and Finance (IOSR-JEF)*, Vol. 3, no. 4. May–Jun. 2014, pp. 61–67, e-ISSN: 2321-5933, p-ISSN: 2321-5925, online source: http://www.iosrjournals.org/iosr-jef/papers/vol3-issue4/J0346167.pdf
95. As defined by economists Bryan Husted and José de Jesus Salazar, online source: https://www.stlouisfed.org/publications/re/articles/?id=1258

5

Responses by Business Sectors

In this section, we will take a look at the responses by various business sectors that cover Transportation, Information and Communication Technology, Pulp and Paper, Cement, Tourism, the Social sector, and others. These responses illustrate sectoral strategies and innovations with examples demonstrating integration of sustainability to an advantage.

5.1 Transportation Sector

The Transportation sector generates about 23% of the total carbon dioxide emissions globally.[1] The sector is dominated by private vehicle use due to increased purchasing capacity of people in both developed and developing countries. Recognizing the impact of the sector, scientists and businesses have been innovating in the areas of sustainable or alternative fuels, efficient design of vehicles, aircrafts, and ships. Boxes 5.1 through 5.4 describe case studies that highlight sustainability initiatives in the Transportation sector.

BOX 5.1 BRITISH AIRWAYS MAKES ADVANCES IN THE AVIATION SECTOR

Aviation is the fastest-growing source of greenhouse gas emissions, responsible for around 2% of the carbon dioxide emissions and only increasing due to increased air travel. The principal emissions of aircraft include greenhouse gases such as carbon dioxide and water vapor. Other major emissions are nitric oxide (NO) and nitrogen dioxide (NO_2), sulfur oxides (SO_x), and soot.

The aviation industry set itself ambitious short-, medium-, and long-term goals as early as 2008. Airports, airlines, air navigation service providers, and the manufacturers of aircraft and engines agreed to these goals. The goals set were:[2]

- Improving aircraft fuel efficiency by 1.5% per year until 2020.
- Stabilizing net emissions from 2020 through carbon-neutral growth.
- Reducing net aviation carbon emissions to 50% of the 2005 levels by 2050.

Let us take an example of British Airways (BA).[3,4]

BA aims to reduce its carbon impact by 50% by 2050. Initiatives by BA to reach this goal include:

- Investing in new aircraft technology and implementing operational fuel saving initiatives, BA's new generation aircrafts include Airbus 380 and Boeing 787 Dreamliner that are some of the most fuel-efficient aircrafts globally.[5]

- Adoption of single engine taxi-out (i.e., only one engine runs when the flight is taxiing) strategy to achieve significant fuel savings during the taxiing time that tends to be long at busy airports. (This strategy was also recommended by aircraft manufacturers like Airbus.)
- Approaching the airport at slow speeds during busy periods. This has a dual advantage. Slow speeds can reduce burning of fuel and reduce time spent circling the airport waiting to land.
- Audits to reduce the weight carried on the aircraft in terms of catering trolleys, seats, water are annually reviewed.
- Surprisingly, around half of their global footprint comes from their catering activity. The recyclable catering waste is channeled appropriately and non-recyclables are sent to waste-to-energy plants for recovering energy.

DISCUSSION QUESTION

- *As a passenger, discuss three ways in which you can reduce your footprint during air travel.*

BOX 5.2 SOLAR AVIATION: CASE STUDY OF SOLAR IMPULSE

The Solar Aviation industry has made multiple attempts to launch solar flights since the 1970s. Several unmanned and manned solar flights successfully took flight for varied lengths of time. One of the most recent and revolutionary advancements in the industry is the flight by Solar Impulse in 2010 and Solar Impulse 2 in 2016.

Solar Impulse is a solar-powered aircraft project co-founded by Swiss psychiatrist and aeronaut Bertrand Piccard and Swiss engineer and entrepreneur André Borschberg. After 13 years of research and development and test runs, they successfully took their maiden flight in Solar Impulse in 2010. Solar Impulse flew from Switzerland to Spain in 26 hours. This included an entire diurnal solar cycle and nine hours of night flying.

In 2016, Solar Impulse 2 completed the first round-the world flight originating from Abu Dhabi and back. The flight took off from Abu Dhabi in March 2015 and touched down in July 2016. The flight had 16 stops with a total flying time of approximately 558 hours (23.25 days) to cover 43,000 kms. The longest flight duration was between Japan and Hawaii covering a distance of 8,924 km in 117 hours 52 minutes (5 days and 5 nights).[6,7]

The Solar Impulse 2 has a wingspan of 72 m and weighs only 2.3 tons. It has 17,248 solar cells. It generated 11,655 kWh of energy during the 23.25 days' flight consuming 0 liters of fuel.[8,9]

The project has several high-profile business partners that include Solvay, Omega, Schindler, ABB, Altran, Bayer, Google, Swiss Re Corporate Solutions, Swisscom, and Moët Hennessy. Masdar, Abu Dhabi's renewable energy company, was their host partner.

DISCUSSION QUESTIONS

- *Discuss the commercial feasibility of Solar Aviation in the coming decade.*
- *Which is the most sustainable mode of transportation (road, rail, or air) and why?*

BOX 5.3 INCREASING COMPETITIVENESS OF ADVANCED BIOFUELS IN THE AUTOMOTIVE INDUSTRY

Biodiesel is a cleaner, renewable, and biodegradable alternate for conventional automotive gasoline or diesel. It is produced from organic substances like vegetable oils, used cooking oil, and animal fats. Carbon dioxide emitted from biodiesel combustion is absorbed via plants producing biodiesel feedstock like soybeans or peanuts. Pure biodiesel blend (B100) emits 74% less emissions when compared to gasoline or petrol. However, most vehicles are not designed to run on B100. Only up to 5% biodiesel is blended with vehicular diesel, but further use requires vehicular design modification. Due to low cost competitiveness, vehicular compatibility and negligible environmental impact, first-generation biofuels did not penetrate the automotive market successfully.[10]

The second-generation fuels or advanced bio-fuels derived from lignocellulosic feed stocks like switchgrass, energy cane and miscanthus, from algae oil and from plant biomass can achieve up to 90% emission reductions. Adoption of advanced biofuels is highly dependent on the cost competitiveness of the fuel that is dependent on factors like enzyme costs, feedstock yields, production scale, and capital expenditure. Owing to improvements in crop yields, agricultural and industrial technology, the overall production cost of lignocellulosic ethanol (LC ethanol) has reduced from $2.39 per gallon to $1.59 per gallon and can compete with conventional gasoline.

New generation vehicles are designed to run on different types of fuels including conventional fuels, alternative fuels, and electricity. However, significant capital investment needs to be made for improved infrastructure to meet the demands of the future. A report by the Boston Consulting Group estimates that advanced biofuels can become economically viable by 2020.[11]

Another emerging technology is the use of electric powered vehicles that use comparatively low carbon electricity and can also reduce emissions by 90% when compared to fuels. Many automotive companies like Toyota, Volkswagon, BMW, Ford, Nissan, and others developed electric plug-in vehicles. BMW launched a sustainable sub-brand, BMWi. BMW i3 and BMW i8 are all electric vehicles that offer emission free vehicles for city driving. The BWM i3 series consumes 11.3–13.5 kWh per 100 kms compared to some models that consume 34 kWh per 100 kms.[12,13,14]

DISCUSSION QUESTIONS

- *Discuss the economic viability of biofuels (like Jatropa or Camelina) for the transportation sector. Are there any significant environmental and social impacts of biofuels?*
- *Are electric vehicles environmentally friendly? Why?*

BOX 5.4 PHASE OUT OF REFRIGERANTS USED IN AUTOMOBILE MOBILE AIR CONDITIONING (MAC)

The automobile manufacturing sector was one of the first sectors to phase-out CFC-12 that had a high Ozone Depleting Potential (ODP) and high Global Warming Potential (GWP). The sector has gradually led the way in transitioning to alternatives that have negligible ODP and a medium to low GWP. In the sector, Mercedes-Benz was the first company to replace CFC-12 with R-134a in 1992.[15] Table 5.1 shows the ODP and GWP values of commonly used refrigerants.

Under the Montreal Protocol, CFC-12 (high ODP, high GWP) was replaced by HFC 134a (near zero ODP, medium GWP) in the vehicle cooling industry. HFC 134a, while having zero ozone-depletion potential, has a very high global warming potential, more than 1,500 times as much warming as an equivalent amount of CO_2. Some countries had filed a petition to amend the Montreal Protocol to also phase out HFC refrigerants, but the petition is still pending as of 2016. HFC 134a is now being replaced by refrigerants like HFO-1234yf, HFC-152a, or CO_2 (near zero ODP, low or near zero GWP).

EU has phased out HFC-134a in 2011 and directed that refrigerants with GWP lower than 150 only can be used in Automobile Mobile Air Conditioning (MAC) systems. US EPA recently issued a rule stating that blends of HFC including HFC-134a now fall under the category of unacceptable refrigerants for MACs as stated in the Significant New Alternatives Policy Program (SNAP). HFO-1234yf and CO_2 are listed as acceptable refrigerants for MAC.[16]

The automobile industry, made up of a relatively few large manufacturers, has worked cooperatively to test a range of alternatives. Honeywell and DuPont developed HFO-1234yf and hold IP rights. German auto manufacturer Daimler is exploring a shift to a CO_2-based MAC system. Concerns about the flammability of

TABLE 5.1

Ozone Depleting Potential and Global Warming Potential of Refrigerants[a]

Refrigerant Type	Refrigerant Name	ODP	GWP
CFC	CFC 12	1	10900
	CFC 11	1	4750
HCFC	HCFC 22	0.04	1810
	HFC 410a	0	2088
	HFC 134a	0	1430
HFO/ HFC blend	HFO 1234yf	0	1
	HFC 152a	0	124
Natural Refrigerants—Other	Carbon Dioxide	0	1
	Ammonia	0	0
	Water	0	0
	Air	0	0
Natural Refrigerants—Hydrocarbon (HC)	HC 290 (Propane)	0	5

[a] Based on Climate Change 2007: The Physical Science Basis, Working Group I, Contribution to the Fourth Assessment Report of the Intergovernmental Panel on Climate Change(IPCC), Table 2.14, Cambridge University Press. Online Source: http://www.ipcc.ch/publications_and_data/ar4/wg1/en/errataserrata-errata.html#table214

HFO-1234yf, high prices, and supply issues are the possible reasons for the switch to CO_2.[17]

General Motors (GM), keen on using HFO-1234yf as the refrigerant, conducted extensive tests to demonstrate that this is a safe automotive refrigerant. Between 2012–2013, GM rolled out around 100,000 vehicles with this refrigerant. Other manufacturers using HFO-1234yf include Hyundai, Subaru, Ford, and BMW.[18]

DISCUSSION QUESTIONS

- *Compare the economic and operations viability of switching to CO_2- or HFO-based refrigerants in MACs.*
- *Discuss the technological advancements of refrigerants in the building air conditioning sector (post 2015).*

5.2 Information and Communication Technology Sector

The Information and Communication Technology (ICT) sector has the potential to incorporate efficiency measures not only in its own sector but also to enable other sectors to become more efficient through development and deployment of smart technologies.

Global e-Sustainability Initiative (GeSI) is a sector initiative by the ICT companies and industry associations to develop and deploy technologies to promote sustainable development. Formed in 2001, GeSI is a great networking platform for businesses. It showcases the sustainable technological innovations of its members to reduce carbon emissions generated through their business-as-usual activities. In a study, GeSI estimated that the ICT sector could potentially reduce carbon emissions by 15% by 2020 that could amount to 600 billion euros.[19] Some examples of smart technologies and innovations by the ICT sector include:

- Smart motors that use variable frequency drives to reduce energy consumption
- Smart logistic technologies for better route and load planning while delivering goods
- Smart building technologies such as building energy systems, HVAC automation systems to enhance energy efficiency
- Smart energy meters to enable consumers to monitor and get real-time feedback of their individual energy use
- Smart lighting systems for buildings and city streets

Vodafone and Philips partnered to execute smart street lighting systems in cities worldwide under the CityTouch program. Philips is a global leader in providing lighting solutions. They innovated LED street lights that offer up to 40% energy savings when compared to conventional lighting systems. Their lights have sensors to monitor parameters like parking availability, traffic density, and air quality. Vodafone's contribution to this joint venture is the Machine-to-Machine (M2M) wireless technology. This technology enables remote monitoring and control of the lighting systems and also eases maintenance. This remote monitoring feature enables over 30% energy savings. This program has made street lighting systems more energy efficient via 530 installations in 33 countries.[20,21]

In Boxes 5.5 through 5.8, we present examples of products and services developed by the ICT sector that have reduced carbon emissions.

BOX 5.5 OPERATIONAL AND LOGISTICS EFFICIENCY AT UPS

The logistics giant UPS adopted technological innovations to enhance the efficiency of their operations and resources. Their *Package Flow Technology* automates and provides an optimal route of all packages to be delivered using a GPS. This helps minimize fuel costs and driving time that tends to be higher due to loading and unloading errors. Their *Delivery Information Acquisition Device* allows the drivers to perform deliveries with optimal routing and minimal errors contributing to fuel and operational efficiency. UPS uses vehicle maintenance and tracking technology to gather data on the vehicle performance in terms of fuel efficiency, tire pressure, usage, and emissions. Collectively, in 2007 all these ICT based systems saved 3 million gallons of fuel and 32,000 million metric tons of carbon emissions.[22]

 **DISCUSSION QUESTIONS**

- *Look up what ICT based systems Amazon uses to make its business operations successful.*
- *Is it more sustainable to buy a product online or in a physical store (Hint: Consider factors like transportation, building energy use, product packaging etc.)?*

BOX 5.6 ERICSSON'S E-HEALTH DELIVERY SYSTEM IN CROATIA

Ericsson is a global leader in providing communications technology services. The Ministry of Health and Social Welfare of the Republic of Croatia engaged Ericsson to incorporate ICT solutions in the national healthcare system.

Ericsson developed a smart phone-based platform to provide healthcare and well-being services. This platform enabled healthcare providers to remotely monitor patients in near real-time and provide them with medical services. The e-health system could maintain digital patient health records, provide video-conferencing options, and e-referral and e-prescription services. The e-referrals of e-prescriptions can be sent to pharmacies, hospitals, or laboratories, thus avoiding emissions due to commutes and high paper use.

Croatia's e-health system could potentially reduce patient trips to hospitals by 50% (approximately 12 million trips per year in Croatia) and paper use for prescriptions by 50%.[23]

DISCUSSION QUESTIONS

- *You may like to read: mHealth—health services via smart phones (http://www.wipro.com/documents/the-mHealth-case-in-India.pdf).*
- *List the social benefits of an e-health system.*

BOX 5.7 SMART ELECTRIC METERING PROGRAM IN CALIFORNIA

The California Public Utilities Commission and the utility company Pacific Gas & Electric Company piloted a smart metering program to test the impact of variable electricity pricing, auto temperature adjustments, and appliance usage adjustments programs. Smart meters allow users to gather real-time data such as hourly use of appliances or an entire facility, current electricity rates, and the ability to manage their electrical appliances. These are particularly useful in places where the electricity prices vary during peak hours, non-peak hours, weekends, and holidays like in California.

Under this program, smart meters were developed for different purposes. They allowed users to:

- View hourly electricity rates (i.e., prices as per time-of-use).
- Make adjustments to the indoor temperature setting or home appliance electric load based on the electricity rates using a smart thermostat (for example, the smart thermostat would increase the indoor air conditioning temperature during peak pricing hours to reduce the energy load).

The pilot results showed that consumers who could view the hourly electricity rates reduced their peak consumption by 13% while the consumers who were equipped with the technology to adjust their residential load, reduced their peak consumption by 27%–43%. Such technological solutions were replicated not only across California but across many other countries.[23]

DISCUSSION QUESTIONS

- *Your 11-story new office building is looking to install a smart meter. Calculate the return on investment (ROI) (in years) of installing the smart meter. (Assume that the electricity demand is mostly for HVAC systems, lighting, and pumps. Make reasonable assumptions to calculate the electricity demand and savings by using smart meters.)*
- *List two technological challenges in executing a smart meter system in the same office building which is over 40 years old.*

BOX 5.8 TELECOMMUTING

Telecommuting is gaining more and more popularity amongst employers and employees on a global level. Traffic congestion, high real estate prices, strain on space availability, the rise of entrepreneurship and start-ups, and the need for space to enhance employee creativity are some of the reasons for this. Telecommuting has a significant environmental impact. It tremendously reduces private vehicle and also public transportation use when it is looked at from a larger perspective. Many companies have been encouraging and incentivizing their employees to telecommute for operational and environmental reasons.

Merrill Lynch introduced the telecommuting program back in 1996 and over 3,500 of its employees worked from home 2–3 times a week. The company helped managers and employees get acclimatized to this style of work that was relatively new then. They provided training and required employees to work for 2 weeks in a telecommuting simulation lab to replicate the telecommuting experience.[24]

Telecommunications company British Telecommunications (BT) conducted a study to assess the impact of employee telecommuting and analyzed whether to encourage or discourage telecommuting as a company policy. They compared the positive impact of reduced vehicle use, both public and private, and also factored in increased energy use at home. They found that telecommuting reduced 1.4 tons CO_2 per employee per year, when compared to the business as usual scenario.[25]

DISCUSSION QUESTIONS

- *List some other employee-centric strategies that can contribute to environmental sustainability.*
- *Quantify the energy consumed (vehicle and building) during your daily commute to office or school in CO_2 per person per year.*

5.3 Pulp and Paper Sector

The pulp and paper industry is one of the most resource intensive and polluting industries. The paper making process involves conversion of fibrous raw materials like wood or waste paper into pulp and paper products. The major processes employed are raw materials preparation, pulping (chemical, semi-chemical, mechanical, and waste paper), bleaching, chemical recovery, pulp drying, and finally paper making. The processes use resources like raw material (from trees), energy (in the form of electricity, fuel, and steam), water, and chemicals (for pulping and bleaching).

The paper making process generates wastewater from various processes including raw material preparation, pulping, screening, washing, bleaching, de-watering, and finishing. Paper industries are required to treat effluents before discharging them out of site. Countries have set standards for effluent discharge.

For example, elemental chlorine was commonly used as a bleaching agent in the paper making process. Use of elemental chlorine in the pulping process generates carcinogenic and toxic pollutants. Due to associated risks, use of elemental chlorine was phased out in many countries. By 2005, over 95% of the pulp production in the United states became Elemental Chlorine Free (ECF). Elemental chlorine was replaced by chlorine dioxide that was then replaced by non-chlorine based chemicals like ozone, hypochlorite, sodium hydroxide, oxygen, and hydrogen peroxide that are more environmentally benign chemicals when compared to elemental chlorine.[26] Similarly, their standards were defined for emission of air polluting substances from the paper industry like sulfur dioxide, particulate matter, hydrocarbons, and carbon monoxide.

There are many innovations and technological advances in the pulp and paper industry targeting efficiency in use of resources, energy and water. Boxes 5.9 and 5.10 present a few examples.

BOX 5.9 RENEWABLE PAPER COATING STARCHES BY CARGILL

The US-based firm Cargill deals with many businesses including paper making. Cargill innovated a range of paper coating products that are renewable, biodegradable, and chemical free.

Paper coating is a process of altering the properties of paper like smoothness, gloss, and weight. Paper coating consists of pigments like chalk or clay that covers the surface of the base paper, natural or petroleum-based synthetic binding agents and water. Cargill has developed three starch based products called C*Film™, C*iCoat™, and C*iFilm™ that can replace synthetic binders like latex and make the paper making process renewable and sustainable on this aspect.[27,28]

Businesses have explored many materials to manufacture tree-free paper:

- *Paper from the kenaf plant*—Vision Paper, based in New Mexico, produces chlorine free, acid free, and tree free paper made from the rapidly renewable kenaf plant belonging to the hibiscus family.[29]
- *Paper from elephant poop*—Indian company Haathi Chaap produces paper by extracting fiber from elephant dung. The raw material is disinfected to ensure it is bacteria free and has no adverse health effects.[30]

DISCUSSION QUESTIONS

- *Discuss the coating material used to manufacture paper for magazines and its environmental impact. Is there any acceptable list or criteria for the use of coating chemicals?*
- *List three other materials that are commonly used to make tree-free paper.*

BOX 5.10 EPSON'S PAPERLAB: COMPACT WATERLESS PAPER MAKING SYSTEM[31]

Seiko Epson Corporation is a Japanese electronics company established in 1942 and is one of the world's largest manufacturers of computer printers. They manufacture electronics including projectors, LCDs, sensor systems, and semi-conductors. EPSON recognized that paper recycling is a very resource intensive process. It involves transportation (fuel use) of used paper to the recycling or paper making center, energy and water use during the recycling process, and manual labor. EPSON developed a technology to localize the paper recycling process and to reduce the resource demand.

EPSON's PaperLab is a compact system that can be installed in offices to recycle waste paper on-site. PaperLab is a dry process (no water required) that uses three technologies, fiberizing of waste paper (converting to thin cotton fibers), binding (to increase strength), and pressure forming.

The output is paper in various forms like A3, A4, or business cards. The system can produce about 14 A4 sheets per minute and 6,720 sheets in an eight-hour day. The prototype was developed in 2015 and production was expected to begin in 2016. This is a good example of how technological innovations by a business can enhance resource efficiency of the sector.

DISCUSSION QUESTIONS

- *Conduct a lifecycle analysis to compare the environmental impact of producing 1 million A4 sheets of virgin paper per year vs the same quantity of recycled paper.*
- *Discuss four strategies to make paper making a more water and energy efficient process.*

5.4 Cement Industry

Globally, cement is the most used construction material for making buildings, roads, highways, etc. Cement production is an energy, water, and material intensive process. In addition to the CO_2 emissions generated from the manufacturing process, the industry produces millions of tons of waste material and gaseous pollutants like cement kin dust, carbon oxides, NO_x, and SO_2 that pollute the environment and cause health issues. The cement industry used fossil fuels for energy supply, but technological advancements have enabled the use of waste materials to fuel the production process. Box 5.11 illustrates how hazardous waste is used in cement plants as a source of energy.[32]

BOX 5.11 CO-PROCESSING OF HAZARDOUS WASTE IN CEMENT PLANTS

Co-processing is a concept where one industry uses the waste from another industry as an alternative fuel for manufacturing processes. Co-processing of high calorific waste in cement kilns provides an environmentally sound resource and an energy recovery opportunity. The high temperatures achieved in cement kilns (which can reach up to 2000°C)[33] make cement kilns an efficient technology for combusting hazardous waste. Since the 1970s, in many countries like Australia, Canada, Japan, and the United States, fossil fuels are being substituted by waste material for use as fuels in cement kilns.

Hazardous wastes used by cement kilns can be in liquid form like industrial solvents from paint and coatings, paint thinners, waste oils and other petrochemical byproducts or in the solid form like plastics, packaging, textiles, tires, or rubber.

The potential benefits of using hazardous wastes as fuel in cement plants include the recovery of energy from waste, conservation of fossil fuels, reduced carbon emissions, and use of existing infrastructure to handle hazardous wastes. Some examples of cement plants where co-processing of waste is done are included below:

CARTAGO, COSTA RICA[34]

In 2004 Holcim, one of the leading cement producers in the world, put a new state-of-the-art cement kiln into operation in Cartago, Costa Rica. Before 2004 co-processing of waste material in cement kilns was not regulated by national legislation. With test runs, Holcim proved that hazardous wastes can be co-processed in cement kilns in an environmentally sound manner. This demonstration helped implement a regulation that legally allowed co-processing of waste in cement plants in Costa Rica. Waste is collected and transported from the waste generating industry. The permitted wastes include used solvents (halogen free), waste oil, waste tires, and rubber scrap, and plastics (except PVC). Emission controlling and monitoring systems are in place to minimize toxic emissions from the co-processing process.

GEOCYCLE VIETNAM[35,36]

Geocycle Vietnam, a business unit of Holcim in Vietnam, offers waste management services including hazardous wastes via co-processing in cement kilns. It follows the EU Waste Framework Directive and Basel Convention that provides guidance on safe management of hazardous wastes. Holcim is expanding its existing cement plants for co-processing to include feeding systems, transfer systems, transportation, testing, and emissions monitoring equipment.

DISCUSSION QUESTIONS

- *Co-processing of hazardous wastes to cement kilns requires transportation. Who bears the risks of waste spill or accidents during transportation?*
- *Is intercountry transportation of hazardous wastes permitted for co-processing as per the Basel Convention on the Control of Transboundary Movement of Hazardous Wastes?*
- *Are there any special standards set for emissions from cement kilns when hazardous waste is co-processed?*

5.5 Tourism Sector

Tourism revolves around the natural environment such as forests, national parks, rivers, lakes, beaches, mountains, and the built environment such as historic monuments and ruins. Both types of ecosystems need to be preserved for the future generations. In addition to being a recreational activity for many, it is a main source of income in many cultures, especially in the Small Island Development States (SIDS).

Tourism can cause overcrowding of popular destinations, over-use of land, clearance of natural vegetation, use of products and services like transportation, use of energy and water, food and beverages, littering, and destruction of the natural and built environment.

Sustainable Tourism was introduced and developed by the World Tourism Organization (WTO) and 2002 was declared as the International Year of Ecotourism. The UN, WTO, national governments, the hotel industry, and NGOs promoted sustainable tourism at an international, national and local level. Certifications and benchmarking standards were also developed to measure the performance of companies and help them move towards sustainable tourism practices.

EarthCheck is a global Benchmarking and Certification program for sustainable travel and tourism. It evaluates the environmental and social performance of tourism businesses including their energy emissions, water, waste, community involvement, paper use, cleaning, and pesticide use and in the process helps businesses improve their performance.[37] Boxes 5.12 and 5.13 give examples of Ecotourism initiatives through industry association and in small islands.

BOX 5.12 INTERNATIONAL TOURISM PARTNERSHIP, ITP (FORMERLY KNOWN AS INTERNATIONAL HOTELS ENVIRONMENT INITIATIVE, IHEI)

In 1992, CEOs of ten multinational hotel groups came together to form IHEI (now International Tourism Partnership (ITP)). The aim was to encourage the hotel industry to improve their environmental performance. It was a non-competitive platform where members could pool resources and share success stories. ITP was a program of the host organization, International Business Forms. In 2013, it has re-located under its new host, Business in the Community. Some of ITP's interesting achievements are:

- In 1993, published a guide to quality environmental management for the industry.
- In 1995, launched the magazine Green Hotelier to share green practices.
- In 2005, published guidelines for Sustainable Hotel Siting and Design.
- In 2010, its first working group was formed on sustainable certification schemes. ITP has another working group on Supply Chain.
- In 2011, launched the Hotel Carbon Measurement Initiative (HCMI) in partnership with the World Travel and Tourism Council. The initiative was to calculate and communicate carbon emissions of hotel stays and meetings.

ITP now has 18 corporate members with a collective reach of over 23,000 properties in over 100 countries worldwide. Execution of ITP's guidelines has resulted in significant water and energy savings.[38]

DISCUSSION QUESTION

- *Refer to the link http://www.thinktur.org/media/Hotel_Carbon_Measurement_Initiative_Methodology_v_-1_1.pdf for a detailed understanding of HCMI Methodology for calculating carbon emissions (see worked example on Page 32).*

BOX 5.13 ECOTOURISM IN SMALL ISLANDS: CASE STUDY OF COUSIN ISLAND SPECIAL RESERVE, SEYCHELLES

Seychelles is a small African Republic, an archipelago in the Indian Ocean. It has small islands over an area of 460 sq. km. with a population of 90,000. Cousin Island is a small island, 34 hectares within the Seychelles. Cousin Island is ecologically very significant, because it is home to globally threatened species like warblers (one of the rarest birds on earth), seabird breeding colonies, the endemic terrestrial birds and nesting hawksbill turtles. In the 1960s, Seychelles has the maximum number of *critically endangered* species.

In 1968, Birdlife International declared Cousin Island as an Important Bird Area. Since 1998, Cousin Island is managed by a non-profit organization, Nature Seychelles, that has a tourism policy and Tourism Code of Practice. Tourism on Cousin Island is guided by the eight principles under International Ecotourism Standard for Certification. These principles include environmentally sustainable practices like limited number of guided tours, no picnicking or overnight accommodation, no picking of specimens or souvenirs, distance to be kept from nesting birds and turtles, use of mooring buoys to protect coral reefs, use of solar power, and use of tourism revenue to fund environmental conservation and education.[39] On the Seychelles islands, there is no urbanization, no main roads, cars, motorized water-sports activities in the national parks, or mass tourism. Hotels are built in accordance with nature and their construction does not affect the natural environment.[40,41]

Multiple initiatives including eco-tourism to preserve the species and biodiversity have transformed the loss-making coconut plantation to an ecologically restored island. As a result, there was a 300% increase in the population of warblers, a healthy population of Seychelles foodies, and successful translocation of magpie robins.

The work done by the organization earned Cousin Island the accolade of becoming the world's First Carbon Neutral Natural Reserve in 2010. It is very interesting to note that when Cousin Island realized that its forest reserves are not sufficient to offset its carbon footprint, it used Seychelles Eco-tourism funds to invest in climate adaptation projects in Brazil and Indonesia. They invested in carbon sequestration projects to offset their carbon footprint. They funded a deforestation project in Brazil, where in agricultural waste is used in place of wood to file ceramic kilns, and in Indonesia they haveve funded a project to upgrade an existing coal-based power plant to a geothermal plant.[42]

DISCUSSION QUESTIONS

- *How does making Cousin island a Carbon Neutral Natural Reserve help preserve its biodiversity? Are these objectives similar?*
- *How can the Polluter Pay principle be applied to the stakeholders at Cousin Island for environmental conservation?*

5.6 Waste Management Sector

Waste Management is a critical issue in every business irrespective of the sector type and the scale of business. Many businesses set-up waste management systems in their manufacturing or business-as-usual processes to divert landfilling of waste generated. They sometimes engage waste management businesses that provide such services.

Waste Management systems are typically handled by the organized sector in developed countries, while they are handled by the unorganized or informal sector in the developing countries. With increasing consumerism and technological innovations, the types and complexity of each waste stream is only increasing. Municipal (domestic), electronic waste (e-waste), bio-medical waste, and construction and demolition waste are the more traditional waste streams. Waste streams like nanoparticles, solar PV, absorbent hygienic products, and disaster waste are the new emerging waste streams. Governments and businesses need to gear up and innovate to manage the vast volumes and complex composition of these waste streams right from the collection to the treatment or disposal process.

One very unique example of businesses collaborating for waste management was when Uber, a global transportation company, partnered with Dell and Goodwill to collect e-waste from the citizens of New York City. This partnership was in honor of America Recycles Day. On that day, folks interested in donating or disposing their e-waste could

BOX 5.14 BELGIUM-BASED UMICORE, E-WASTE RECYCLING EXPERTS

Umicore is a producer of rechargeable batteries, solar PV systems, mobile phones, electrified vehicles, and also a global leader in extracting and recycling precious metals from electronic and electrical waste.

Umicore has global expertise in extracting metals from complex electronic waste products that are supplied to them from global e-waste suppliers. They use cutting edge technologies to maximize resource extraction efficiency. Umicore claims that recycling of metals from e-waste is far more lucrative and efficient when compared to natural mining. For instance, they can extract 200–250 grams of gold per ton from printed circuit boards (PCB) and 300–350 grams of gold per ton from cell phones when compared to ~5 grams of a gold ton of ore.[44] Their precious metal refining process can extract 99.95% of fine metals from electronic scrap.[45]

Umicore complies with strict environmental standards, even for products with complex product composition. Their plants are ISO 9001 and 14001 and OSHAS 18001 certified. They use efficient off-gas and wastewater treatment systems and dispose non-recyclable material in a safe manner.[46]

DISCUSSION QUESTIONS

- *Umicore is a global leader in PCB recycling. Learn more about the recycling process and technology used.*
- *Write a note on the market of recycling used mobile phones. Read about the Mobile Phone Partnership Initiative (MPPI) by accessing http://www.basel.int/Implementation/TechnicalAssistance/Partnerships/MPPI/Overview/tabid/3268/Default.aspx.*

use the Uber application on their smart phones to arrange a pick-up. All the e-waste was sent to the Goodwill centers for refurbishing or recycling.[43]

In Boxes 5.14 through 5.17, we present some innovative business models in the waste management sector.

BOX 5.15 WONGPANIT CO. LTD., THAILAND'S PIONEERING RECYCLING LEADER[47,48]

Among the many waste recycling business models, Wongpanit, a Thailand based waste dealer, stands out for many reasons. They believe that most types of waste in waste dumps are valuable, and that the community plays as much part in waste management as governments and businesses.

Operating since 1974, they have a very transparent waste trade system, wherein the waste purchasing prices are publicly listed. It not only invests in trading recyclables for monetary benefits, but also significantly contributes to social development and environmental conservation. The company has engaged and empowered stakeholders like waste-miners or scavengers, scrap traders, municipal services, schools and temples into their business model.

Wongpanit has set up a franchisee system and widened its reach and collection coverage to nearly 1,000 stores in all Thai sub-districts. The company has its quality standards and offers training for the community and its franchise owners. They incentivize bodies like schools, temples, and communities to set up waste banks that can make money by collecting and selling waste to the company. Some of its successful business initiatives are:

- One-stop shop for collection of all types of waste.
- Collaborating with local governments to promote recyclable waste separation at source for sale.
- Increasing public interest through slogans like *waste is gold* and *waste separation for society and the environment.*
- Increasing value of recyclables through fine segregation of waste to meet recyclers' requirement.
- Public announcement of real-time recyclable prices: board, webpages, and SMS.

Wongpanit has found a recycling system for all kinds of waste such as coconut shells into activated carbon for wastewater treatment, odor absorbents, pencil making; food into compost, sending back used fish and alcohol bottles to the factories, recycling iron, steel, wood, waste oil, cloth, old batteries, battery acid, and many other waste products.

 **DISCUSSION QUESTIONS**

- *Discuss how Wongpanit benefits from setting up a network of franchisees.*
- *Wongpanit offered you a franchise to set up shop in town. Would you take up the offer? What waste collection strategies or operations would you put in place prior to signing up?*

BOX 5.16 DISASTER WASTE RECOVERY

Natural disasters generate large quantities of debris and owing to the unexpected nature of disasters, it is hard to predict and prepare for waste arising from disasters. For example, the Great East Japan Earthquake of 2011 in Ishinomaki City is estimated to have generated 6.16 million tons of waste that is more than what the city would generate in over 103 years.[49] Disaster waste can be comprised of soil, vegetation, trees, building debris, hazardous wastes like batteries, oils, asbestos, plaster, furniture, vehicles, and biomedical waste. The complex nature of disaster waste poses a high risk to human health, soil, ground water, and river bodies, and this make restoration and waste management very challenging.[50]

Disaster Waste Recovery (DWR) is a UK-based non-profit company dedicated to managing waste due to natural and man-made disasters like earthquakes, droughts, flooding, hurricanes, tsunamis, and areas inflicted with terrorism or international conflict. DWR collects debris and disaster waste, recycles debris and rubble, does cleanups and restores infrastructure and coastal areas, provides training in waste and debris management and helps countries prepare for disaster waste management. They use innovative solutions like use of debris for rehabilitation of roadways or to make paver blocks or masonry blocks, to generate power or fuel, and to re-use plastics or styrofoam.

DWR has undertaken a host of projects in disaster-stricken countries like Syria, Serbia, Haiti, Gaza, Lebanon, Turkey, Pakistan, Indonesia, Sri Lanka, Kosovo, Bosnia, Croatia, the Maldives, New Orleans, and Burundi.[51,52]

DISCUSSION QUESTIONS

- *Read UNEP and OCHA's Disaster Waste Management Guidelines from—https://docs.unocha.org/sites/dms/Documents/DWM.pdf.*
- *Discuss Disaster Waste Management in Haiti and the Fukushima waste cleanup.*

Manufacturers were vested with Extended Producer Responsibility (EPR), which required them to take back their used products for recycling. Some of the challenges faced to comply with EPR were collection and separation of used or discarded products for their recovery and recycling. Research centers, small and medium sized companies, and large corporations have been working together to address this issue of traceability and separation of waste products. They formed Consortiums to develop technological solutions to trace and recover waste products. We describe the EU funded consortium, Dibbiopack project, and the PRISM initiative in Box 5.17.

BOX 5.17 TRACEABILITY AND RECOVERY OF WASTE PLASTIC

An EU funded consortium, the Dibbiopack project, made several innovations to promote circular economy of plastics. They developed compostable and biodegradable packaging material made from renewable sources that are capable of prolonging product freshness. A promising feature developed was including an RFID tag on products that allows for manufacturers to not only trace their products in the supply chain but also to recover it once it's been used and discarded.

Another approach to recovering waste plastic materials is by using fluorescent materials on product labels to help identify the packaging product composition that will aid sorting and, hence, recovery. Labels made of fluorescent materials will act as an invisible bar code on packaging products that will be sorted based on their product composition when scanned at material recovery or recycling facilities. This technology was developed by another consortium, the Plastic Packaging Recycling using Intelligent Separation technologies for Materials (PRISM) project, and can help sort materials such as polypropylene packaging, high density polyethylene milk bottles, and sleeved polyethylene terephthalate.[53,54]

5.7 Cosmetic Sector

The cosmetic sector uses a mix of natural and synthetic materials to manufacture personal care products like shampoo, body wash, sunscreens, and body lotions. The chemicals and synthetic materials used in cosmetics and personal care products are dangerously harmful to the marine ecosystems. For example, the micro-beads present in our body or face washes are polyethylene substances that find their way to oceans and lakes. Chemicals like BHA (butylated hydroxyanisole), BHT (butylated hydroxytoluene), Dibutyl Phthalate (DBP), and Triclosan found in lipsticks, moisturizers, and hand sanitizers are a threat to aquatic animals and plants.[55]

The government and industry is taking measures to reduce their environmental impact. For example, the US government introduced a law in 2016 to phase out the use of micro-beads in all personal care products by July 2017.[56] Companies like Unilever, Colgate-Palmolive, Procter & Gamble, Avon, Crest, The Body Shop, Bath & Body Works, Boots, and many others pledged to phase out micro-beads as early as 2012.[57]

Water pollution during manufacture, product use, and disposal is one of the key impacts of the cosmetic industry. Biotherm, a L'Oreal company, attempted to create awareness and engage consumers towards this issue. They launched the Water Lovers initiative to allow consumers to calculate and understand their water footprint, which is the first step to reducing it.[58]

Box 5.18 includes examples of cosmetic companies that have adopted practices and innovated alternatives to alleviate their environmental impact.

BOX 5.18 PACKAGING INNOVATIONS BY COSMETIC BUSINESS[59,60]

- Lush and M-A-C (an Estee Lauder Company) offer their consumers' incentives like free face mask or lipstick for returning empty cosmetic containers. The used containers are recycled into new ones.

- Similarly, cosmetic retailers like Sephora and Nocibe incentivize their customers by offering discounts when empty perfume bottles are returned.
- Lush has innovated chewable tablets as a replacement of traditional toothpaste, which eliminates the use of toothpaste tubes, which are complex to recycle.[61]
- The Body Shop partnered with Newlight Technologies to package their products in the carbon-negative material, AirCarbon™. This carbon sequestering polymer is manufactured by combining air with methane-based carbon emissions from landfills or power facilities.[62]

DISCUSSION QUESTIONS

- *Watch the video on Newlight's carbon-neutral plastics—https://www.youtube.com/watch?v=LwoqJj2brLQ.*
- *Taking the cue from Lush's chewable tablets example, think of one of your personal care products that can be re-invented on similar lines to reduce the adverse impact of the use of plastic.*

5.8 Food & Agriculture Sector

This sector is a key to human health and nutrition. As the world moves towards a more affluent economy, people's food and beverage choices tend to move towards packaged and processed food products that are proving to cause food related diseases like Type 2 diabetes and obesity. Sugary beverages and increasing meat consumption are some of the obvious causes of these diseases. Red meats are also considered to be one of the foods that have the highest carbon footprint. The availability of food, land degradation, food quality, and waste management are some of the prime challenges facing the industries in the food and agricultural sector. Governments and businesses are taking measures to address these challenges.

Beverage companies like PepsiCo and Coca-Cola faced public pressure to the environmental and health impact of their products. These companies pledged to reduce the calorific content and sugar content of their beverages. The Chinese government intends to reduce its meat consumption by 50% in an effort to reduce its impact on the climate. The Danish government is considering imposing a tax on red meat.[63]

Farming practices for food production are one of the key aspects of the food sector that should be made sustainable. Uses of pesticides, excessive use of land and water, and soil degradation are some key issues to be addressed. Sustainable agricultural practices are essential for the survival of a healthy human race. Through the entire cycle of food production and consumption, businesses have attempted to incorporate sustainable practices. Boxes 5.19 through 5.22 discuss examples of some businesses that are contributing towards sector sustainability.

BOX 5.19 SUSTAINABLE COCOA AND COFFEE FARMING BY KRAFT FOODS

Kraft Foods is a 100-year-old company that produces and sells food products like cheese, coffee, cereals, cookies, crackers, chocolates, candies, nuts, Jell-O, beverages, salad dressing, etc.

Kraft Foods launched its first sustainable coffee farming program in 1983 and extended it to sustainable cocoa farming in 2005. Kraft has had a multi-dimensional approach to sustainable farming. They focus on getting their products certified under sustainable sourcing programs like Rainforest Alliance, Fair Trade, and UTZ Certification. As pioneers in developing sustainable farming programs, they support their industry partners on this and also their sub-brands.

In 2003, Kraft started working with Rainforest Alliance for helping coffee farmers adopt sustainable practices and in 2005 extended this partnership to benefit their cocoa farmers.

In 2007, they started supplying sustainably sourced coffee to over 500 McCafes and 6,400 McDonalds across Europe. They launched two coffee varieties made from Rainforest Alliance Certified™ beans. In 2010, Kraft purchased over 50,000 tons of Rainforest Alliance Certified™ coffee from suppliers.

Adopting practices under these certification programs helped farmers boost crop productivity, abated environmental impacts of farming, conserved bio-diversity, wildlife, land and water bodies, and improved the farmers' standard of living due to premium sale prices of certified products. In the West African country of Côte d'Ivoire, the cocoa farmers benefited by learning better farming practices like how to take care of plants, pest and disease control, plant breeding, etc. In 2008, over 2,000 farmers achieved the Rainforest Alliance Certification that earned them an additional $250,000 from the sale of premium certified cocoa. Kraft is the world's largest purchaser of Fairtrade Certified cocoa and Fairtrade Organic cocoa. As of 2012, Kraft was purchasing 20,000 metric tons of Fairtrade Cocoa per year.[64,65,66]

DISCUSSION QUESTIONS

- *Discuss the objectives of the five principles provided under the Sustainable Agriculture Network 2017 Standards—https://dl.dropboxusercontent.com/u/585326/2017SAN/Certification%20Documents/SAN-Standard-2017.pdf.*
- *Cocoa production globally is under distress. Discuss the reasons behind this downfall. Factor issues like skewed supply-demand and the impact of climate change.*

BOX 5.20 FURTHER WITH FOOD: CENTER FOR FOOD LOSS AND WASTE SOLUTIONS[67]

The United States spends a significant amount of resources, energy, water, labor, and money in growing, harvesting, processing, transporting, and selling food that is never bought or consumed by consumers. This unconsumed food ends up in landfills that emit greenhouse gas emissions.

In 2015, the US Environmental Protection Agency (US EPA) and the United States Department of Agriculture (USDA) announced the first national goal to halve the food loss and wastage in the country. The goal is in line with the UN Sustainable Development Goal 12.3.

Twelve organizations including Rockefeller Foundation, the USDA, US EPA, World Resources Institute (WRI), the World Wildlife Fund (WWF), the Natural Resources Defense Council (NRDC), the Grocery Manufacturers Association, the Academy of Nutrition and Dietetics, the Innovation Center for U.S. Dairy, Feeding America, the Food Marketing Institute, the National Consumers League, and the National Restaurant Association came together to launch the Further Food initiative to reduce food wastage, Further with Foods. Set to be launched in 2017, the forum will create awareness and educate stakeholders on best practices to prevent, recover, and recycle food before letting it go to waste. The portal would allow stakeholders to share information and initiate new projects to reduce food wastage.[68]

BOX 5.21 EGYPT-BASED ORGANIC FOOD PRODUCER

Operating since 1977, Sekem is the first organic foods producer in Egypt that also exports in Europe. They produce health foods, dairy products, honey, fruits and vegetables, herbal tea, medicines, and textiles that are natural and organically produced. Their focus is on biodynamic farming, which means to preserve soil and biodiversity while cultivating products.

In an area where the desert land was covering more than two-thirds of the land area in the Arab Region, Sekem reclaimed part of the desert land by cultivating organic farms to produce food for the local market. They reduced the use of synthetic pesticides in their in-house cotton production and cultivated microorganisms that replaced chemicals used in farming. This had multi-fold advantages:

- The yield of cotton was 30% higher.
- The elasticity and quality of cotton was better than conventional cotton.
- The ability of soil to absorb atmospheric carbon dioxide was higher.
- About 20%–40% less water was required for cultivation.

The company worked with the national government to ban airborne pesticide distribution for cotton cultivation, a high polluting application technique. The company's philosophy is not to maximize profit but to help the environment and community through its business. Its non-profit Egyptian Biodynamic Association (EBDA) offers capacity and skill development program for its workers, provides health services to its employees and community members, and shares profit with the farmers whose lands have been certified under the EU organic standards. EBDA also spreads knowledge and know-how of biodynamic farming in Egypt and Africa.[69] Sekem's cotton is Global Organic Textile Standard (GOTS) and Fairtrade certified.

DISCUSSION QUESTIONS

- *What is Biodynamic Farming?*
- *Discuss the environmental, social, and technical criteria that businesses need to comply to get their product GOTS certified.*
- *Discuss ways in which consumer behavior can be cultivated towards healthier food choices and less wastage.*

BOX 5.22 FLORIDA ICE & FARM, BEVERAGE COMPANY'S WATER EFFICIENCY INITIATIVES

Florida Ice & Farm is a food and beverage company based in Costa Rica that produces beverages and food for global brands like PepsiCo, Tropical juices, Heineken, Corona, and Gatorade. The company decided to adopt sustainable practices in 2008, even when the entire world was reeling under the global financial crisis.

As a beverage company, the company's focus was to tackle water efficiency as huge volumes of water was used not only in its products but also in the production process. In 2003, the company used 14 liters of water for every liter of beverage produced. In efforts to reduce water consumption, it resulted in reduction to 8 liters per liter of water by 2008 then to 4.72 liters per liter of water and the aim is to reach to 3.5 liters, a world benchmark. The company further *off-sets* its water usage by compensating communities to conserve water through watersheds and its employees help build infrastructure for water delivery to underprivileged communities on company time.

DISCUSSION QUESTIONS

- *What is Integrated Watershed Management (IWM)? What role can a private company play in IWM?*
- *Florida Ice & Farm was hailed as a sustainability champion in 2011. Find other organizations that have won such an accolade.*

5.9 Mining Sector

Globally, the mining industry is gaining more significance due to high demand. Overexploitation of minerals, disparity between demand and supply, and monopoly of minerals by select countries due to unequal geographic distribution of minerals are some of the key sustainability challenges.

Rapid industrialization and technological advances including *Green* technologies have further increased the demand of minerals. For example, batteries require lithium, tantalum is needed for mobile phones' hybrid card, and wind turbines use neodymium, a rare-earth metal; the electric car model Prius requires two pounds of neodymium and

22–30 pounds of lanthium. But the existing supply is unable to meet the rapidly increasing demand.

Mining activities have a huge environmental impact due to exploitation of natural ecosystems like mountains, forests, oceans; pollution of water bodies and ground water; transportation needs, pre-and post-extraction of minerals; air (dust) pollution; and communities are also impacted. Some countries have strict environmental policies, while providing mining authorization, while some countries are more lax.

In 2001, mining companies established the International Council on Mining & Metals (ICMM) with the objective of promoting safe, fair, and sustainable practices in the industry. ICMM members include Rio Tinto, Xstrata, Anglo American, BHP Billiton, Vale, Sumitomo, Barrick, LonMin, Goldcorp, MMG, and JX Nippon. ICMM works towards promoting reporting and assurance, environment, health and safety, socio-economic development, and materials stewardship. In 2003, ICMM made a commitment to not conduct any mining activities in World Heritage Properties.

Mining of precious metals can be an environmental and social issue. Box 5.23 presents an example to showcase an initiative taken towards responsible and conflict-free mining.

BOX 5.23 RESPONSIBLE COBALT INITIATIVE

Lithium-based batteries have established their superiority over lead-acid batteries due to their lighter weight and ability to pack more energy for recharging. Electronics like smart-phones, cars, and laptops that contain lithium-based batteries have fueled the demand for cobalt. Cobalt is extracted from cobalt mines in countries like the Democratic Republic of Congo, which caters to 55% of the global mine supply.

Apart from the environmental hazards associated with unsustainable mining practices, one of the health hazards in the hand-dug mines are those that employ children as young as seven years of age.[70]

A majority of cobalt from Congolese mines is processed by Congo DongFang International Mining, which is a subsidiary of Zhejiang Huayou Cobalt, that sells it to battery component manufacturers, battery manufacturers and electronics and vehicle manufacturers like Apple, HP, Samsung, Sony, HP, and Lenovo. See the global map which represents the global movement of cobalt to smelting and refining centers to component and battery manufacturers and to electronic or vehicle manufacturers in the reference as referred above.[71]

In interviews conducted by media houses and by Amnesty International, a human rights organization, most manufacturers in the supply chain were unaware of the exact source of the cobalt they purchased and the health risks of crude mining procedures done without protective equipment.[71]

Chinese Chamber of Commerce for Metals, Minerals and Chemicals Importers and Exporters, which is a unit of Ministry of Commerce of China, is playing a lead role in setting up the Responsible Cobalt Initiative (RCI) which has been joined by companies like Apple, Samsung, HP, SDI, and Sony. Businesses that partake in this Initiative follow OECD guidelines for mining supply chains where cobalt is extracted, transported, manufactured, or sold. Companies had to set up supply chain management systems so they are able to trace the source of the cobalt purchased and the conditions and mining procedures that are followed. Any health or environmental risks identified should be assessed, and a strategy should be developed to respond to the identified risks. Protection of artisanal cobalt miners from unhealthy working environments, unfair labor prices and exploitative practices were responsibilities of the RCI members.[72]

DISCUSSION QUESTIONS

- *Read paper on The Emerging Cobalt Challenge by RCS Global—https://www.rcsglobal.com/wp-content/uploads/2016/11/rcsglobal-cobalt-briefing-paper.pdf.*
- *Discuss Electronic Industry Citizenship Coalition (EICC)'s Responsible Raw Materials Initiative and how can they address issues of unsustainable and untraceable cobalt mining problems.*

5.10 Social Sector

More and more businesses are looking to address social and environmental challenges as part of their company's social responsibility or philosophy. Large corporates and small-scale social enterprises or non-governmental organizations are partnering to address these issues on a local or national scale. Such partnerships offer multiple advantages, including using the local knowledge and network of the social enterprise or NGO and technical know-how and operational efficiency.

The Grameen Group is one such social enterprise that works for the uplifting of the financially challenged strata of society in Bangladesh in various areas like healthcare, financial uplift, and agriculture. Grameen has partnered with numerous multi-national corporations (MNCs) to set up and operate social businesses that focus on problem areas like health, nutrition, and clean water. Box 5.24 briefly discusses such social businesses initiated by Grameen with MNCs.

BOX 5.24 EXAMPLES OF SOCIAL BUSINESSES

GRAMEEN VEOLIA WATER

Around 83% of the groundwater in the rural areas of Bangladesh was contaminated with arsenic that was affecting the health of the population. To tackle this issue, Grameen Healthcare and Veolia Environment partnered in 2008. They set up a water treatment plant that provided safe drinking water to 7,000 people who live in villages with contaminated groundwater. By 2010, the plant was able to provide treated water that was more than the demand.[73]

GRAMEEN DANONE FOODS

Grameen Group partnered with Danone Asia PTE Ltd. to reduce child malnutrition by selling fortified yogurt at affordable prices.

BASF GRAMEEN

In 2009, BASF and Grameen partnered to improve the health of people in Bangladesh through improved nutrition and protection from insect borne diseases like malaria. They sell insecticidal nets and sachets containing nutrients at affordable prices.[74]

GRAMEEN SHAKTI

About 70% of rural Bangladesh has no access to electricity. Grameen Shakti is a non-profit village renewable energy scheme, under which small loans are provided to villagers to install solar systems in their homes to replace kerosene-based lanterns.[75]

Note that none of the above social businesses are charity organizations, but they actually sell products and strive to become self-sustainable while providing job opportunities to the locals.

DISCUSSION QUESTIONS

- *Refer to this link for more examples of Grameen's social businesses—https://www.bcgperspectives.com/content/articles/corporate_social_responsibility_poverty_hunger_power_social_business/.*
- *Discuss how Grameen has utilized the concept of micro-financing for uplift of economically challenged sectors of society.*

In Chapter 6, we will look at more examples of how communities and businesses have partnered to contribute to environmental sustainability and creation of green jobs.

5.11 Collective Industry Responses

While sustainability stems from individual business initiatives, they need to be supported and driven by collective industry associations. Working collectively creates a better and more sustainable impact and in return makes the sector more competitive and responsible. In this section, we will see some examples of initiatives by industries as a collective entity. Box 5.25 through 5.31 provide such initiatives.

BOX 5.25 SWISS PRODUCER RESPONSIBILITY ORGANIZATIONS[76]

Switzerland is a mature consumer market for electronics and electrical (EEE) products. It is the first country to implement an organized industrywide E-waste management system. The primary E-waste management mechanism in Switzerland is through the Producer Responsibility Organization (PRO). PRO is a co-operative industry body to collectively meet the Extended Producer Responsibility (EPR) obligations of its member organizations. The Swiss EEE producers took a proactive approach and set up PROs in 1990, well before the EPR became a mandatory legislative requirement.

The Swiss PROs are not-for-profit organizations that have the operational responsibility of e-waste management including collection, transportation, and financing operations. The member organizations collectively make decisions on the operational and financial instruments.

The financial model follows the "Consumer Pays" principal. The producers charge the consumers Advanced Recycling Fees (ARF), via the retailers and distributors. The ARF fund is passed on to the PROs, and it funds the collection, transportation, and recycling activities. The Swiss PRO collection system is an all-inclusive, collective takeback system, that is, it does not differentiate between brands while taking back from consumers making it more convenient for them.

KEY FEATURES AND TAKEAWAYS

- The Swiss PROs use retailers and public areas such as railway stations and community as their collection points.
- They do not differentiate between historical (products sold before EPR legislation) or orphan (whose producers do not exist) and collect all E-waste.
- The SWISS ARF model shows when the disposal fee is more than the recoverable fee, then advanced recyclable fees fills in the gap. While in countries where recoverable fee is higher than disposal fee due to lower collection costs, etc., then that eliminates the need for ARF.
- Some factors that facilitated active producer participation are legislative reporting responsibility, peer pressure, low membership fee, and ease of the compliance process.
- Certain manufacturers who did not participate in the PROs in the beginning joined when they faced high collection and recycling costs.
- For consumers, one of the big incentives of recycling E-waste is that they avoid high waste disposal fees as the Swiss household waste management system follows the pay-per-use system.
- SWISS PROs ensure there is competition between recyclers and avoid monopoly by granting rights to the lowest bidding recycler.

 **DISCUSSION QUESTIONS:**

- *Discuss the origin and operating model of the two Swiss PROs, SWICO and S.E.N.S.*
- *Give an example of a PRO that works on another stream (for example packaging waste such as Tetra packs, etc.).*

BOX 5.26 ELECTRONIC INDUSTRY CODE OF CONDUCT (EICC)[77]

A group of electronic companies came together in 2004 to form the EICC to develop industry-wide practices and standards for health and safety, social, environmental, and ethical issues in the companies and their in the electronics industry supply chain. The EICC has standards to ensure the supply chain is safe and humane with standards such as no forced or bonded employment, no child labor, non-discrimination, occupational safety, emergency preparedness, sanitation, and food and housing for the labor force. Primarily, volunteer members ran the EICC for the first nine years.

EICC initially had eight members in 2004 and now that has increased to over 100 electronics companies with combined annual revenue greater than $4.5 trillion and with an employment of over 6 million people.

EICC members require their Tier I suppliers to also follow the EICC Code of Conduct. This is critical because over 3.5 million people from over 120 countries contribute to the manufacture of EICC members' products.

DISCUSSION QUESTIONS

Discuss the four major Standards listed in EICC's Code of Conduct—http://www.eiccoalition.org/media/docs/EICCCodeofConduct5_English.pdf.

BOX 5.27 US GREEN BUILDING COUNCIL (USGBC)

In 1993, Rick Fedrizzi, David Gottfried, and Mike Italiano devised a standard to promote sustainability practices in the building and construction industry. The establishment of this standard, USGBC, was a participative process wherein 60 firms and nonprofits gathered for the founding meeting.

With open discussion and sharing of knowledge and idea a rating system for green buildings was established that is now known as LEED. This rating system has gained international acclaim and acceptance and has been adapted and adopted by many countries. Using this system, hundreds of thousands of square feet per day are designed in an environmentally sound manner.

USGBC works with government, member businesses and allied organizations to support policies and programs that advance greener buildings and communities. Figure 5.1 shows a marked increase in the Green Building projects across the globe.

Organizations are increasingly opting to green their buildings because of government regulations, corporate commitments, market demand, lower operating costs, and branding.

FORMATION OF LEED BY USGBC

LEED, or Leadership in Energy and Environmental Design, is a green building certification program that guides the design, construction, operations and maintenance of buildings and large estates toward sustainability. It is based on prerequisites and credits that a project meets to achieve a certification level of Certified, Silver, Gold, or Platinum. This is a classic example of a whole business sector evolving collectively to meet environmental challenges. It has created an involuntary competitive spirit amongst building owners who are vying for the best possible certification for their premises for operational, environmental, and branding reasons.[78]

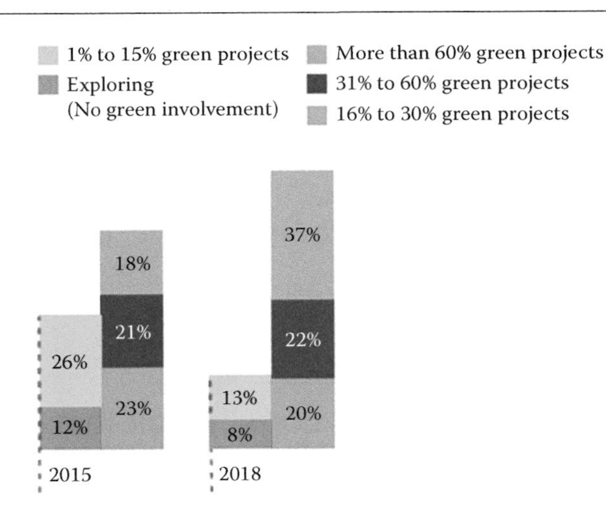

FIGURE 5.1
Increasing green building activity across the globe. (From SmartMarket Reports, World Green Building Trends, 2016, Dodge Data and Analytics. Reference link: https://www.construction.com/toolkit/reports/world-green-building. p.10. With permission, online Source: http://www.czgbc.org/Download/World%20Green%20Building%20Trends%202016%20SmartMarket%20Report%20FINAL.pdf.)

DISCUSSION QUESTIONS

- *Discuss five strategies listed under the LEED Building Design + Construction (BD + C) Rating System for the categories such as Water Efficiency, Energy and Atmosphere.*
- *Is Indoor Air Quality adequately reflected and positioned in the LEED system?*

BOX 5.28 TROPICAL FOREST ALLIANCE 2020

Tropical Forest Alliance 2020 is a global public-private partnership between governments, businesses, and civil societies that work together to ensure deforestation-free sourcing of commodities like palm oil, soy, beef, paper, and pulp through the supply chain. The overarching objectives of TFA 2020 is to conserve natural resources, protect ecosystems, reduce greenhouse gas emissions, and improve livelihoods of small-scale farmers that supply commodities to large companies.

The Alliance partners are committed to work together to share best practices, expertise, and knowledge to help commodity production while conserving tropical forests and encourage use of degraded lands and reforestation activities. The Alliance works with investors, multilateral organizations, and donors to tackle investment related issues that will help generate economies of scale and increase competitiveness within the supply chain. The Alliance has undertaken the deforestation issue in countries like Africa, Brazil, and Indonesia.[79]

The Alliance Civil Society Partners include the Climate Policy Initiative, the Forest Stewardship Council, Rainforest Alliance, UNDP, the Resources Institute, and others. Governments of the nations like Colombia, Ghana, Norway, Democratic Republic of the Congo, Netherlands, UK, and the United States are its partners, while its private sector partners include Cargill, General Mills, Kellogg's, Marks & Spencer, Mars, McDonald's, Nestle, and many others.

DISCUSSION QUESTION

- *How can we assess the performance of the Tropical Forest Alliance? What could be the indicators?*

BOX 5.29 WORLD BUSINESS COUNCIL FOR SUSTAINABLE DEVELOPMENT AND THE CEMENT SUSTAINABILITY INITIATIVE

The World Business Council for Sustainable Development (WBCSD) is a global organization that aims to help businesses collaborate and work towards developing sustainable business practices. It facilitates a network of businesses; over 200 global companies that wish to improve their environmental impact. Some of its member companies are 3M, ABB, Accenture, Acer, Loreal, Michelin, and many more. Their focus areas are promoting energy efficiency (clean fuel, efficient buildings, and transportation systems, and renewables), sustainable food, and land-use (increased food production, forests, and land protection, with soil and water management), cities, and mobility.

An example of one such industry association initiated by WBCSD is the *Cement Sustainability Initiative* (*CSI*). This initiative brought together 24 major cement producers that account for 30% of the world's cement production. The purpose is to help companies identify and execute steps for sustainable business development and develop a framework for other cement companies to get involved. The environmental issues dealt with are employee health and safety, CO_2 and climate protection, responsible use of fuels and raw material, emissions monitoring and reduction, local impacts on land, and biodiversity, and water management in the cement industry.[80]

DISCUSSION QUESTIONS

- *Discuss the following programs by WBCSD, their objectives, current initiatives and success potential:*
  - *Below 50*
  - *Circular Water Management*

BOX 5.30 BUSINESS CALL TO ACTION[81]

Business Call to Action (BCtA) was launched at the United Nations in 2008 to accelerate progress towards the Sustainable Development Goals (SDGs) by challenging companies to develop inclusive business models. The idea was to showcase how people at the base of the economic pyramid (BoP) can be engaged to improve their livelihoods. The BCtA is also a unique multilateral alliance among donor governments.

Over 170 companies, ranging from multinationals to social enterprises, and working in 65 countries, have responded to the BCtA. BCtA member companies are market leaders that provide examples of successful, profitable and scalable models for reaching poor communities and contributing to global development. The BCtA members have committed to improve the lives and livelihoods of millions in developing countries through access to markets, financial services, affordable healthcare, water and sanitation, education, and other critical services.

DISCUSSION QUESTION

- *The BCtA website has a large repository of case studies. Read these case studies and place them under a typology of interventions and business models. Draw conclusions or interpretations from this exercise.*

We discussed Eco-Industrial Park in Chapter 2 where governments played the key role. We describe in Box 5.31 examples where industry association or group of businesses have taken the lead.

BOX 5.31 ECO-INDUSTRIAL PARKS

The concept of Eco-Industrial Parks (EIP) emerged when businesses realized the economic and environmental benefits of collaborating with each other. In an EIP, a network of service or manufacturing businesses collaborates to share resources like energy, water, and raw and recycled or scrap materials. They share best operating and management practices, environmental management infrastructure, waste materials or by-products of one industry that are of value to another industry. Such synergies help improve the individual and collective economic and environmental performance of the member industries, enhances industrial competitiveness, and increases access to technology and finance.

The Kalundborg Industrial Ecosystem was the first modern EIP set up in the seaside industrial town of Kalundborg, Denmark, in 1959 when private businesses and plant managers developed an intelligent resource exchange system. Four major industries in town came together to exchange and utilize each other's resources. The industries included a 1,500 MW coal-fired power plant (Asneas Power Station), a large oil refinery (Statoil Refinery), a plasterboard manufacturer (Gyproc), and a pharmaceutical manufacturer (Novo Nordisk). Other industries and communities are also the recipients of by-products of the EIP. The power plant sells steam to Novo Nordisk, waste heat to homeowners and fisheries, sulfur dioxide scrubber sludge containing gypsum to Gyproc, fly ash and clinker for road building, and cement production.

Statoil supplied cooling water and wastewater to Asneas, where it is treated and used as boiler-feed water and supplied fuel gases to Gyproc and Asneas. Novo

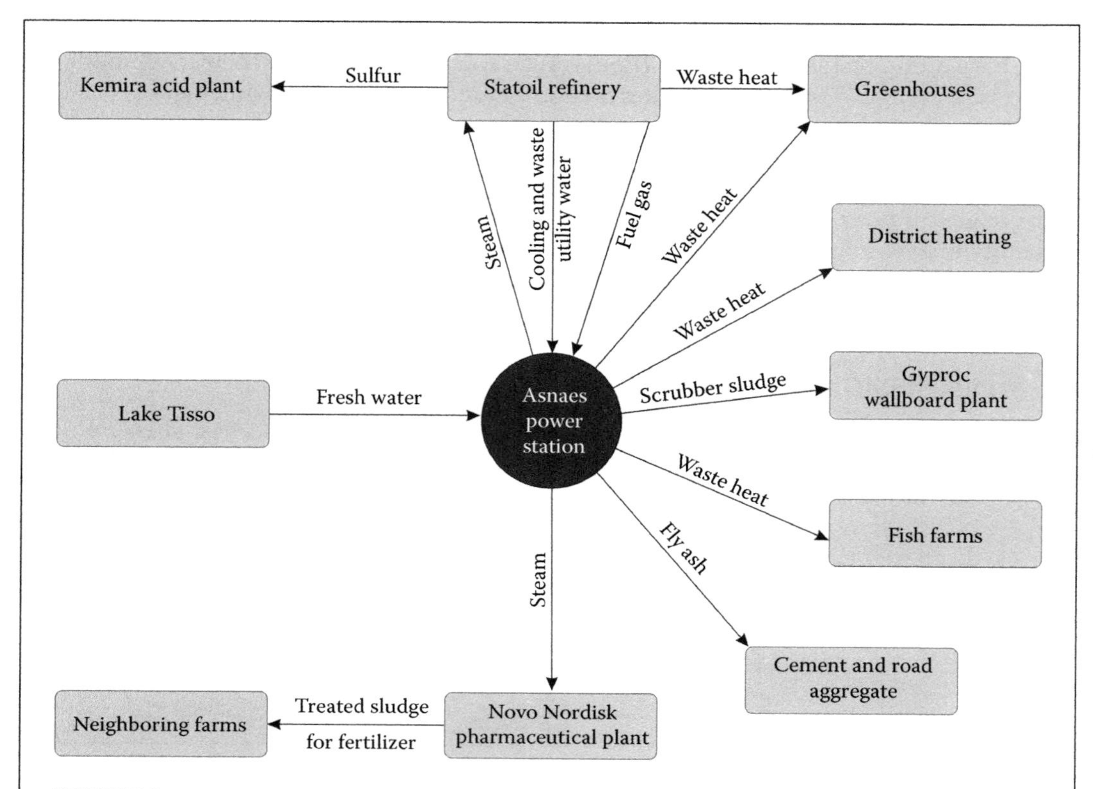

FIGURE 5.2
Industrial symbiosis at Kalundborg. (From Ehrenfeld J. and Gertler N., Journal of Industrial Ecology, Volume I, p. 70, 1997. With permission, http://dnr.wi.gov/topic/smallbusiness/documents/casestudies/industrialecologyinpractice.pdf.)

Nordisk supplied over 3,000 cubic meters of sludge per day to neighboring farms to be used as fertilizers. To facilitate this resource exchange, a network of pipe to supply water, fuel gases, and steam was built. See Figure 5.2 that presents the industrial symbiosis in Kalundborg.[82]

The industrial symbiosis had noteworthy environmental and economic impacts. Roughly 2.9 million tons of material is exchanged per year, the collective water demand went down by 25%, and the power station's water use went down by 60% due to use of recycled water. It is estimated that the industries in town realize annual cost savings of about $15 million.

DISCUSSION QUESTION

- *Cooperation and presence of the companies in the ecosystem of material flows is important for a successful symbiosis. If there are premature, early, or short-term exits by the founding members, then how will the consortium deal with the situation? How can the risks of such situations be managed?*

5.12 Leadership Examples in Sustainability

Many businesses embraced sustainability early on. These businesses saw opportunity in incorporating sustainability into their practices. Transitioning to more sustainable practices has its own set of technological and economic challenges. Some businesses took leadership to overcome these challenges and converted them into opportunities through innovative solutions. We provide examples of such leadership initiatives across different sectors. See Boxes 5.32 through 5.38.

BOX 5.32 THE WORLD'S BIGGEST RETAILERS SUSTAINABILITY INITIATIVE, THE WALMART SUSTAINABILITY INDEX

WALMART SUSTAINABILITY INDEX (WSI)[83]

Walmart, as a company, has focused on sustainability, both environmental and social. They developed the WSI to help their consumers easily understand how sustainable a product is. WSI can be used as a tool for Walmart to understand their suppliers' approach to monitoring and managing its social and environmental impact on the full lifecycle of its products.

Launched in 2009 in collaboration with The Sustainability Consortium (TSC), the WSI helps retailers and their suppliers to:

- Integrate sustainability into the business of buying and selling merchandise.
- Reduce cost, improve product quality and create a more resilient supply chain.
- Strengthen customers' trust in retailers and the brands they carry.

With over 100,000 suppliers in its network, Walmart has the ability to use the Sustainability Index program to move entire industries to perform better in terms of what the law requires, benefiting consumers, workers, and the planet. More than 700 Walmart product categories are covered under the WSI.

DISCUSSION QUESTION

- *Illustrate with two examples how Walmart has influenced the re-design of products manufactured by its suppliers.*

BOX 5.33 PUMA'S ENVIRONMENTAL PROFIT & LOSS STATEMENT

PUMA has always been an industry leader when it comes to environmental sustainability. Towards the end of 2009, PUMA decided to assess their environmental impact in monetary terms. This initiative came along when PUMA realized that it pays the regulatory bodies for services like electricity, water, and wastewater treatment, but it does not pay nature for its services like fresh water, clean air, healthy bio-diversity, and land use. So, they measured how much compensation they would have to pay nature for PUMA's manufacturing process and business operations. The results revealed that PUMA's in-house operations had done damage worth EUR 8 million to nature in 2010 alone. When the supply chain of external partners was factored in, PUMA would owe an additional EUR 137 million to nature.

This study covered all significant activities starting from production of raw materials, refining, manufacturing, offices, and logistics through to the point of sales. This impact highlighted that PUMA would have to work with its external supplier factories and raw material producers to minimize its environmental impact. In addition to tightening their commitment to minimize their environmental impact, PUMA published their analysis results. This transparency display of the environmental damage done by PUMA not only improved their brand perception amongst consumers and industry peers, but also set an industry standard for measuring corporate environmental impact.[84]

DISCUSSION QUESTIONS

- *Research and discuss the key environmentally damaging activities of PUMA that was claimed to result in an environmental damage of EUR 8 million in 2010.*
- *Read about Patagonia Don't Buy this Jacket campaign where they requested their customers to make environmentally sound purchasing decisions at the risk of losing business—http://patagonia.wpengine.com/wp-content/uploads/2016/07/nyt_11-25-11.pdf.*

BOX 5.34 APPLE: 2014's MOST ENVIRONMENTALLY FRIENDLY COMPANY

Greenpeace 2014 Green Gadgets report[85] evaluated electronic companies and products with the least environmental footprint. Apple has topped the electronics sector in addressing its environmental footprint in 2014. Some of the more interesting sustainability initiatives by Apple are included below:

- In 2009, Apple was the first company to eliminate Polyvinyl Chloride (PVC)/ Brominated Flame Retardants (BFRs) in all computer products, excluding external cables.
- In 2014, Apple was the only company to eliminate the use of PVC and BFRs in all mobile phones, PC components, including external cables. It is working without companies like HP, Dell and Acer, to lobby with the EU to restrict the use of all BFRs and PVC, to create a level playing field and elimination of hazardous substances across the whole sector.
- Apple achieved a worldwide recycling collection rate of 85% of the total weight of the products it sold between 2003 and 2010.

- Apple is incorporating renewable energy use in its supply chain. It is planning to power its new US factory for iPhone glass screens by 100% renewable energy from solar and geothermal power.
- Apple publicly discloses the carbon footprint of all its products.

DISCUSSION QUESTIONS

- *Discuss what strategies Apple uses to achieve a high (85%) recycling rate? (Hint: Read the report at https://www.apple.com/environment/reports/docs/apple_environmental_responsibility_report_0714.pdf).*
- *Trace Apple's environmental performance over the last five years based on a few key indicators. How did you decide these indicators?*

BOX 5.35 UNILEVER SET UP THE MARINE STEWARDSHIP COUNCIL

In the early 1990s, the impact of overfishing was a huge concern for bio-diversity conservationists, businesses, and coastal communities who depend on fish for their staple diet and livelihoods. Overfishing disrupts the ecological balance of fish species, eradicates threatened fish species, and damages the marine ecosystem like algae and coral reefs. Unilever also felt the impact of overfishing when its frozen fish business line was at risk.

In 1995, Unilever and the World Wildlife Fund (WWF) came together to establish the Marine Stewardship Council (MSC) to address the problem of unsustainable fishing and protect ocean health. The developed a certification program for fisheries that are well managed and sustainable. The fisheries under this program do not over-exploit fish species, do not cause harm to seabirds while fishing, and maintain the productivity and diversity of marine ecosystems that follow international laws and standards so that they are certified by MSC.

Big global businesses like IKEA, Unilever, and McDonald's have pledged to purchase and sell only MSC certified seafood. This incentivizes fisheries across the globe to become MSC certified if they wish to continue supplying to large corporations.[86]

DISCUSSION QUESTIONS

- *Is aquaculture a sustainable farming technique?*
- *If not, what measures can be taken to make it more sustainable?*

BOX 5.36 AEON ECO PROJECT FOR RETAIL OUTLETS

Based in Japan, AEON Co. Ltd. is the largest retailer in Asia. Operating since 1758, the company has expanded tremendously and now operates about 625 general merchandise stores and 274 malls in Japan and overseas.

The Environmental Foundation was established in 1990 with the aim to conserve biodiversity and prevent global warming. Under its AEON Hometown Forests Program it has planted 11,170,000 trees as of February, 2016.

In 2012, as part of its commitment towards environmental sustainability, the company launched the AEON Eco Project. It was based on three key strategies that were to be implemented by 2020:

- Reduce its energy use by 50%
- Renewable energy generation of 200,000 kW
- Establish 100 disaster prevention bases in Japan, which tends to be affected by natural disasters frequently

As of 2014, AEON has achieved 22% reduction in its energy use when compared to its 2010 energy use. It has installed renewable systems with generation capacity of 55,868 kW and has set up 21 disaster prevention locations across Japan.[87]

Under this project, AEON is working towards establishing or upgrading its existing malls into Eco malls. Some of the sustainable features of AEON malls are[88]:

- Installation of chargers for electric vehicles and plug-in hybrid cars
- Installation of Solar Power generation systems for electricity supply to the malls
- Use of Ice Thermal Storage System to power the air conditioners
- Installation of LED lighting
- Greening of the roofs and walls to reduce energy demand and improve the micro-climate
- Promote the use of public transportation
- Recycling of the waste generated at the malls
- Promoting recycling by incentivizing consumers to return used paper products in exchange for coupons that can be used towards purchase of merchandise

BOX 5.37 NEW BELGIUM BREWERY'S INTERNAL ENERGY TAX

New Belgium Brewery is committed to reducing the environmental impact of their beer production process. The company invested in a local wind energy program that supplied 100% of their electricity demand in 1999. They purchased Renewable Energy Certificates (RECs) and installed an on-site cogeneration plant.

The company introduced an Internal Electricity Tax scheme in 2013. Under this scheme, the company charges itself 2.4 cents for every unit of electricity it consumes. The revenue collected under this scheme is used for renewable or energy efficiency projects. In 2015, they used the revenue from their internal tax to install a solar photovoltaic system of capacity of 96.5 kW worth $250,000 at one of their facilities. This system in addition to the cogeneration plant and existing solar system had enabled the company to generate 19.5% of their total electricity demand through renewable sources on site.[89,90]

BOX 5.38 ENERGY EFFICIENT INITIATIVES BY SHREE CEMENT, INDIA

Based in India, Shree Cement is among the largest five cement plants in the country and has incorporated multiple energy efficiency strategies. It is the first cement company in the world to have an EN 16001 certification, a European standard for Energy Management Systems that includes continuous monitoring of the plants energy use. The captive power plants use biomass as fuel and use bottom ash as fuel for production of cement clinker material, which is a highly energy intensive process due to the high temperatures required for heating clinker. In addition to this, use of clinker (clinker is lumps of limestone and clay) is reduced with increased content of fly ash in the cement. This has a significant impact on the reduction of carbon emissions because clinker production is the main source of CO_2 emission in cement production. Due to this initiative, Shree Cement obtained 0.45 million Certified Emission Reduction (CER) units for "optimal utilization of clinker."[91]

 **DISCUSSION QUESTIONS**

- *Discuss the PDCA (Plan-Do-Check-Act) cycle of the EN 16001 certification.*
- *Discuss the advantages of use of fly ash in the cement industry.*

Endnotes

1. Based on: IEA Statistics, CO_2 Emissions from Fuel Combustion—Highlights, International Energy Agency/OECD, p. 10, 2015, online source: https://www.iea.org/publications/freepublications/publication/CO2EmissionsFromFuelCombustionHighlights2015.pdf
2. For further reading: The Global Flightplan: Reducing Emissions from the International Aviation Sector through carbon-neutral growth from 2020, *UNFCC COP19 Climate Talks,* Warsaw, An information paper presented by the Air Transport Action Group, 2013.
3. Based on: British Airways, 2009/10 Annual Report and Accounts, pp. 37–38, online source: https://www.britishairways.com/cms/global/microsites/ba_reports0910/pdfs/Environment.pdf
4. Online source: http://responsibleflying.ba.com/wp-content/uploads/RFFE_2014_Summary_online.pdf
5. Based on: Fleet Facts: Boeing 787-9, British Airways, online source: http://www.britishairways.com/en-gb/information/about-ba/fleet-facts/boeing787-9
6. For further reading: The first ever round-the-world solar flight at a glance from March 2015–August 2016, Solar Impulse: Around the World in a Solar Airplane, 2016, online source: https://www.solarimpulse.com/img/pdf/SI_RTW_FACTSHEET_PRESS.pdf
7. Based on: Our Adventure, *Solar Impulse: Exploration to Change the World,* 2016, online source: https://www.solarimpulse.com/adventure
8. Based on: Zero-Fuel Aircraft, *The Airplane of Perpetual Endurance, Solar Impulse: Exploration to Change the World,* 2016, online source: https://www.solarimpulse.com/adventure#zero-fuel-aircraft
9. Based on: What's Next?, *Solar Impulse: Exploration to Change the World,* 2016, online source: https://www.solarimpulse.com/
10. Based on: Alternative Fuels Data Center (ADFC), Biodiesel Benefits and Considerations, U.S. Department of Energy: Energy Efficiency & Renewable Energy, 2016, online source: http://www.afdc.energy.gov/fuels/biodiesel_benefits.html

11. For further reading: Balagopal, B., Paranikas, P. and Rose, J., *What's Next for Alternative Energy*, The Boston Consulting Group, 2010, online source: https://www.bcgperspectives.com/content/articles/energy_environment_sustainability_whats_next_for_alternative_energy/
12. Based on: Born Electric: An Overview of the BMW I Models, BMW, online source: http://www.bmw.com/com/en/newvehicles/i/overview.html
13. Based on: Sustainability is the Answer that calls everything into question: The BMW i3—sustainability thought through from being top end, BMW, online source: http://www.bmw.com/com/en/newvehicles/i/i3/2016/showroom/sustainability.html
14. Based on: Alternative Fuels Data Center (ADFC), Charging plug-in electric vehicles at home, U.S. Department of Energy: Energy Efficiency & Renewable Energy, 2016, online source: http://www.afdc.energy.gov/fuels/electricity_charging_home.html
15. Based on: Mercedes Demonstrates First Automative Use of CFC-Free Air Conditioning System New Refrigerant to Replace Ozone-Depleting CFCs, PR Newswire Association LLC, 1992, online source: https://www.thefreelibrary.com/MERCEDES+DEMONSTRATES+FIRST+AUTOMOTIVE+USE+OF+CFC-FREE+AIR...-a011742964
16. Based on: SNAP Regulations, Significant New Alternatives Policies (SNAP), EPA: United States Environmental Protection Agency, 2016, online source: https://www.epa.gov/snap/snap-regulations
17. For further reading: Seidel, S., Ye, J. and Andersen, S.O., Technological change in the production sector under the Montreal Protocol, C2ES: Center for Climate and Energy Solutions, 2015, online source: http://www.c2es.org/publications/technological-change-production-sector-under-montreal-protocol
18. For further reading: Sciance, F., The Transition from HFC-134a to a Low-GWP Refrigerant in Mobile Air Conditioners—HFO-1234yf, General Motors Public Policy Center/EPA, 2013, online source: https://www.epa.gov/sites/production/files/2014-09/documents/sciance.pdf
19. Based on: Jung, P., Evaluating the Carbon-Reducing Impacts of ICT, *bcg perspectives*, 2010, online source: https://www.bcgperspectives.com/content/articles/technology_software_sustainability_evaluating_carbon_reducing_impacts_of_ict/
20. Based on: Philips Lighting takes light beyond illumination, Light + Building, 2016, online source: http://www.philips.com/a-w/about/news/archive/standard/news/press/2016/20160314-Philips-Lighting-takes-light-beyond-illumination.html
21. Based on: Philips Lighting launches new LED family of street lights future proofed by design for Internet of Things age, Philips, 2016, online source: http://www.philips.com/a-w/about/news/archive/standard/news/press/2016/20160310-Philips-Lighting-launches-new-LED-family-of-street-lights-futureproofed-by-design-for-Internet-of-Things-age.html
22. Based on: Smart 2020: Enabling the low carbon economy in the information age—United States Report Addendum, Global e-Sustainability Initiative, pp. 22–30, 2008, online source: https://www.bcgperspectives.com/Images/BCG_Report_Enabling_low_carbon_Economy_inthe_Information_Age_Nov2008[1]_tcm80-23231.pdf
23. Based on: E-health Croatia: Life cycle assessment of ICT enablement potential, Ericsson, 2017, online source: https://www.ericsson.com/article/e-health_croatia_1633598732_c
24. For further reading: Hamilton, E., Bringing Work Home: Advantages and Challenges of Telecommuting, Boston College, p. 27, 2016, online source: http://www.bc.edu/content/dam/files/centers/cwf/research/publications/pdf/BCCWF_Telecommuting_Paper.pdf
25. For further reading: Evaluating the carbon-reducing impact of ICTs: An assessment of methodology, Global e-Sustainability Initiative and The Boston Consulting Group, p. 41, 2010, online source: https://www.bcgperspectives.com/Images/Evaluating_the_carbon_reducing_impacts_of%20ICT_Sept_2010_tcm80-59352.pdf
26. For further reading: Kramer, K.J., Masanet, E., Tengfang, X. and Worrell, E., *Energy Efficiency Improvement and Cost Saving Opportunities for the Pulp and Paper Industry—An ENERGY STAR Guide for Energy and Plant Managers*, Energy Analysis Department, Environmental Energy Technologies Division, Ernest Orlando Lawrence Berkeley National Laboratory, U.S.

Environmental Protection Agency, 2009, online source: https://www.energystar.gov/ia/business/industry/downloads/Pulp_and_Paper_Energy_Guide.pdf
27. For further reading: Case Studies of Market Making in the Bioeconomy, EC- Europa, online source: https://ec.europa.eu/research/bioeconomy/pdf/13-case-studies-0809102014_en.pdf
28. Based on: Coating Starches, Cargill—Helping the World Thrive, 2017, online source: http://www.cargill.com/products/industrial/papermaking/coating/index.jsp
29. Based on: Kinsella, S., Tree Free Papers: Vision Paper, Conservatree, 2003, online source: http://www.conservatree.org/paper/PaperTypes/Rymsza.shtml
30. Based on: Pooper Making, haathi chaap—elephant poo paper, online source: http://elephant-poopaper.com/making1.html
31. Based on: Epson Develops the World's First Office Papermaking System that Turns Waste Paper into New Paper, EPSON, 2015, online source: http://global.epson.com/newsroom/2015/news_20151201.html
32. For further reading: Stajanca M. and Estokova A., Environmental Impacts of Cement Production, Technical University of Kosice, 2012, online source: http://ena.lp.edu.ua:8080/bitstream/ntb/16692/1/55-Stajanca-296-302.pdf
33. Based on: Innovative Waste Management, Holcim Leadership Journey: AFR, 2015, online source: http://www.holcim.com/fileadmin/templates/CORP/doc/Sustainable_Dev/Extract_from_AR_-_AFR.pdf
34. Based on: The GTZ-Holcim Public Private Partnership, *Guidelines on Co-Processing Waste Materials in Cement Production*, The GTZ-Holcim Public Private Partnership, Holcim Support Ltd and Deutsche Gesellschaft fur Technische Zusameenarbeit (GTZ) GmbH.
35. Based on: Co-processing, 2011, Geocycle, Holcim, 2006, online source: http://www.geocycle.vn/solutions/co-processing.html
36. Based on: Hayes, P., Geocycle & Waste Processing in Cement Kilns, Holcim, 2011, online source: http://vietnamsupplychain.com/assets/upload/file/publication/1325666888479-215.pdf
37. Based on: Products & Services, Earth Check, 2017, online source: https://earthcheck.org/products-services/
38. Based on: International Tourism Partnership: About us, online source: http://www.ihei.org/about-us/
39. Based on: NOAA: National Oceanic and Atmospheric Administration, Ecotourism Principals for Cousin Island Special Reserve, Seychelles, Handout 10.1—Case Study: Addressing Carrying Capacity, Sustainable Tourism, online source: http://sanctuaries.noaa.gov/management/pdfs/sustain_tour_mod10_ho_10_1.pdf
40. Based on: Seychelles Islands: Ecotourism, a Success Story!, Indian Ocean, 2014, online source: http://www.indian-ocean.com/seychelles-islands-ecotourism-a-success-story/
41. Based on: Cousin Island, Nature Seychelles, 2014, online source: http://www.natureseychelles.org/visiting/cousin-island
42. Based on: Shah, N., From Coconuts to Conservation, Cousin Island Special Reserve—An Innovative Approach to Conservation and Ecotourism, *Ashoka Changemakers: Geotourism Challenge 2010: Places on the Edge—Saving Coastal and Freshwater Destinations*, 2010, online source: https://www.changemakers.com/coasts/entries/coconuts-conservation-cousin-island-special-reserve
43. Based on: Mikaela, Recycle Your E-Waste Responsibly with #TechTakeback, UBER Newsroom, 2015, online source: https://newsroom.uber.com/us-new-york/recycle-your-e-waste-responsibly-with-techtakeback/
44. Based on: Caffarey M., Umicore Precious Metals Refining: A Key Partner in Closing the Life Cycle of EEE (Electrical and Electronic Equipment), Umicore, USA: SERDC, 2012, online source: https://www.serdc.org/Resources/Documents/Summit%20Presentations/SERDC%20Summit%20Presentation%20-%20Mark%20Caffarey.pdf
45. Based on: Recycling, Umicore, online source: http://www.umicore.com/en/industries/recycling/

46. Based on: Precious Metals Refining, Umicore, online source: http://www.preciousmetals.umicore.com/recyclables/eScrap/process/
47. For further reading: Yamamoto M. and Hosoda E., *The Economics of Waste Management in East Asia*, Routledge, p. 108, 2016.
48. Based on: The P&O Nedlloyd Award for Infrastructure, The Winner: Wongpanit Garbage Recycle Separation Plant, WorldAware, 2004, online source: http://www.worldaware.org.uk/awards/awards2004/wongpanit.html
49. For further reading: Recovery Status Report: The Great East Japan Earthquake 2011—Case Studies, International Recovery Platform, 2013, online source: http://www.recoveryplatform.org/assets/irp_case_studies/ENGLISH_RECOVERY%20STATUS%20REPORT%20JAPAN_revised%202014.3.27.pdf
50. Based on: Nepal: DWR Starts Demolition Training of Trainers, Disaster Waste Recovery, 2015, online source: http://www.disasterwaste.org/nepal-dwr-starts-demolition-training-of-trainers/
51. Based on: DWR in Nepal, Disaster Waste Recovery, 2014, online source: http://www.disasterwaste.org/dwr-in-nepal/
52. For further reading: Berg P., Bjerregaard M. and Jonsson L., *Disaster Waste Management Guidelines*, UNEP/OCHA, United Nations Office for the Coordination of Humanitarian Affairs, Environmental Emergencies Section, 2011, online source: https://docs.unocha.org/sites/dms/Documents/DWM.pdf
53. Based on: Furlong, H., Trending: New Materials, Traceability Could Help Mould Circular Economy for Plastics, *Sustainable Brands: The Bridge to Better Brands*, 2016, online source: http://www.sustainablebrands.com/news_and_views/next_economy/hannah_furlong/trending_new_materials_traceability_could_help_mould_circ
54. Based on: Embree, K., Project gets funding for invisible markers to improve plastics recycling quality, *Plastics Today: Community for Plastics Professionals*, 2016, online source: http://www.plasticstoday.com/packaging/project-gets-funding-invisible-markers-improve-plastics-recycling-quality/8630306024196?cid=nl.x.plas08.edt.aud.plas2day.20160226
55. Based on: Aragon, B., 7 Cosmetic ingredients that are bad for the environment, *Cinco Vidas: Setting the Standard for Safe Self-Care*, 2016, online source: http://cincovidas.com/7-cosmetic-ingredients-that-are-bad-for-the-environment/
56. For further reading: Putrich, G.S., Microbeads to be phased out in US starting in 2017, *Plastic News*, 2015, online source: http://www.plasticsnews.com/article/20151229/NEWS/151229895/microbeads-to-be-phased-out-in-us-starting-in-2017
57. Based on: Selier, J., Industry Responds to US Microbead Ban, *Beat the Micro Bead*, 2016, online source: https://www.beatthemicrobead.org/en/industry
58. Based on: Water lovers, the Biotherm initiative to preserve the planet's water resources and aquatic biodiversity, *L'Oreal: Sharing Beauty with All*, 2016, online source: http://www.sharingbeautywithall.com/en/living/raising-consumer-awareness-about-living-sustainably/water-lovers-biotherm-initiative-preserve-planets-water-resources-and-aquatic-biodiversity
59. Based on: Ahlquist, M. Recycling our Black Pots, *Lush*, 2017, online source: http://www.lush.ca/en/Stories-Article?cid=article_recycled-black-pot
60. For further reading: Beauty Faces up to Sustainability, *The Sustainability Consortium*, Quantis, 2016, online source: https://www.sustainabilityconsortium.org/tsc-news/quantis-special-report-beauty-faces-up-to-sustainability/
61. Based on: Teeth: For fresh breath and a healthy smile, *Lush*, online source: http://www.lushusa.com/face/teeth/
62. Based on: *Air + Greenhouse Gas → Thermoplastic*, Newlight Technologies, 2017, online source: https://www.newlight.com/
63. For further reading: Brackley, A. and York, B., What's Next for Business? Sustainability Trends for 2017, *SustainAbility*, 2017, online source: http://radar.sustainability.com/annual-trends-report/
64. For further reading: McGrath, C., Why Kraft Food cares about fair trade chocolate, *MIT Sloan: Management Review*, 2012, online source: http://sloanreview.mit.edu/article/why-kraft-foods-cares-about-fair-trade-chocolate/

65. Based on: Kraft Foods Cocoa Sustainability Overview, *kraft foods*, 2009, online source: http://phx.corporate-ir.net/External.File?item=UGFyZW50SUQ9MTkyMDR8Q2hpbGRJRD0tMXxUeXBlPTM=&t=1
66. Based on: Coffee sustainability: Kraft Foods and Rainforest Alliance, *Crossroads' Global Hand*, 2017, online source: http://www.globalhand.org/en/search/success+story/document/28858
67. Visit https://furtherwithfood.org/
68. Based on: Rockefeller Foundation, USDA, EPA to Create Center for Action Against Food Wastage, *Sustainable Brands: The Bridge to Better Brands*, 2016, online source: http://www.sustainablebrands.com/news_and_views/collaboration/sustainable_brands/rockefeller_foundation_usda_epa_lead_creation_center
69. Based on: Refining the Future of Growth: The New Sustainability Champions, *World Economic Forum*, 2011, online source: http://www3.weforum.org/docs/WEF_GGC_SustainabilityChampions_Report_2011.pdf
70. Based on: Ross, A., Congo cobalt mined by children may be in your mobile phone—Amnesty, *Reuters*, 2016, online source: http://uk.reuters.com/article/uk-electronics-congodemocratic-idUKKCN0UX1B1
71. Based on: Exposed: Child labor behind smart phone and electric car batteries, *Amnesty International*, 2016, online source: https://www.amnesty.org/en/latest/news/2016/01/Child-labour-behind-smart-phone-and-electric-car-batteries/
72. For further reading: *OECD Due Diligence Guidance for Responsible Supply Chains of Minerals from Conflict-Affected and High-Risk Areas*, second edition, OECD Publishing. 2013, online source: https://www.oecd.org/corporate/mne/GuidanceEdition2.pdf
73. Based on: Grameen Veolia Provides Safe Drinking Water for All, Grameen Viola Water Ltd, 2017, online source: http://www.grameenveoliawaterltd.com/index.php?p=action
74. Based on: Grameen BASF, 2011 Yunus Center, online source: http://www.muhammadyunus.org/index.php/social-business/grameen-basf
75. For further reading: Grameen Shakti (Bangladesh), 2012 Green Economy Coalition, online source: http://www.greeneconomycoalition.org/glimpses/grameen-shakti-bangladesh
76. Based on: Khetriwal D. et al., A comparison of electronic waste recycling in Switzerland and in India, 2005, online source: http://www.ewasteguide.info/files/Sinha-Khetriwal_2005_EIAR.pdf
77. Based on: The EICC is the world's largest industry coalition dedicated to electronics supply chain responsibility, EICC, 2017, online source: http://www.eiccoalition.org/about/members/
78. Taken from: SmartMarket Reports, World Green Building Trends, McGraw-Hill Construction, 2013, online source: http://www.gbcsa.org.za/wp-content/uploads/2013/06/WGBC-Trends-Report_2013.pdf
79. Based on: Objectives, Tropical Forest Alliance 2020, World Economic Forum, 2017, online source: https://www.tfa2020.org/about-tfa/objectives/
80. Based on: Cement Sustainability Initiative, Key Sustainability Issues, *Cement Sustainability Initiative, World Business Council for Sustainable Development*, 2015, online source: http://www.wbcsdcement.org/index.php/key-issues
81. Visit http://www.businesscalltoaction.org/
82. Taken from: Ehrenfeld J. and Gertler N., Industrial ecology in practice, the evolution of interdependance at Kalundborg, *Journal of Industrial Ecology*, Vol. I, p. 69, 1997, online source: http://www.johnehrenfeld.com/Kalundborg.pdf
83. Based on: Sustainability Index, Walmart, online source: http://corporate.walmart.com/global-responsibility/environment-sustainability/sustainability-index-leaders-shop
84. Based on: PUMA's Environmental Profit and Loss Account for the Year Ended 31 December 2010, PUMA, 2010, online source: http://about.puma.com/damfiles/default/sustainability/environment/e-p-l/EPL080212final-3cdfc1bdca0821c6ec1cf4b89935bb5f.pdf
85. Based on: Greenpeace, Green Gadgets: Designing the Future—The Path to Greener Electronics, 2014, online source: http://www.greenpeace.org/international/Global/international/publications/toxics/2014/Green%20Gadgets.pdf

86. Based on: 2010-2015: New Horizons, Marine Stewardship Council: Certified Sustainable Seafood, 2010–2015, online source: https://www.msc.org/about-us/our-history/2010-2015-new-horizons
87. For further reading: Aeon Environmental and Social Report 2015, 2015, online source: https://www.unglobalcompact.org/system/attachments/cop_2015/189821/original/AEON_Environmental_and_Social_Report.pdf?1442556777
88. Based on: AEON Mall's Approaches to CSR, AEON, online source: http://www.aeonmall.com/en/csr/environment.html#po01-1
89. Based on: New Belgium Brewing Case Study, *Business & Sustainable Development Commission,* 2016, online source: http://report.businesscommission.org/case-studies/new-belgium-brewing-case-study
90. Based on: New Belgium Adds 50 Percent More Onsite Solar Capacity, *BrewBound: Craft Beer News, Events & Jobs,* BevNET.com, 2014, online source: http://www.brewbound.com/news/new-belgium-adds-50-percent-more-onsite-solar-capacity
91. For further reading: Refining the Future of Growth: The New Sustainability Champions, World Economic Forum and Boston Consulting Group, 2011, online source: http://www3.weforum.org/docs/WEF_GGC_SustainabilityChampions_Report_2011.pdf

6

Community Response

6.1 Roles Played by the Community

Putting sustainability in practice requires a multi-level and multi-stakeholder approach. We need to involve the government, financing institutions, business, and communities, to develop partnerships. In Chapters 2, 3, and 4, we saw several examples where these partnerships have resulted in better outcomes leading to innovation. These examples also show how we could collectively move towards the attainment of Sustainable Development Goals (SDGs).

It is important that communities are recognized as an important foundation to demonstrate, up-scale, and replicate the changes we want to see in society, in the way that we live and govern. If sustainability is to be mainstreamed in practice, then communities must be empowered to play a central role. Figure 6.1 shows levels and types of interaction between community and key stakeholders towards sustainability.

The community plays a unique role in ensuring the linkages between various stakeholders. As an entity that is directly impacted by the damage to the ecosystem and depletion of resources, the community acts as an *informal regulator* by putting pressure on the government, businesses, and financing institutions. In many ways, communities ensure that their activities are in compliance, in harmony, and sustainable. Apart from the *watchdog* function, the community plays an important role as a *facilitator* as well as *developer* on a proactive basis. Both these roles are important. In this chapter, we will illustrate these roles with examples.

The term community can include entities like NGOs, Community Based Organizations (CBOs) or the local citizen groups, individuals or activist movements that support a cause, students, social entrepreneurs and professionals that belong to a neighborhood, town, city, or region.

NGOs/CBOs are one of the key entities of the community. NGOs act like *nodes* that interact with other stakeholders like the government, businesses, and financing institutions as well as the informal and under-privileged sectors of the community.

Through formal and informal mechanisms like city/village hall meetings, volunteer organizations, petitions and protests and demonstration pilots (to lead by example), the community influences government policies, business activities, and investments. It seeks transparency, participation, environmental or social justice, and an inclusive growth.[1]

6.1.1 Community and Government

As discussed in Chapter 2, national governments started factoring environmental sustainability as an important perspective in the overall economic and social development.

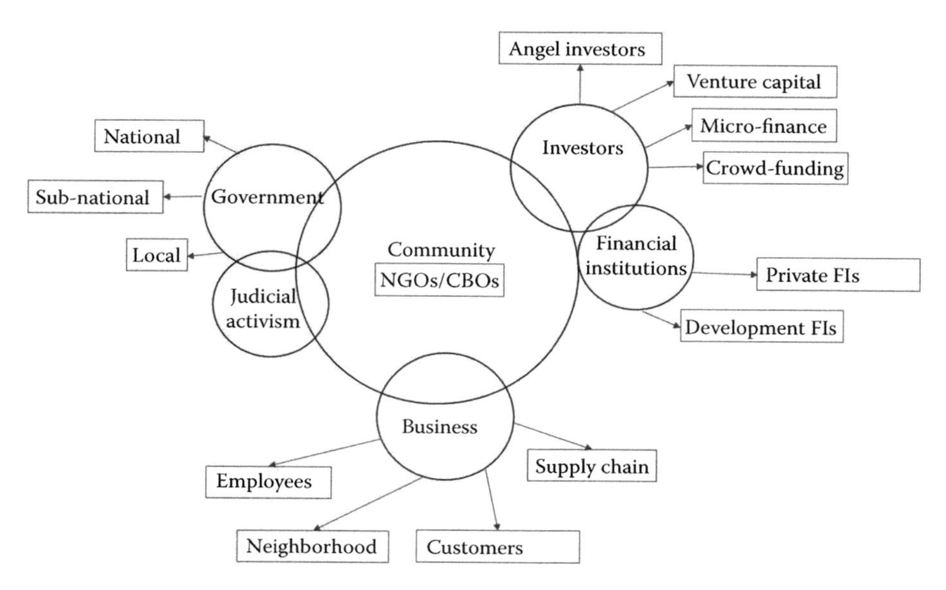

FIGURE 6.1
Levels of interaction between community and key stakeholders towards sustainability.

Efforts were undertaken to increase the per capita income of the country along with a better quality of life for communities on an inclusive basis. In this endeavor, it became very important that communities and the government work together.

The government has the authority to formulate, introduce, and amend policies and regulations. It has access to funds, infrastructure, and human resources and has the experience to execute programs at various scales. The community, on the other hand, is knowledgeable about the local issues, the situation on the ground, and importantly has access to the people. Communities can also give the government feedback on the effectiveness of the sustainability programs. If the communities find the policies amiss or not in line with the priorities, then they appeal to the government to introduce, modify, or revoke such policies. Community response is one of the cornerstones in putting environmental policy into practice and investing in infrastructure.

Several governments have supported programs in the interest of communities by giving them a platform to participate, as well as to voice their opinions and suggestions. These programs include information sharing to help involve and empower the community for making decisions.

In Chapter 2, we looked at how the Government of India established the National Green Tribunal (NGT) as a platform for various stakeholders including the community to seek environmental justice. Indonesia's PROPER program was also discussed in Chapter 2 as an example of an information-based instrument to enhance public transparency regarding the environmental performance of businesses. An example of market-based instruments was the introduction of a Green Credit Card scheme in Korea by the Ministry of Environment. The Green Credit Card incentivized consumers with rewards for saving on resources like tap water, electricity, and gas heating.[2] The EU established a Directive 2002/14/EC as a general framework for informing and consulting employees at an enterprise level on issues like the enterprise's activities, undertakings, economic situation, management structure, any changes in organization, and contractual relations. This empowered employees to ensure that their rights and interests are protected.[3]

6.1.2 Community and Financing Institutions

In Chapter 3, we introduced two types of Financing Institutions (FI): viz., Private Sector FI (PSFI) and Development FI (DFI). DFIs such as the World Bank and Asian Development Bank (ADB) have been providing support to governments for development by offering grants or concessional loans and providing technical assistance. DFIs use a partnership approach through Country Partnership Strategies and Country Partnership Frameworks[4] over five years that often involves community consultation at national level. DFIs follow Environmental Social Safeguards where they compare *global standard* or *best practices* with the national standards, identify gaps, and support required reforms and capacity development.

In Chapter 3, we saw that the DFIs have influenced the environmental and social governance of developing nations in this process. The PSFIs, on the other hand, have influenced the commercial banks and the businesses to ensure that sustainability considerations are incorporated in making investment decisions.

The community came into the picture when the DFIs started looking for development effectiveness. As a first step, resettlement and rehabilitation of the affected community members were considered important factors for mitigating the adverse social impacts. Introduction of measures for social development on an inclusive basis was the next step focusing on poverty reduction and sustainable livelihoods. Insulating communities from climate change related risks through adaptation measures is an important consideration today.

The PSFIs feared community backlash and came up with measures that ensured that the community was well engaged in their transactions and were provided with adequate compensation and employment. In some cases, the PSFIs provided support for entrepreneurship through micro-finance. To PSFI, credit and reputational risks were important considerations. The Equator Principles (EP) discussed in Chapter 3 included the required social dimension. PSFIs following EPs started engaging with the communities, improving labor conditions on health and safety, supporting supply chains and improving conditions of the neighborhoods. More recently, the era of impact investments has begun, where the impact investors did not mind a lower Return on Investment (RoI) if there was a good social impact. Opportunities for Conservation Finance have also arisen that look at conservation of nature along with profits from ecosystem services. We have discussed in Chapter 3 both impact funds and conservation finance.

6.1.3 Community and Business

Businesses primarily focus on making profits and rightly so. However, in the process of operating successful businesses, there are instances of irreversible damage to the ecosystems and risks to the human population, especially to vulnerable groups.

Businesses are now paying more attention to working for and with the community. One of the mechanisms developed to promote business and community interaction is Corporate Social Responsibility (CSR) (also discussed in Chapter 5). Under CSR, businesses make an attempt to incorporate sustainability into their core business practices. In Chapters 4 and 5, we illustrated several examples of business and community partnerships that helped communities, labor, and supply chains.

To empower consumers to make well-informed choices and decisions, the market developed eco-labels as information-based tools. Eco-labels promoted greener products and services as well as served as a driver towards branding and innovations. Here, the business, retailers, and consumer-based organizations took the lead with the help of national governments. In Chapters 2 and 4, we presented eco-labels for sectors such as textiles, food, and everyday home products such as electronics. Apart from the eco-labels, policies like Fair Trade, Fair Wages, and *Codes of Conduct* were adopted to protect the interests of employees, and local suppliers as well as the consumers.

Such sustainable business initiatives did not come about so easily. It was public pressure and increasing consumer and investor demands that brought about this change. This led to a gradual shift towards greener products and services. At the same time, many national governments developed mechanisms for regulating green products and services with incentives and disincentives. A section of communities showed their willingness to pay a higher or premium price for the green products in the interest of sustainability.

6.2 Community and Government

We introduce in this section various roles played by communities towards sustainable development in partnership with the government. We begin with examples of innovative models used in engaging with the communities. We then illustrate examples of community involvement in urban planning followed by the role played by the communities in the conservation of nature. Finally, we showcase examples of communities getting involved in environmental monitoring and end the section with examples of judicial activism.

6.2.1 Innovative Mechanisms for Involving the Communities

There could be several innovative mechanisms to involve communities apart from traditional approaches of holding public seminars, organizing events, and conducting workshops. We illustrate in Box 6.1 three interesting and innovative case studies on community participation and involvement in the UK, United States, and Canada.

Creative ways of communication have been gaining popularity owing to the rise of social media and a creative line of thoughts along with environmental awareness. Artists are using their art forms to create awareness amongst the community, local and international, regarding issues that are posing a threat to our sustainable future. Graffiti and street art are such powerful platforms used to reach out to the public in urban and semi-urban areas via physical or online sources.

Dr. Love, a famous Georgian artist, is known as the crusader of communicating on urban air pollution. Air pollution in cities is quite an important issue nowadays, since more and more green areas have disappeared and have been substituted with buildings. Moreover, rapid increase in the number of cars has worsened the situation significantly, practically leaving the city without fresh air. The artist has expressed his protest toward air pollution in the city through his paintings. Many have appreciated Love's work to confront air pollution as well as to make our society think about the problem through his art.[11]

In July 1969, photographer Mark Edwards was lost on the edge of the Sahara Desert, and was rescued by a Tuareg nomad. This changed Mark's life. Bob Dylan's songs "A Hard Rain's A-Gonna Fall," "Sad Forests," "Dead Oceans," "Where the People are Many and their Hands are All Empty" had led a touching impression to the World. The World seemed to be on fire.

Edwards thought of illustrating each line of the song "A Hard Rain's A-Gonna Fall." In the years that followed, Mark traveled around the world on assignments that allowed him to capture the photographs. The result became *Hard Rain*, an outdoor exhibition, book and DVD that brought global challenges alive in a moving and unforgettable way. See Box 6.2 for details.

Dr. Leah is an award-winning composer, sound artist and creative producer working at the intersection of art, science, and technology. Dr. Leah realized the opportunities for hydrophone recordings as a measure to understand the *health* of the river. The soundscapes of rivers generated by a hydrophone can expose many insights on the active aquatic life. She found that the polluted and stagnant waterways were silent, often with a hum of anthropogenic sound from boats and machinery on the riverbanks. The healthy waterways exhibited sounds of species ranging from dolphins, fish, and turtles to shrimp and insects.[12]

Box 6.2 describes the Edwards' exhibition *Hard Rain* and Dr. Leah's work on River Listening.

BOX 6.1 EXAMPLES ILLUSTRATING INNOVATIVE MECHANISMS FOR INVOLVING THE COMMUNITIES

POP-UP PUBLIC PARTICIPATION WORKSHOPS ON THE THAMES RIVER IN THE UK

The Your Tidal Thames (YTT) Project has taken an innovative approach to stakeholder engagement, with *pop-up* workshops, where the engagement is taken to the community.[5] YTT was an initiative by the Department of the Environment, Food, and Rural Affairs (Defra) in the UK and ran for one month, between January and December 2012. This was a joint project between Thames21 and Thames Estuary Partnership (TEP), Thames Strategy Kew to Chelsea, and Thames Landscape Strategy.[6]

YTT was based on a community approach towards river management. Instead of the traditional workshops, the community was engaged on their own terms, in a new and entertaining way. The YTT team set up pop-up workshops at local events being held in the area, such as volunteering days and local community fairs. Through posters and leaflets about the purpose of the river catchment plan, community members were invited to ask questions, make suggestions, or fill out short forms with their ideas about what issues should be addressed. These ideas were then taken into consideration, mapping the issues for making decisions.

NYC CHANGEBYUS IN NEW YORK

ChangeByUs is an online platform, where New Yorkers can submit their ideas to make the city more sustainable.[7] The Program was launched in 2011 and is run by the New York City's Office of the Mayor, with participation from the Citizens Committee for New York City, Pratt Center for Community Development, ioby, and the United States Forest Service.[8]

Based on the ideas submitted, the website connects with the community members, and invites them into various project groups. These project groups are then connected to the city's existing resources and community organizations to help them achieve their goal of making NYC a greener and better place to live.

TALK GREEN TO US IN VANCOUVER, CANADA

The *Talk Green to Us* campaign was launched as a means to meet the city's goal of being the *Greenest City in the World* by 2020.[9] An online forum was created, where residents could submit their ideas of how Vancouver can reach its targets, and the community can vote for their preferred ideas.

The idea that had the most number of votes was to make the city pedestrian friendly, as well as to increase the travel done by foot and bike. It was proposed that a complete cycling network is to be established that is safe and attractive to all, including children, seniors, and novice cyclists. The Greenest City Planning Team (GCPT) acknowledged this idea. Another idea that was suggested was to ban bottled water and install public water fountains instead, where people can easily refill their own water bottles. Only four months after this idea was suggested, the GCPT launched a free app which showed the locations of drinking water fountains and bottle refill stations throughout the city.[10]

 **DISCUSSION QUESTION**

- *Think of issues faced by your city such as river pollution, waste management, air quality, etc. Prioritize these issues. Come up with ideas on how to engage the community in an innovative and effective manner to address these issues.*

BOX 6.2 EXHIBITION *HARD RAIN*[13]

The *Hard Rain* exhibition is an example of fusion between music and visuals. It has been seen by over 15 million people at over 50 venues on every continent. It is a very successful photographic display that attracts huge public and critical acclaim, along with the support and endorsement of political and environmental leaders across the world. Mark's photographs illustrate every line of Dylan's prophetic songs, setting the scene for a moving and unforgettable exploration of the state of our planet at this critical time. Hard Rain puts the puzzle together to show that there is *one* problem: aligning human systems with natural systems.

RIVER LISTENING[14]

Sound is so closely associated with the state of the environment. According to Dr. Leah, bioacoustics and acoustic ecology have emerged as extremely valuable technologies for non-invasive environmental monitoring. The technique essentially involves auditory recordings of the aquatic environment. Scientists have now developed advanced software tools for species recognition, based on the hydro-acoustics. There is a need to consolidate these software tools for recognition and explore the value of *listening to the data* in innovative ways. The data collected can also be made available, especially to the tourists and citizen scientists through digital technology, social networking, and creative collaborations.

DISCUSSION QUESTIONS

- *Research other examples of creative communication on the environment as practiced by communities.*
- *Prepare a compendium of such efforts as a class or group assignment.*

6.2.2 Participatory Urban Planning

The Town of Antigonish in Nova Scotia, Canada, set sustainability as a vital goal for the next 25 years and beyond. Community leaders recognized that Antigonish could be considered as a pilot community in Nova Scotia for implementing a comprehensive sustainability framework. Box 6.3 shows an example of an Integrated Community Sustainability Plan (ICSP) that was developed for Antigonish. The ICSP demonstrated how communities can play a proactive role to develop an action plan on sustainability.

BOX 6.3 INTEGRATED COMMUNITY SUSTAINABILITY PLAN AT THE TOWN OF ANTIGONISH, NOVA SCOTIA, CANADA[15]

In May of 2007, the Town of Antigonish began an extensive process to further sustainability initiatives in the Antigonish area. The initiative was intended to provide the community with the tools needed to consider the future and to engage businesses, institutions, planners, developers, public administrators, and the community to develop a plan based on principles, community values and achievable options for action.

The activities under ICSP were identified through geographical and digital outreach, two annual community conferences, and the dedicated work of approximately 36 community volunteers representing a broad base of stakeholders as a steering group. Over 500 people representing the community, public, private, and not-for-profit sectors were engaged in the public consultation process.

The ICSP process resulted in the identification of 38 key goals/objectives in collaboration with the community and other stakeholders for the period 2010–2025. The focus areas of ICSP were: economy, environment, governance, infrastructure, services, and quality of life.

DISCUSSION QUESTIONS

- *Have you come across examples like ICSP in your country or city?*
- *Would you call ICSP a Government supported or a Community led initiative?*
- *Do you think that ICSP can happen or be put in practice only through political will or leadership?*

Similar to the example of the Town of Antigonish, there are examples where communities have worked closely with the local bodies to complement the governance. We present in Box 6.4 an illustration of the proactive role played by communities in Mumbai and Pune in India.

BOX 6.4 COMMUNITY RESPONSES SUPPORTING THE GOVERNANCE AT MUMBAI AND PUNE, INDIA

ADVANCE LOCALITY MANAGEMENT (ALM) IN MUMBAI[16]

In 1997, the ALM scheme was launched by MCGM (Municipal Corporation of Greater Mumbai) in partnership with the community. ALM was initiated in 1998 which led to the formation of 658 ALMs in 24 wards of Mumbai.

ALM formed a strong citizen's group working with MCGM to improve the civic amenities in their local area. A sustainable waste management program for the neighborhood was one of the focal areas. Other areas were promotion of rainwater harvesting, solar heating, improving vigilance and safety, and eviction of illegal hawkers.

PARTICIPATORY BUDGETING IN PUNE[17,18]

Participatory Budgeting (PB) is an innovative process that empowers citizens to participate in the public spending decisions. This mechanism was first introduced in 1989 in Porte Alegre in Brazil and is now practiced in over 250 cities in Europe, Peru, Colombia, and India. In PB, local municipal authorities collaborate with NGOs and citizens to identify development priorities and develop these priorities into actionable and concrete projects to be then implemented by the authorities. This collaborative approach ensures that the government is providing the citizens with what they want, and, in turn, citizens provide their insights and support needed during implementation.

In the Indian city of Pune, PB was initiated by the Pune Municipal Corporation (PMC) in 2005 on an informal basis. During this pilot, hundreds of citizens participated and offered suggestions for projects within the municipal budget. Encouraged by the response, PB became a formal process in the city in 2006–2007. Civic societies like the Urban Community Development Department and Community Development Society facilitated PB workshops and also trained community members to conduct such meetings within the neighborhood.

In PB, the citizens were responsible for surveying their neighborhoods, identifying priority tasks, and submitting them in a formal manner to be discussed at project prioritization and budget allocations meetings. Citizens could suggest projects in the areas of footpaths, cycle tracks, road, street lights, traffic signals, bus stops, public parking, public toilets, solid waste management, water (supply), storm water, gardens, public buildings, signage, and others. In 2014–2015, of the 4645 citizen-suggested public work projects, 846 were included in the annual budget. This was a steep increase from the 600 suggestions in 2012–2013 and 3300 suggestions in 2013–2014. The PMC increased the budget allocated for PB projects from ~\$35 million to ~\$57 million.

PB in Pune brought to light many issues in the area of solid waste management. Waste workers involved in the PB workshops requested more space for decentralized waste management solutions. The PMC supported these ideas and built more than 30 waste sorting centers and 16 decentralized waste processing facilities.

SWACH, PUNE[19]

At a time when waste pickers experienced poor social and economic status in the country, Pune was the first city in India to formally register waste pickers way back in 1995. The organization of waste pickers was formulated into a union, which later led to the formation of the workers cooperative. This cooperative is the key to the success of this model, which SWaCH refers to as a *pro-poor public private partnership.*

The model has evolved as a result of more than 20 years of consistent dialog and deliberations between the Pune Municipal Corporation (PMC) and the Kagad Kach Patra Kashtakari Panchayat (KKPKP) that organized 9000-plus rag pickers/waste collectors to work with dignity and created a sustainable de-centralized, waste management model in the country.

This cooperative has been working for the recognition and uplift of waste pickers in the city. Their focus is to improve the working conditions and represent the waste pickers in front of the government for protection of their rights. The organization collaborated with the PMC to integrate the waste pickers into the local waste management system.

SWaCH is authorized by the PMC to provide door-to-door waste collection and other allied waste management services. The scope of SWaCH includes collection, resource recovery, trade, and waste processing. In addition to developing a sustainable waste management system, SWaCH aims to provide a decent livelihood in the recycling industry.

Through its 2300 members, SWaCH services over 400,000 households across 76 zones or *prabhags* in 15 municipal administrative wards of the PMC and the number is growing.

DISCUSSION QUESTIONS

- *The ALM program in Mumbai has had a fairly uneven experience. The program suffered due to poor support from the MCGM for a while and not all ALMs were actively engaged. Research more on the current situation of ALMs in Mumbai and identify the barriers and challenges faced. Look for similar examples in other cities. Discuss how the ALM concept can be evolved in your city and hold a discussion with city officials.*
- *What could be the barriers to introduce Participatory Budgeting? What may be the pre-conditions for its success? Research additional case studies like PB in Pune.*
- The *agreement between SWaCH and Pune Municipal Corporation (PMC) is available at http://www.swachcoop.com/pdf/Agreementforwebsite.pdf. Please review and suggest improvements, if any.*

6.2.3 Conservation of Nature

Most communities in rural and coastal areas depend on natural resources and are sensitive towards their protection. We present here three examples of the proactive role played by the communities in the conservation sector. These examples come from Kenya, Cambodia, and South Africa.

There are many initiatives taken by the Kenyan government and the community to protect and conserve black rhinos. A Wildlife Conservation and Management Act issued in 2013 provides guidance on facilitating community based wildlife conservation initiatives.

Despite the regulatory framework, the black rhino population in Kenya was reduced from 20,000 in 1970 to less than 600 in 2013. Natural death, poaching, or illegal killing were leading this rhino population to the brink of extinction.[20] Box 6.5 illustrates the case of community led Sera Wildlife Conservancy in Kenya.

In order to protect forests and conserve the ecosystems for the future, the monks of Samraong Pagoda in Cambodia adopted an 18,261-hectare area of forest lands in the north-west part of the country. Their aim was to protect the forests from illegal logging and land encroachment, and to promote community livelihoods. This forest is now referred to as Monks Community Forest (MCF). We describe efforts taken by the communities to protect the MCF in Box 6.6.

Protecting forests and at the same time promoting tourism is often a difficult balance to achieve. However, if *eco-tourism* is well designed and operated, it helps to achieve the objectives of sensitization, creation of livelihoods as well as protection of forests. Box 6.7 describes community based eco-tourism in Chi Phat, Cambodia, and the Khomani San area in South Africa.

In the later sections of this chapter we will be looking at examples of partnerships between business and community. However, it may be worthwhile to illustrate an example

BOX 6.5 COMMUNITY BASED CONSERVATION IN AFRICA[21]

Sera Wildlife Conservancy, a community based initiative, was formed by the Samburu communities. Under this conservancy, the Samburu community worked with two other neighboring communities to establish Sera Rhino Sanctuary in 2015, the first community operated sanctuary in East Africa aimed at protecting the endangered species.

The Government of Kenya demonstrated immense trust in this local community-based initiative and translocated two rhinos from Lewa, Nakuru, and Nairobi National Parks to the Sera Conservancy. The community rangers were trained by the Kenya Wildlife Service in tracking the animals through satellite-based transmitters, anti-poaching operations, and patrolling. These activities also boosted tourism and job opportunities in the community.

DISCUSSION QUESTIONS

- *Discuss the perils of animal poaching, specifically, its impact on ecosystems as well as on livelihoods.*
- *Discuss some of the rhino reintroduction and conservation strategies in the African continent and various ways the local communities have been involved, especially the tribals.*

BOX 6.6 MONKS COMMUNITY FORESTRY, CAMBODIA[22]

MCF in Cambodia enhances local livelihoods by providing resources, shelter, subsistence crops, and commercial products. To protect the MCF, the village sub-committees and volunteers came together. In addition to patrolling, volunteers raised awareness within the community about the importance of conservation. These measures reduced the illegal activities significantly. The forest houses, today, many endangered animal species. Villagers collect or harvest permitted resources from the forests like old timber, bamboo, wild ginger, fruit, mushrooms, and fishing. This provides an income of $150–$200 per month per person, which further incentivizes the communities to participate in the initiative.

DISCUSSION QUESTIONS

- *How does one decide on the limits to extract resources from forests that we want to conserve?*
- *Do you perceive any risks in involving communities in conservation of nature? Are there examples where you see that community involvement in conservation has failed? What can be the reasons?*

BOX 6.7 EXAMPLES OF COMMUNITY BASED ECO-TOURISM

Since 2007, the Wildlife Alliance and the Chi Phat community from the Southwestern part of Cambodia have been working together to preserve the Cardamon Mountain ecosystem by providing alternative and sustainable livelihoods to the illegal poaching and logging to the local villagers. In 2008, the community officially opened to tourism, and the project is now run and owned by elected members of the community. Since then, more than 15,000 international tourists have visited Chi Phat, and the community has received more than $600,000 in revenue, with 80% going to the families providing the services and 20% going to the CBET community fund for equipment and infrastructure maintenance. This conservation model has resulted in *reforesting 733 hectares of degraded tropical areas, the cancellation of 36 land concessions, provided sustainable livelihood alternatives to 5,000 local community members, and protected 720,000 hectares of tropical forests from illegal loggers and commercial and industrial encroachment*. Furthermore, forest burning has stopped by 100% and hunting by 80%.[23]

Khuin Kwa Kalahari Experience (KK) offers authentic eco-tours within and around the Kgalagadi Transfrontier Park in South Africa. What sets KK apart from other tours is that the enterprise is led by the Khomani San community, and the tours embrace the traditions of the Khomani San culture. This offers visitors not only the opportunity to see the extraordinary landscapes and biodiversity of the region, but also to experience life through the eyes of the Khomani San descendants, one of the oldest groups of people on the planet. Through its inclusive approach, KK provides the community with employment opportunities, builds local knowledge and skills in conservation and tourism, and aligns the needs and values of the community with conservation.

DISCUSSION QUESTIONS

- *List the guiding and operational principles for undertaking eco-tourism.*
- *Check whether your country has a community oriented eco-tourism policy.*
- *What could be some strategies to recognize and promote eco-tourism industry?*
- *Research success stories where objectives of conservation, sensitization, and creation of livelihoods are achieved.*

of a business that spearheaded the Save the Whale Shark Campaign in India. Box 6.8 describes the case of Tata Chemicals in India.

We introduced in the opening section of this chapter the reactive role played by the communities in the form of a watchdog. There have been numerous cases where communities have protested against the business and the government in the interest of the protection and conservation of our environment and natural resources. We highlight in Box 6.9 the case of the Chipko movement in India.

BOX 6.8 SAVE THE WHALE SHARK CAMPAIGN

The *Save the Whale Shark Campaign* is a community-led conservation project that was launched in 2004 by Tata Chemicals Ltd (TCL), in partnership with Wildlife Trust of India (WTI) and the Gujarat State Forest Department and with support from the fishing communities of the Dhamlej, Veraval, and Sutrapada regions.

The whale shark is the largest fish in the world and was once slaughtered along the Gujarat coast in large numbers for its oil and meat. This led to a steep decline in its population; hence, it was declared an endangered species in 2001. The idea of the Save the Whale Shark campaign was to create conservation awareness amongst the coastal communities and garner their support for protecting this endangered fish. In the span of 14 years, 601 whale sharks[24] have been caught in the fishing nets in the coast of Gujarat. Taking inspiration from this campaign, the fishermen are now voluntarily cutting their nets to release the sharks. They have now become protectors of this species and not the hunters.

To date, a total of 585 whale sharks, which were incidentally caught in the fishing nets of the locals, have been rescued and released. The forest department of Gujarat provided monetary relief to fishermen towards fishing net damages incurred during the whale shark rescue operation.

Research on the elusive whale shark is also being carried out on its feeding, breeding, and migratory patterns to strengthen the available scientific knowledge. This is helping the conservation efforts. The research activities include studying the whale sharks through photo-identification, genetic analysis, and satellite tagging.

To take this campaign forward, this project has been taken from the west to the east coast, with the support from the Andhra Pradesh Forest Department, EGREE Foundation, and IUCN. A study has been undertaken by the WTI on the traditional ecological knowledge (TEK) of the whale shark distribution along the Andhra Pradesh and Odisha coasts of India.[25]

 **DISCUSSION QUESTIONS**

- *What could be the role of whale sharks in the marine ecosystems?*
- *Apart from fishing, what are the other threats to the population of whale sharks?*
- *Similar to whale shark conservation programs, explore reading about the sea turtle conservation programs in coastal countries where communities are involved with the support of business as well as government.*

BOX 6.9 CHIPKO MOVEMENT IN INDIA[26–28]

The Chipko (tree hugging) movement in India dates back to the fourteenth century to protect trees and forests. In the seventeenth century a woman-led group in the Bishnoi district of Rajasthan protested the felling of trees and even sacrificed their lives.

In the 1970s, the government of India allotted a locally important forest area in the State of Uttar Pradesh, India, to a sports company. This decision was not accepted by the communities as the government prohibited use of forest wood for making agricultural tools and instead permitted the company to use timber to make sports goods on a large scale. This was expected to increase soil erosion leading to floods.

Between the 1970s and 1980s, the communities launched the *Chipko movement* against contractors and to prevent the forest auctions. To stop the cutting of the trees, many women hugged the trees.

The Chipko movement was lauded and supported by the activists in India. Taking inspiration from the success, communities in one of the more affected regions in the State of Uttar Pradesh could bring a 15-year-long stay on felling of trees.

 **DISCUSSION QUESTIONS**

- *Similar to the Chipko movement, discuss some of the more recent citizen or community-led campaigns against unsustainable development.*
- *To what extent have these campaigns been effective?*
- *Some argue that some of these campaigns could be politically motivated or supported by rival businesses. Do you see such cases?*

The non-violent method followed in the Chipko movement has spread to many other regions including East Africa, West Africa, Somalia, Lebanon, Tasmania, Papua New Guinea, and Mexico. In 2004, in Tasmania community members became *tree sitters* who climbed onto eucalyptus trees to protect Tasmania's old-growth Styx forest from being logged. Many individuals who protested for this cause were threatened, imprisoned, tortured, or killed in the process.[29]

6.2.4 Participatory Environmental Monitoring

Communities can play an important role in environmental monitoring. Until recently, monitoring activities like measuring air pollution was a task that could be performed only by trained scientists using very sophisticated and expensive equipment. Environmental sensors are now getting advanced, miniaturized, and cheaper, opening new methods of collecting environmental data.

Environmental data capture is, therefore, not part of the regulatory agencies today. It is now led by the citizens. Citizens can *sense* the environment using readily available sensor devices with smart phones, and share this information using existing cellular and Internet communication infrastructure.

Democratization of technology and Do-It-Yourself (DIY) hardware platforms have the potential to enable citizens to sense. It is estimated that by 2020, *citizen environmentalists* will have more personal sensors, measuring air and water pollution, energy consumption, health , etc., than the governments. Citizens are now interested to know more about the state of the environment they are living in and not solely depend on the monitoring reports of the regulators.

The US EPA is challenging communities across the country to collect data using hundreds of air quality sensors as part of the Smart City Air Challenge. In 2016, the agency offered up to $40,000 apiece to two communities to help them develop and implement plans for collecting and sharing data from air quality sensors. The award money only covers part of the program costs, so communities are expected to partner with sensor manufacturers, data management companies, or others to get resources and expertise to implement their plans.[30]

EPA Victoria in Australia launched the Victoria citizens' science program in 2014[31] to involve the community in important environmental issues. They empowered citizens by involving them in real environmental project activities like data collection, analysis and monitoring and assessment, tasks which are usually done by trained professionals. This program gives the citizens a sense of ownership of the issues and also creates awareness. Examples of the citizen science program projects included training the citizen volunteer to estimate the levels of smoke in the air using visual landmarks, training them to identify the types of micro-plastics in freshwater and marine waters, analyzing the potential sources, and to train them to collect water quality data for improved management of one of the creeks in the region.

Box 6.10 provides a few interesting examples of participatory air quality monitoring and the Do It Yourself (DIY) approach. The idea of crowdsourcing environmental monitoring is more towards knowledge creation rather than compliance monitoring. Community participation are in monitoring builds sense of ownership on the data and guides on how to use data for better understanding and decision making.

The accuracy and reliability of many environmental sensors is yet to be established as they are in the nascent stage of development. Many stakeholders have voiced their concerns about this.

Scientists have been devising innovative mechanisms for environmental monitoring. Sensors are now being developed made from gallium nitride, a material that can perform in extreme heat and at high power levels.[37]

Plants have been found to have good sensing capabilities. For instance, each single root apex can simultaneously and continuously monitor many chemical and physical parameters. A digital network and a powerful algorithm transforms each tree into an environmental informer. A group of Italian, British, and Spanish researchers are working on developing a network of microsensors that can be embedded in plants, sending us information on how plants respond to changes in temperature, humidity, air pollution, and many other changes in their environment. A project called PLEASED (PLants Employed As SEnsing Devices)[38] has been launched with €1.07 million ($1.46 million) funding by the EU.

BOX 6.10 EXAMPLES OF ENVIRONMENTAL MONITORING BY CITIZENS

AIR QUALITY MONITORING

Air Quality Egg[32] is a network of about 1,300 CO_2 and NO_2 sensors. The sensors typically cost less than $250 each. The Air Quality Egg was developed by a community effort, born out of groups from the Internet of Things Meetups in NYC and Amsterdam and New York City. The data monitored by the Egg is sent to the cloud that stores and provides free access to the data. A website is operated where one can see graphs and other visualizations of the data. Conversations thus begin on the status of air quality.

Another popular device is the Smart Citizen Kit.[33] Here data is uploaded to the Smart Citizen website which shows about 800 kits deployed across the world, with more than half deployed in Europe. The basic kit cost less than $200, before tax and shipping. There are dozens of such sensor packs and gateways now available, and the number is constantly growing.

Smart Citizen is a platform to generate participatory processes of people in the cities. Connecting data, people, and knowledge, the objective of the platform is to serve as a node for building productive and open indicators with distributed tools.

The Kids Making Sense[34] program empowers youth to drive positive change and improve public health by collecting credible air quality data around their neighborhoods. In this program, students participate in hands-on science-based tasks, discuss their findings with an air quality scientist, and share their data with the global air quality community. They can even use their data to identify local sources of air pollution and take actions to be part of the solution.

THE DO-IT-YOURSELF MANTRA

EnviroDIY[35] is a community of enthusiasts sharing Do-It-Yourself Ideas for environmental monitoring in the United States. This project is an initiative of Stroud™ Water Research Center that has the objective of advancing what seeks to advance knowledge and stewardship of freshwater systems through global research, education, and watershed restoration.

All EnviroDIY members can showcase their environmental sensing gadgets or describe their own homegrown approaches to monitoring, sensor calibration, installation hardware, radio communication, data management, training, or any number of other topics. Members can pose and answer questions and can network within interest groups to collectively develop new devices, tutorials, or other useful products.[36]

 **DISCUSSION QUESTIONS**

- *Can community-based monitoring complement monitoring carried out by regulatory agencies?*
- *Do you think community based monitoring can ever substitute for regulatory monitoring carried out to check compliance?*

BOX 6.11 MULTIPARTITE ENVIRONMENTAL MONITORING IN THE PHILIPPINES[39,40]

To ensure compliance with the Environmental Clearance Certificate (ECC) and the Environmental Management Plan, the Multipartite Monitoring Team (MMT) was established by the Department of Environment and Natural Resources (DENR) of the Philippines.

The concept of MMT was developed after protests by the local communities to developmental projects. The MMT is comprise of stakeholders like local government units, NGOs, the community, a woman's sector, and relevant governmental agencies.

For sensitive projects (Category A), a MMT is formed immediately after the issuance of an ECC. Proponents required to establish an MMT are required to put up an Environmental Monitoring Fund (EMF) not later than the initial construction phase of the project.

The MMT is tasked to undertake monitoring of compliance with ECC conditions as well as the EMP. The MMT is required to submit a semi-annual monitoring report within January and July of each year. The Environment Management Bureau (EMB) has formulated guidelines for operationalizing area-based or cluster-based MMT.

The MMT Report is more credible and unbiased being an output signed by a multisectoral group duly represented by a broad spectrum of concerned stakeholders who joined and took part in the monitoring activity.

For projects whose significant environmental impacts do not persist after the construction phase or whose impacts could be addressed through other regulatory means or through the mandates of other government agencies, the operations of MMT are terminated immediately after construction or after a reasonable period during implementation.[41]

 **DISCUSSION QUESTIONS**

- *In your opinion, should the community members of MMT necessarily be environmental scientists and professionals?*
- *Should MMT be established for a project or for a region covering several projects that may lead to cumulative impacts?*

The community can play an important role in monitoring the environmental compliance as well. There are cases where the regulator has partnered with the communities to jointly oversee the compliance. We illustrate in Box 6.11 the concept of Multipartite Monitoring followed in the Philippines.

6.2.5 Judicial Activism

Judicial activism is a term given to court rulings that are based on the judge's personal or political opinions instead of the law. In some cases, judicial activism refers to instances when the courts adopt an activist stance in cases that require rulings when there is no precedence in laws or previous judgements or when citizens use judicial routes to seek environmental justice.

In India, the incident of the Bhopal Gas tragedy resulted in reforms of statutory provisions regarding the role of citizens in environmental laws. Post the Bhopal incident, Indian citizens who wished to act against a pollutant in their individual or public interest could challenge the polluters legally. This mechanism was termed as Public Interest Litigation (PIL) under which citizens could challenge environmentally harmful practices on behalf of society. The example of Taj Trapezium discussed in Chapter 2 is the result of judicial activism by M. C. Mehta, an Indian Supreme Court judge and an environmental activist. More recently as discussed in Chapter 2, a National Green Tribunal was set up by the government of India to tackle environmental cases.[42]

Judicial activism has been an influencing factor to identify gaps in the policies and influence formulation of required legislation. Box 6.12 shows two examples where judicial activism followed by pressures from the communities led to introduction and transformation of policies on waste management at the national level.

BOX 6.12 EXAMPLES OF INFLUENCE ON POLICIES AND REGULATIONS DUE TO JUDICIAL ACTIVISM IN WASTE MANAGEMENT IN BANGLADESH AND INDIA

BANGLADESH ENVIRONMENTAL LAWYERS ASSOCIATION[43,44]

The Bangladesh Environmental Lawyers Association (BELA) is a non-governmental and not-for-profit organization established in 1992 by a lawyer, Dr. Mohiuddin Farooque. The primary objective of BELA is to use law as a tool to develop and monitor laws and regulations for protection of the environment. It was established at a time when the state of environmental laws and justice in Bangladesh was rather dismal.

BELA's activities include Public Interest Environmental Litigation (PIEL), research, and increasing public awareness in environmental matters and capacity building. Using PIEL, BELA has filed over 250 cases revolving around issues like river pollution, industrial pollution, vehicular pollution, labor welfare, compensation for losses due to development projects, encroachment and derogation of critical wetlands, relocation of industries, prevention of hill cutting, conservation of forests, defending forest rights, fishermen's rights, farmers' rights, and others.

BELA has been consistently working on improving the waste management scenario in Bangladesh. One of the waste streams targeted by BELA was the shipbreaking wastage. Chittagong Shipbreaking Yard is one of the world's largest shipbreaking yards, where retired ships are sent over by developed country ship owners to reduce associated breakage costs in regions that are less stringent in environmental and safety requirements. Hazardous waste resulting from shipbreaking operations was indiscriminately disposed. The workers were underpaid and faced high health risks due to injuries, noxious fumes, and the handling of the asbestos. The local fishing industry was also affected by this polluting activity.

BELA filed several cases against the shipbreaking industry for the rights of workers and to ban ships into Bangladesh that contain poisonous substances. In 2009 BELA got the Supreme Court to ban all ship recycling that did not meet environmental standards. Consequently, the Government of Bangladesh introduced national policies to protect the environment and occupational health and safety of workers in the ship breaking yards. However, challenges like poor enforcement still remain.[45]

AN ENVIRONMENTAL POLICY ADVOCATE DRIVES MSWM REVOLUTION IN INDIA[46–48]

Almitra Patel, an environmental policy activist, pioneered key milestones in the Indian Municipal Solid Waste Management (MSWM) sector. Her involvement in this area started when she protested against poor waste management processes in her own neighborhood in the city of Bangalore in 1991. She surveyed over 80 Indian cities and found that the MSW systems are extremely poor and neglected in most of India in 1994–1995. Subsequently, she filed a Public Interest Litigation (PIL) in 1996. In response, in 1998, an expert committee on MSWM was set up by the government and the Municipal Solid Waste (Management and Handling) Rules were drafted and issued between 1999 and 2000.

Unfortunately, the overall scenario of MSWM has left a lot to be desired in India but some cities have made significant progress. Ms. Patel continues to pursue efforts to bring about change at a government and city level. She filed another PIL with the Supreme Court of India for non-compliance of the MSW rules in 2004 urging immediate action. The pressure on the government towards compliance thus continued.

 **DISCUSSION QUESTIONS**

- *Several interesting community initiatives have been launched to clean up the cities. Study the case of the Warangal city clean up in India that shows a collective positive action by the community in partnership with the local government.[49] Do you think such community driven campaigns on cleaning of the cities last?*
- *Similar to Chittagong, Alang in the coast of Gujarat has been a place for the ship salvaging industry. You may like to research on the impacts due to operations at Alang.[50] Japan and the Gujarat government have joined hands to upgrade the existing Alang shipyard. The two parties have signed a Memorandum of Understanding, focusing on technology transfer and financial assistance from Japan to assist in the upgrading of operations at Alang to meet international standards. You may like to learn about these international standards.*

6.2.6 Compliance Review and Inspection Panels

In Chapter 3 we discussed the important role played by the Development Financing Institutions (DFIs) such as the World Bank in assisting the national governments and building the local capacities. The Community Driven Development (CDD) approach highlighted in Chapter 3 illustrated partnerships between the DFIs and the communities towards positive and sustainable outcomes.

Communities have played a role of a watchdog to the DFIs by monitoring the projects and programs sponsored by the DFIs towards the compliance and commitments made. In response, DFIs have set up compliance review and inspection panels to address community concerns, assess them independently, and resolve them in an amicable and transparent manner.

The World Bank created the World Bank Inspection Panel in 1993 to serve as an independent mechanism to receive and resolve complaints on World Bank funded projects. The Inspection Panel handles issues like forced relocation of affected communities, adverse impact on the livelihoods, rights of indigenous people, environmental harm, etc. Any citizen or community member can register a complaint with the Inspection Panel.

The Inspection Panel is an impartial body, which operates independent of the World Bank Groups. Its prime aim is to hold World Bank Groups accountable for its actions and to serve as a forum for affected people to voice their concerns or problems.[51]

Box 6.13 described a few cases that were dealt with at the Inspection Panels.

Similar to the World Bank, the ADB established the Accountability Mechanism (AM, an independent forum for affected persons (APs) or communities to address their concerns and seek solutions about ADB-assisted projects, including noncompliance with ADB policies and procedures. The AM covers Office of the Special Project Facilitator (OSPF), the Compliance Review Panel (CRP), and Office of the Compliance Review Panel (OCRP). OSPF holds informal consultations with APs to hear their complaints and seek solutions, the problem-solving function. The CRP looks into noncompliance with ADB's operational policies and procedures leading to direct and material or likely harm to APs.

Using such panels as a platform, the community has played the role of a watchdog and has sought environmental justice. Between 2012 and 2016, ADB received 65 complaints, out of which 25 complaints were found eligible.[55]

BOX 6.13 EXAMPLES OF CASES DEALT AT THE WORLD BANK'S INSPECTION PANEL[52]

THE NARMADA LESSON, INDIA[53]

The World Bank was working with the Government of India on two key development projects on the Narmada river, the Narmada River Development (Gujarat) Sardar Sarovar Dam and Power Project, and the Narmada River Development (Gujarat) Water Delivery and Drainage Project. The projects involved construction of a large concrete dam, a 1200 MW powerhouse, transmission lines, a water conveyance system via a 460 km long canal, branch and distributary canals, and a drainage network to serve an irrigation area of about 2 million hectares across multiple States. The project activities were found to lead to adverse environmental impacts and required resettlement of 70,000 inhabitants. Protests were made by local and international NGOs garnering attention worldwide.

The World Bank commissioned a first-ever independent review of the World Bank funded project in 1992 to assess the resettlement and rehabilitation impact of the affected population and the environmental impact of the projects. The review team visited the sites, and it was found that the project failed to incorporate the Bank's procedures and policies. In 1993, the World Bank support was withdrawn upon the Indian Government's request that committed to follow social and environmental standards thereafter. This milestone project served as a key learning experience for the World Bank to factor in social and environmental dimensions into their projects.

TRANSPORTATION SECTOR DEVELOPMENT PROJECT, UGANDA

The World Bank was funding a 66-km road upgrade project in Uganda. In 2014, the Bigodi and Nyabubale-Nkingo communities that were living along the road complained about the serious harm that was caused to the communities and children with cases of child sexual abuse, sexual harassment of female employees, and teenage pregnancies caused by road workers. There was also an increased presence of sex workers leading to spread of HIV/AIDS. The resettlement plan was not satisfactory, and there were inadequate road and occupational health and safety measures.

Upon visit and assessment, these concerns were found valid. The Bank's oversight and supervision was found to be inadequate. The Bank was asked to strengthen oversight of the projects and take corrective measures. However, this particular project was later shelved in 2015.[54]

 **DISCUSSION QUESTIONS**

- *What process does the World Bank's Inspection Panel follow when they receive a complaint from the affected party or their representatives?*
- *Compare this with ADB's CRP process.*
- *Investigate World Bank funded or developmental projects in your country where the Inspection Panel or CRP intervened due to complaints. What was the impact of this intervention?*

The major subjects of AM complaints since 2003 are: (i) resettlement, compensation, and land acquisition; and (ii) adverse environmental impact. These subjects both require effective and active stakeholder consultation during project design and implementation. If meaningful consultations, which include improved information sharing, regular involvement, and communication with project beneficiaries, and resolving conflicts quickly, are carried out in a timely fashion, the probability of complaints should be less. Even if there is a complaint, chances are that it will be resolved faster because of the trust and respect built from regular consultation. With meticulous efforts to plug these deficiencies, the objectives of meaningful consultation will be achieved in the coming years.

The AM cases suggest that close relations of ADB with NGOs/CSOs in the borrowing countries will have benefits for the functioning of the AM and GRMs and may well reduce the likelihood of safeguard complaints materializing. NGOs/CSOs have played critical roles in the implementation of the AM Policy. Of the 25 eligible AM complaints, NGOs/CSOs were involved in bringing complaints to the AM in more than 50% of the cases. Some took on major roles as complainants' advisors, facilitators, representatives or intermediaries, or merely as observers to ensure fairness in the process. Therefore, ADB needs to enhance its dialogue with NGOs and CSOs to promote trust and partnership building. They are often the conduits for channeling urgent concerns of the APs. ADB's AM has increased its outreach activities to build relations with international organizations, the independent accountability mechanisms of other multilateral development banks and bilateral development agencies, and local and regional NGOs/CSOs. This has resulted in increased internal and external partnerships during the AM Policy implementation period. ADB needs to increase NGOs/CSOs partnership going forward.

6.3 Microfinance to Empower the Community

Microfinance has been an important intervention to help empower communities. It has been found that access to microfinance often leads to reduced economic vulnerability as well enhances sustainable development. In this section, we will introduce examples of community oriented microfinance.

BOX 6.14 EXAMPLE OF A ROTARY INITIATIVE TO SUPPORT AND EMPOWER THE COMMUNITY [56–58]

The Rotarian Action Group for Microfinance (RAGM) and Community Development is a group that was formed to help clubs and districts to develop and deploy effective Microfinance and Community Development programs.

For example, RAGM worked with local partner organizations to provide microfinance services in the Honduras region in Central America. Affordable loans were provided to the Hondurans to support schools, provide food, shelter, health care, and to empower women by facilitating entrepreneurial roles in the communities like chicken farming, food product supply, candle making, and others.

The program has supported 8,000 beneficiaries to advance their businesses like shoemaking, moto-taxis, tortilla making, and sun baked bricks by procuring technologies, more raw materials, or adding repair services.

The Rotary Foundation is one of the global philanthropic organizations operating across many countries. Economic and community development is one of the Rotary's six areas of focus. The Rotary Clubs of Quito Occidente, Ecuador, and Marin Evening, California, the United States, along with 64 clubs from seven districts, together raised money, built partnerships, and mobilized local community members to invest in skills training in Esmeraldas, Ecuador. Box 6.14 describes the details.

In Brazil, waste collectors (called catadores) collect recyclable products like paper, cardboard, plastics, glass, metals, and other products and sell them to a recycling depot or middlemen. They face many problems like unsafe and unhygienic working conditions for them and their children, social stigma and harassment, delayed payment from the depots, or lower rates from middlemen. Because of the minimum quantity requirement by depots and poor payments by the middlemen, the collectors have formed or joined cooperatives to sell recyclables collectively to depots. Box 6.15 provides the details.

BOX 6.15 FUNDING WASTE MANAGEMENT COOPERATIVE TO EMPOWER WASTE COLLECTORS IN BRAZIL[59]

A cooperative of waste collectors was formed in Brazil supported by the Participatory Sustainable Waste Management (PSWM) project funded by the Canadian International Development Agency in partnership with educational institutions in Canada and Sao Paulo. The objective was to increase income received by collectors and provide a structure to collectively increase sales of their merchandise.

The cooperative needed organizational and financial management capabilities in addition to having required quantities and capital to transport material to the depots. To facilitate the cost of transportation, the PSWM provided microfunds to transport materials, expand business or cover expenses in case of delayed payment from depots. Through these cooperatives, the collectors improved their incomes and in addition learnt to collaborate and participate in meetings and share their opinions. The PSWM project however largely depends on public donations that are limited. It also faces the challenges of creating awareness in the sponsors to appreciate the importance of this project from an environmental and social perspective.

Use of solar energy plays an important role in providing energy access, especially to the rural poor. Many programs have been initiated across the world that have been promoting solar cookers and solar lamps with the help of government and philanthropic organizations. In some cases, financial assistance is provided through microfinance with contributions made by the beneficiary communities.[60] Many of these programs are community focused with an interest on building the capacities of the youth to stimulate entrepreneurship and lead to sustainable solutions.

Similar initiatives have been taken up to promote water and sanitation infrastructure supported by microfinance. Under these programs drinking water connections and in-house toilets are provided to lower income communities in urban and rural areas. We describe in Box 6.16 two such initiatives in India.

BOX 6.16 MICROFINANCE INITIATIVES

THE SOUL PROJECT AT IIT BOMBAY, INDIA[61]

In India, 65 million of the 76 million homes that have no access to electricity use carbon-emitting kerosene lamps. The lack of electricity hampers their personal and economic activities. Further, use of kerosene is hazardous to health and safety, because it is flammable and releases harmful emissions. The other fuel used is firewood that not only pollutes the indoor air quality in the poorly ventilated kitchens but depletes the forest produce and reserve. In addition, women have to walk miles to collect firewood for cooking.

The Indian Institute of Technology (IIT) in Mumbai, India, came up with the Solar Urja Lamps (SoUL) project in which solar lamps are assembled, distributed, and repaired by the villagers. More than 700,000 SoUL have already been distributed across States of Maharashtra, Madhya Pradesh, Rajasthan, and Odisha covering nearly 8,000 villages. There are about 76 million students studying in blocks of India where more than 50% of the households use kerosene as a main source of lighting. The SoUL project has helped these students to study at night in addition to saving kerosene and reducing the air emissions.

The solar lamps are assembled locally, used by local people, and serviced by the locals. By transferring technical skills and knowledge and training locals to use technology, they cease to become dependent on anyone. Local assembly not only results in employment generation, but also allows the locals themselves to repair and maintain the solar products in the future. The outreach and training activities are conducted by NGOs who closely coordinate with IIT Bombay.

The funds for this project are contributed by three partners: 36% from the National Clean Energy Funds (NCEP), Ministry of New and Renewable Energy (MNRE), Govt. of India; 40% from various philanthropic partners including state governments, trusts, corporate social responsibilities, and individuals; and 24% comes directly from the student beneficiaries. The total cost of the Solar Urja Lamp (SoUL) is Rs. 500, but a student gets this lamp at a student discounted price of Rs. 120. Thus, the lamps are heavily subsidized, but there is the owner contribution.

THE GUARDIAN STORY[62]

Gramalaya Urban and Rural Development Initiatives and Network (Guardian) is the first and only MFI in India that finances water and sanitation infrastructure projects like drinking water connections and in-house toilets for the lower income communities in urban and rural areas. For these projects, Guardian has financed around $154 million and has provided loans between $75 and $215 for projects like new toilet construction, new water connection, renovation of existing water and toilet facilities, installation of water purifiers, rain water harvesting structures and bio gas plants. As Guardian works in both rural and urban areas, they offer loans accordingly with an interest rate of 21% per annum with a loan tenure of 18 months.

In addition to providing infrastructure, Guardian brings about a culture change and sensitizes communities on hygienic practices like using toilets and washing hands with soap. Guardian has a goal to reach 100,000 households in the state of Tamil Nadu in India with this objective by 2020. In 2015–2016, Guardian has provided loans for 3818 households with water connected and to 12,892 women for building toilet facilities within the house.

DISCUSSION QUESTIONS

- *You may like to learn more about the repayment of loans to the Guardian.*
- *Has Guardian established any congruence with the state and national schemes ongoing to provide water and sanitation infrastructure?*
- *Who maintains the infrastructure that is provided?*
- *Does the Guardian's annual or bi-annual evaluation of the impact of the program to understand the benefits relate especially to health?*

6.4 Community and Business Organizations

Many business organizations have recognized the importance of doing good to the communities, workers, and neighborhoods. In Chapter 5, we described the efforts taken on for Corporate Social Responsibility (CSR) by business. More recently, businesses have been blending the CSR thinking in their core business and building a brand. We illustrate in Box 6.17 the Buy-One-Give-One model of Tom Shoes.

For large corporations and those operating complex supply chains, employees become a critical component of the community. Many business organizations have, therefore, come up with interesting engagement models where sustainability is pitched and innovations are simulated and supported as an outcome.

One of the early and well-known examples of such an engagement model is by 3M called 3P, that is, Pollution Prevention Pays. 3M pioneered the concept of pollution prevention with the creation of the 3P program in 1975. The 3P program is based on the reality that pollution prevention is more environmentally effective, technically sound, and economical than conventional pollution control. The 3P program came with a slogan—*Contribute an idea–we will reward you*, focusing on shop floor innovations. The innovations came from product reformulation, process modification, equipment redesign, and the recycling and reuse of waste materials. We describe the 3P program in Box 6.18.

BOX 6.17 THE BUY-ONE-GIVE-ONE MODEL OF TOMS SHOES

Since 2006, the TOMS Shoes business operates a One for One™ model, in which, for every pair of shoes purchased, a new pair is given to someone in need for free. This buy-one-give-one model has given away over 35 million pairs of shoes across six countries. TOMS partnered with Save the Children, in 2011, and this partnership has led to donation of over 600,000 pairs of footwear to children in Australia, China, El Salvador, Jordan, Lithuania, Tajikistan, and the United States.

However, the program has also been faced with criticism. Some have called this program the *worst charity in development* as it does the bare minimum to address the root cause of poverty. It also undermines the demand for locally-produced products. Blake Mycoskie, the founder of TOMS, has taken steps to address these concerns, and has launched a shoe manufacturing operations in Haiti that is staffed only by locals.[63]

TOMS Shoes has inspired many businesses to adopt a similar business model. One such not-for-profit company is Mealshare.[64] It is a Canadian company which provides one meal for free to youth in need for every meal that is purchased from a select *mealshare* partner restaurant. When a customer goes to the restaurant, they order their meal like normal, and for each meal sold, the restaurants contribute $1.00. Mealshare then forwards most of these funds to their partner charities, which are responsible for providing groceries and ingredients for the needed meal.

 **DISCUSSION QUESTIONS**

- *Research the impact of the TOMS Shoes model on its sales and the brand reputation.*
- *Do you think criticism made of the Buy-One-Give-One Model are fair and justified?*
- *Compare the Mealshare and Buy-One-Give-One Model in terms of environmental and social benefits.*

BOX 6.18 3M: POLLUTION, PREVENTION PAYS (3P) PROGRAM[65]

The multinational corporate 3M runs an innovative and inspiring program called the 3P program. They invite voluntary participation of their employees at all levels. Over the last 32 years, 3M employees have developed over 6,300 3P projects which are reviewed by a multi-disciplinary team from the engineering, manufacturing, laboratory, and Environmental, Health and Safety departments. To qualify, the projects must meet the following fundamental criteria:

- Elimination or reduction of a pollutant
- Improving energy and resource efficiency in terms of manufacturing materials and other raw materials used
- Reducing costs associated with pollution control equipment costs, reducing operating and materials expenses, or increasing sales of an existing or new product

In 2014, this program celebrated 39 years of successful completion. In this span of time, over 2 million units of pollutants have been prevented from entering the atmosphere, which has saved $1.9 billion (USD) over that time frame.

The 3P program's success and longevity can be attributed to its simple, inclusive program design.

DISCUSSION QUESTIONS

- *Investigate some of the 3P projects that have been successfully implemented. How have these projects benefitted the company in terms of launching new or improved products and increased profits or reduced costs.*
- *Are there other examples where companies have empowered their employees or associates to innovate for environmental, social, and economic benefits?*
- *Does 3M pass to the innovator profits made out of the ideas contributed? How are the issues like Intellectual Property Rights (IPR) addressed?*

Google *20% time* program[66,67] is another example of an innovative mechanism targeted at the employees. The mechanism encouraged an employee of Google to spend 20% of their time on a project which is not a part of their job description. The idea was to inspire employees to innovative and improve their work. The innovation mechanism could generate 50% of Google projects that resulted in some of the company's most successful products such as Gmail, AdSense and Google Talk, Google Teacher Academy, etc. These products have influenced consumer behavior across the world.

In addition to the *20% time* program, Google also operates Google Green Initiatives Program that is promoted for employees to influence their consumptive behaviors. This strategy has led to creation of communities that not only understand the importance of sustainability but practice it in their day-to-day life. We present some details in Box 6.19 on Google's Green Initiative.

Sustainable products are also made by following crowdsource-based innovation processes. One prominent example is Unilever's Foundry IDEAS platform, which acts as a digital hub for consumers and entrepreneurs to work together on tackling global sustainability challenges. Another example is LEGO's Sustainable Materials Center, which encourages all employees of the company to come up with alternatives to substitute fossil-based raw materials to manufacture LEGO toys and packaging.

In Chapter 4 we discussed the increasing sustainability considerations in the operations of the supply chains. Many business organizations have started addressing this interest by developing codes of conduct or guidelines on sustainable business operations. The idea often has been to minimize the business risks (especially reputational) and make the operation of the supply chain more robust and participatory. We show in Box 6.20 the example of Starbucks C.A.F.E. (Coffee and Farmer Equity) Practices.

There are examples where business organizations have established an explicit Green Supply Chain Management Policy and have embedded sustainability considerations in the supply chains. We cite in Box 6.21 the example of Mahindra Live Spaces in India.

BOX 6.19 GOOGLE GREEN INITIATIVE

Google employees are provided the option to use a green transportation system that includes biodiesel shuttles and electric vehicles. There is a car-pooling program created for Googlers called GFleet. It includes the newest generation of plug-in vehicles and shuttles. This energy efficient transportation system has resulted in net annual savings of more than 29,000 metric tons of CO_2 which is equivalent to 5,700 cars off the road or avoiding 87 million vehicle miles every year.

Google encourages it employees to get to work by foot, bike, cycle, unicycle, skateboard, scooter, and even kayak. In 2016, over 3,700 Googlers in 26 offices participated in the Google Bike to Work Day.

DISCUSSION QUESTIONS

- *Discuss the importance of the role of a business organization to foster a behavioral change in its employees.*
- *Do you think innovations like Gmail, AdSense and Google Talk, and Google Teacher Academy will actually reduce the carbon footprint? Can these savings in carbon emissions be computed on an on-line basis? (Hint—don't miss the carbon emissions arising from data centers).*

BOX 6.20 STARBUCKS C.A.F.E. (COFFEE AND FARMER EQUITY) PRACTICES[68]

Starbuck's C.A.F.E. Practices is an exemplary example of a worldwide organization using its clout and network to promote sustainable practices within its vast supply chain. The C.A.F.E. Practices is a program that focuses on sourcing of coffee grown sustainably and at the same time recognizing and awarding their suppliers who support their endeavors. Starbucks launched this program in collaboration with Conservation International, a non-profit environmental organization and SCS a third-party evaluation and certification body to help certify green suppliers. Supplier practices are verified by trained organizations to ensure they are compliant with social, economic, and environmental indicators listed in the C.A.F.E. Practices scorecard. The scorecard includes 200 indicators under the categories of Environmental Leadership—Coffee Growing, Social Responsibility, and Economic Accountability.

The C.A.F.E. Practices is estimated to have benefited over one million workers employed by participating agricultural farms.[69]

DISCUSSION QUESTIONS

- *Do you think that C.A.F.E. Practice's sourcing guideline could be an imposition or barrier to the coffee growers and processors?*
- *Is there any technical and financial assistance offered to meet the sourcing requirements? Visit https://www.starbucks.com/responsibility to read more about Starbucks community based initiatives.*

BOX 6.21 GREEN SUPPLY CHAIN MANAGEMENT POLICY AT MAHINDRA LIVE SPACES IN INDIA[70]

Mahindra Live Spaces in India has made a commitment to build a sustainable value chain as one of the primary sustainability goals. During the reporting year 2014–2115, Mahindra Live Spaces identified 150 of their biggest suppliers and started mentoring and motivating them towards becoming responsible partners in their sustainability journey.

Mahindra Live Spaces developed a special Green Supply Chain Management Policy to encourage contractors, suppliers, and vendors to ensure total compliance for better health, safety, and environment. This policy has been cascading across all their projects after incorporating the concerns of the suppliers.

Mahindra Live Spaces organized the suppliers and contractors meetings that were attended by 225-plus suppliers/individuals representing 165-plus companies. The participating suppliers were given presentations on sustainable operational practices as well as dialogue sessions was held. Through these engagements, the suppliers gained knowledge on various sustainability related aspects such as energy conservation, water efficiency, waste management, use of alternate energy, local sourcing, operational safety, and biodiversity. Mahindra Live Spaces shared their Sustainability Roadmap 2020 during the meetings. The company also encouraged their suppliers to use the globally recognized tools for GHG accounting and water foot printing.

 **DISCUSSION QUESTION**

- *What could be the business strategy of Mahindra Live Spaces to implement the Green Supply Chain Management Policy? Compare this strategy with Starbucks C.A.F.E. Practice.*

6.5 Eco-Entrepreneurship

Social enterprises are today considered as effective platforms to meet the objectives of innovation and employment generation, creating the intended development impact and realization of profits. These enterprises are often supported by the business organizations as well as by government.

Business Call to Action (BCtA)[71] was launched at the United Nations in 2008 to accelerate progress towards the Sustainable Development Goals (SDGs) by challenging companies to develop inclusive business models that engage people at the base of the economic pyramid (BoP). People with less than $8 per day in purchasing power, as consumers, producers, suppliers, distributors of goods and services, and employees, were considered at the BoP.

The BCtA is a unique multilateral alliance supported by donor governments that are supported by the United Nations Development Program that hosts the secretariat. Over 170 companies, ranging from multinationals to social enterprises, and working in 65 countries, are members of BCtA. BCtA member companies are market leaders that provide examples of successful, profitable, and scalable models for reaching poor communities and contributing to global development. We describe in Box 6.22 case studies of two members of BCtA while Box 6.23 illustrates Eureka Forbes water entrepreneurship initiative.

BOX 6.22 NATURA IN MEXICO[72,73]

Natura Cosméticos of Brazil is one of Latin America's largest cosmetics firms. As a member of BtCA, Natura pledged to provide new skills training to low-income sales representatives in Mexico. It integrated all of its nearly 75,000-plus beauty advisers or consultants into a specialized job training program that was aimed at optimizing and upgrading job skills and self-development over three years.

Some 98% of Natura's beauty advisers in Mexico are women who lack formal education or prior work experience. Natura, as part of its focus on sustainability, empowered these advisers with entrepreneurial skills that will serve them well in the future.

The training modules, designed by Natura, increased productivity as well as earning potential by teaching essential business skills. Some of the training offered includes strategic planning, direct sales, computer skills, customer service, accounting, and project management. There was also a focus on creating self-awareness and helping the consultants become agents for change in their own communities.

With a strong focus on human development and basic business skills training, the program also aims to transform sales representatives into inspired entrepreneurs, aligned with the Natura brand. By promoting such entrepreneurship, Natura could increase its sales in Mexico by 134%.

ACCESS AFYA IN NAIROBI[74]

Within Nairobi's slums, health is a serious concern: the majority of residents suffer from preventable diseases and diarrhea as a result of poor sanitation, contaminated water, overcrowding and malnutrition. The limited healthcare options available to slum residents include informal pharmacies with sub-standard medications, and public facilities that lack proper supplies or qualified staff.

Access Aya (Access Health in Swahili) is a Kenyan social enterprise that was established to provide access to some 15,000 residents of Nairobi's slums to quality healthcare, following a commitment to BCtA.

Access Afya was founded on the principles of customer engagement, planning for growth, working through trusted community partners, and utilizing low-cost technologies to provide reliable, affordable healthcare, in low-income urban areas. The company innovated a high-tech healthcare system that provides consistent, quality service, and medication to low-income Kenyans in both micro-clinic and school settings.

To tap into the health market Access Afya developed the micro-clinic concept, which integrates quality clinical care into a community's daily lives. This novel delivery model relies on small community spaces, rigorous quality and customer care, digital information systems, and critical community engagement and education to create demand for products and services. It has adapted the traditional clinic concept to areas where space is a challenge, education is limited, and the trust of communities is crucial; partnerships with local leaders ensure positive working relationships.

Another of Access Afya's innovations is marketing healthcare directly to schools: The Healthy Schools program provides check-ups for children and teachers, health report cards, deworming, hand-washing education, and nutrition counseling within schools. This program not only improves child health, it increases school attendance; a powerful tool for lifting children out of poverty.

BOX 6.23 EUREKA FORBES: WATER ENTREPRENEURSHIP[75]

Eureka Forbes started an initiative captioned water entrepreneurship among slum communities under its Community Fulfilment division to address the challenge of poor water supply and sanitation. Under this initiative Water Shops were designed and implemented to provide drinking water solutions and in this process empower and encourage local people to become stakeholders. The Water Shops created awareness about the need to purify water and create in this process self-sustainable communities.

Water Shops are essentially water dispensing systems that provide quick and easy access to purified drinking water. The water supplied meets all standards and requirements of the World Health Organization, at a price as low as 1–5 Rs (0.014–0.07$) per m^3.

To make the Water Shops work, water scientists and field experts test the water conditions. Eureka Forbes has developed seven unique technologies to combat 17 key water pollutants. Depending on the level of contaminants present in water, the company customizes the technology and installs a community water purification plant. Buyers can get the drinking water anytime (24 × 7), without any operator dependency at any of the dispensing units using their unique identification card called a Water Card.

The Water Shop model is based on cross-sector partnership between *gram panchayats* (local bodies), State governments, NGOs and local communities which aim to promote entrepreneurship in rural areas. Water Shops give villagers the opportunity to not only run the plants but also sell water and generate a business model while working closely with the city-based water distributors and state and central Governments.

DISCUSSION QUESTION

- *The Eureka ForbesWater Entrepreneurship model is fraught with challenges as there has been a reluctance to pay by the communities (despite the low tariff) requiring therefore support of external subsidies to keep the plants running.*[76] *What could be the strategies to overcome such challenges?*

There are examples where businesses have encouraged entrepreneurship in the communities marrying the interest of socio-economic development as well as market penetration. An example of Natura in Box 6.22 illustrates how capacity building of the consultants could increase the sales turnover of Natura's products. We describe a similar case study on Eureka Forbes in India that promoted water entrepreneurship.

Many business organizations have launched or supported the concept of community-based centers as distributed enterprises. These community centers provide a host of services to the people, especially in the rural areas such as electricity, pure water, health and education, marker information, and a wireless Internet connection. Business models are then developed to make the community centers financially sustainable. We describe in Box 6.24 the case of Greenstar's Eco-business.

There are examples where businesses have taken the lead in not just the *supply side* but have also influenced the *demand side*, that is, consumption. Some of these examples have been exemplary. Communities or the user of the services have played an important role in making these efforts successful. We cite in Box 6.25 the case of Opower's Data Analytics

BOX 6.24 GREENSTAR'S ECO-BUSINESS THROUGH COMMUNITY CENTERS[77,78]

Greenstar builds a solar-powered community center that delivers electricity, pure water, health and education information, and a wireless Internet connection, to villages in the developing world. They record art, music, photography, legends, and storytelling in traditional communities, and bring these unique, priceless products to global markets both directly to the consumer, and through licensing to businesses. Products offered on the Web can also include coffee, dried fruits, and spices, as well as handicrafts, art and cultural works like ceramics, brassware, musical instruments, tapestries, etc.

Greenstar returns the revenues from this *digital culture* to the villages to support their ongoing development. The revenue is also used to repay the investors who provide the startup capital necessary to launch Greenstar's Community Center. Villagers own the Greenstar Village Center and become the shareholders.

Greenstar was started in 1998 by a small group of people with varying backgrounds, including a high-tech executive, the head of the US Department of Energy's National Renewable Energy Lab, and a member of the White House Office of Media Affairs. The company presently has a decentralized network of offices in Los Angeles, Boston, Washington DC, Cairo, and Hyderabad, India, and operates across the West Bank, Jamaica, India, Ghana, Brazil, and Tibet.

DISCUSSION QUESTIONS

- *Understand the financial viability of the Community Centres of Greenstar by visiting http://www.greenstar.org/.*
- *Develop an impact assessment framework to assess the environmental and social benefits of the Greenstar program.*
- *Do the community centers of Greenstar interact with each other and to what extent is the cross-pollination of ideas? Do you think it is a good idea to establish a knowledge network across the centres?*

and Knowledge Sharing efforts that has led to reduction in the energy consumption and emissions of GHG.

Many times, communities launch initiatives by setting up cooperatives and drawing support from the local government, businesses, and trusts. The cooperatives operate for the interest of its members and help them towards sustainable livelihoods. The activities undertaken by the cooperatives often lead to improving the sustainability of the neighborhood. We illustrate in Box 6.26 examples of such cooperatives in India and South Africa that provide income generation opportunities to the women while addressing the problem of waste management.

Business and community linkages can take a form of a social enterprise when structured on the basis of a win–win situation. There are examples where business organizations make offerings to the community and involve them to participate, build capacities and take over part of the business. Such business models have shown sustainable outcomes and on a scale where technology is used to its advantage. Renewable energy is a sector where we see considerable experience and success. We cite two interesting cases in

BOX 6.25 OPOWER'S DATA ANALYTICS AND KNOWLEDGE SHARING TO REDUCE ENERGY CONSUMPTION[79]

Opower (acquired by Oracle in 2016) is a software solutions company that is empowering its consumers to manage their own energy efficiently by providing them with knowledge and tools. To implement this project, Opower has partnered with its utility providers around the world, such as PG&E, Exelon, and National Grid.

Opower's big data platform stores and analyzes over 600 billion meter reads from 60 million utility end customers. This helps the utilities to proactively meet regulatory requirements, decrease the cost to serve, and improve customer satisfaction.

By empowering consumers with knowledge about their energy consumption, and by leveraging proven behavior-changing techniques, the company changes the way people think about their energy use. It encourages them to save energy, save money, and reduce carbon emissions. Opower could thus boost energy security, help the environment, and save consumers money. See https://www.oracle.com/industries/utilities/products/opower-energy-efficiency-cloud-service/index.html to view a video on the information driven behavioral change.[80]

DISCUSSION QUESTIONS

- *Why should a utility be interested to use Opower's demand side management program?*
- *Does it make a sense for a utility provider to reduce energy consumption?*

BOX 6.26 STREE MUKTI SANGHATANA (SMS) IN MUMBAI, INDIA[81]

SMS was established in Mumbai with a commitment to improve the situation of marginalized women by imparting training and building expertise to explore sustainable livelihoods and vocations. SMS developed environmental entrepreneurship among urban poor women through training and building effective stakeholder partnerships.

In order to institutionalize its efforts in four cities, SMS has organized 3,000 women into three federations of Self-Help Groups (SHGs) and ten cooperatives. Through this, SMS has encouraged women to take up contracts for waste management projects with local companies and residential associations. This ensures a steady income generation to sustain their families as well as their cooperatives, and also creates a healthy environment, besides providing dignity of labor to this much-maligned community.

SMS has also entered into partnership for technical know-how with the Bhabha Atomic Research Center (BARC), Government of India, and started bio-gas plants for small hotels, canteens, etc. SMS has initiated an experiment of a barter system with companies by collecting dry waste from offices and supplying recycled stationery and consumer items.

More recently SMS has established an Environmental Training Center to provide training in rainwater harvesting, water recycling, and sanitation under one roof for both urban and rural women.

ALL WOMEN PLASTIC RECYCLING IN SOUTH AFRICA

All Women Recycling (AWR) is a small business, based in the Southern Suburbs of Cape Town, that has created a unique product to address the growing problem of plastic waste. The enterprise recycles discarded plastic 2-L PET bottles, which they source from dumpsites, community centers and schools, into greeting cards, and into kliketyklikboxes™.

The kliketyklikbox™, a unique versatile and trendy eco-friendly gift box, is now sold all around the world through agents and distributors. By recruiting and training young women that have been unemployed for two years or more, AWR not only addresses plastic waste pollution, but also tackles unemployment amongst one of the most vulnerable groups of South Africa's society.[82]

Waste to Food (W2F),[83] located in Philippi, decided to address the problem of widespread disposal of organic waste to landfills in South Africa. Their strategy was to contribute to the employment and address the challenge of food security at the same time. Through its partners, W2F collected food waste from large producers, such as retailers and hospitality group, and from markets.

With their innovative processing system, the waste is subsequently recycled into commercial high-quality vermicompost. The organic compost is then sold to commercial clients such as garden centers, seedling producers, and farmers. The compost is also directly applied in community food gardens and in the W2F greenhouse to increase the production of organic vegetables. These vegetables are then consumed by the local community and sold to the retailers. By employing people from the Philippi Township and training them to become independent compost entrepreneurs and gardeners, W2F could offer new livelihood opportunities to address poverty in the community.

Box 6.27 from India on use of rice husk at decentralized power generating units and application of smart micro-grid technology. Both the examples underscore the importance of building local capacities and promotion of eco-entrepreneurships.

6.6 Education and Sustainability

6.6.1 Greening of the Universities

Education is central to the United Nations Environment Program's mandate of *inspiring, informing and enabling nations and peoples to improve their quality of life without compromising that of future generations*. Universities, as the pinnacle of formal, organized education, have a particular responsibility to teach both principles and practices towards sustainability. To assist university administration, teachers, researchers, and students, UNEP came up with a Toolkit.[84] This Toolkit provides strategies, tools, and resources, gleaned from the literature to inspire, encourage, and support to transform both education and campus towards sustainability. Specifically, the Toolkit can be used to establish green, resource-efficient, and low carbon campuses. It was also hoped that the *green campus* will help develop the *green curriculum*, extending beyond institutional boundaries and reach the business and communities.[85]

BOX 6.27 A ELECTRIFYING RURAL INDIA WITH HUSK POWER[86,87]

Today more than 300 villages and hamlets with over 200,000 villagers residing in the rice belt of India are the proud beneficiary of HPS (Husk Power System).

HPS's technology cost-effectively converts agricultural residue (rice husk, mustard stems, corn cobs, certain grasses, etc.) into electricity. A typical plant can serve two to four villages within a radius of 1.5 km, depending on size and population. The plant employs local villagers who are trained by HPS to conduct plant operations. HPS provides end-to-end renewable energy solutions by installing 25-kW to 100-kW mini power plants and then wiring villages and hamlets of up to 4,000 inhabitants to deliver electricity on a pay-for-use basis.

The interested household needs to give a token installation charge of Rs. 100 ($ 1.5). Once installation is done, the consumers pre-pay a fixed monthly fee ranging from $2 to $3 to light up two fluorescent lamps and one mobile charging station. This offers consumers savings of at least 30% over competing kerosene and diesel energy sources (annual savings of up to $50) and a lighting package that can serve the whole household.

HPS takes due care to ensure smooth supply of the husk to its plants. For ensuring this, HPS has evolved a strong relationship with the rice husk suppliers beyond just a buyer–seller relationship. HPS, for example, helps to get insurance for the family members of rice husk suppliers, provides a technician free of cost for the maintenance of the de-husking machines, and gets them a contract agreement for regular purchase of husk at a fixed cost, which is subject to annual revision.

HPS manages a network of decentralized power plants across wide geographies using a cloud-based, real-time monitoring system. Electricity is distributed through micro-grids made from bamboo or other sustainable materials instead of concrete. A custom made pre-paid meter helps the company ensure that consumers utilize no more than the wattage they have paid for. Low-cost transformers developed by the company ensure consistent voltage flow and theft free distribution.

HPS has institutionalized the standard procedures and practices like commissioning of the power plant, and its operation and maintenance schedules. A trained pool of skilled manpower has also been created along with developing a strong network of vendors. There are more than 75 plants in five districts and the organization is expecting to meet the target of electrifying more than 5,000 villages and hamlets by 2017.[88]

To enhance livelihoods in local communities, HPS is currently piloting programs to train local electricians to manufacture simple electronic products, such as pre-paid meters or innovative products such as small refrigeration devices for rural microenterprises.

Bio-char is a waste product of the gasification process that can be used to make incense sticks and char briquettes. HPS employs a group of women from the local villages for this activity providing them with income generating opportunities. HPS is also currently working on additional ways to monetize this waste through silica precipitation.

HPS's solution displaces use of kerosene lamps and diesel-based solutions. Each plant helps avoid 125–150 tons of CO_2 per year, which can be monetized through Certified Emission Reductions (CERs).

GRAM POWER INITIATIVE

Gram Power, a for-profit company that seeks to provide sustainable electricity to rural India, offers a flexible, modular, and reliable smart micro-grid system that can create access to prepaid grid-level electricity for nearly 40% of the world population at costs lower than their current expenses on kerosene. Gram Power has identified that distribution of power is the key challenge in scaling up rural electrification models. Their solution is based on two key components: renewable-based captive generation of energy customized for local energy demands, and the utilization of multiple energy sources and even the utility grid integrated for energy supply.

The Gram Power micro-grid system works along simple lines. Gram Power's Smart Prepaid Meter (patent pending) is installed for every consumer. Thus, power is paid for before consumption. The proprietary grid communication monitors and collects data from all meters. The grid can be managed remotely. There is a configurable load limit on every meter and consumers are charged by hours of consumption. Overload detection and self-recovery on meters are available. Community meters can be provided for communal power usage. The entire system is set up with community support and labor. Gram Power trains the community on system maintenance and safe electricity usage practices. A local entrepreneur is appointed and trained to sell energy. The prepaid meter charges consumers by the hour, just like pre-paid cell phones. A Rs 50 recharge (less than 1$) buys 200 hours of CFL lighting or 50 hours of a fan. The meter constantly displays the number of hours left, and as the load increases, the number of hours keeps reducing. This makes users aware of their power consumption and incentivizes them to conserve power and use energy-efficient appliances.[89,90]

 **DISCUSSION QUESTIONS**

- *A supply of rice husk is critical for distributed generation of biomass-based power in rural areas. Many power generating plants in India and in other countries have suffered and faced losses and, in some cases, even closure. Discuss the strategies employed by Husk Power to overcome this challenge.*
- *Learn more about the micro-grid systems and especially the technological improvements that have taken place in the last decade focusing on energy storage devices.*

The Toolkit addresses the gap between preaching and practicing, where educational institutions not only speak about green, but also practice it. Educational institutions as a stakeholder can influence not only the academic community but the larger community outside. This they can do through not only their curriculum but also through implementing environmental friendly projects on their campuses and in their buildings. The universities can take the lead in bringing change as well as influence the community and business outside their campuses.

The UNEP Toolkit became a part of a wider Greening Universities Initiative established through UNEP's Environmental Education and Training Unit, in collaboration with other UN agencies, under the umbrella of the recently formed Global Universities Partnership for Environment and Sustainability (GUPES). GUPES brings together over 100 Universities from across Africa, Asia and the Pacific, Latin America and the Caribbean, West Asia, Europe, and North America.

There are associations formed to foster and network those interested in sustainability education at the universities. We present one such initiative called Advancing Sustainability in Higher Education in Box 6.28.

6.6.1.1 Green Schools

Green schools have a common goal that they all share, which is their commitment towards environmental sustainability. We present in Box 6.29 two examples, one international initiative by the World Wildlife Fund (WWF) and one by the U.S. Department of Education Green Ribbon Schools. Today, a large number of programs operate on green schools across the world.

6.6.1.2 Sustainability Literacy[91]

Sustainability education is essential today. It is required at all levels but more so in schools, colleges, and at the institutions of higher learning. Greening of Universities and Green Schools are examples where sustainability is getting embedded in education. In addition to the universities and schools, we also need continuing education programs on sustainability for professionals, regulators, and financing institutions. All must become sustainability literate.

Under the efforts to reach Sustainable Development Goals (SDGs), the Higher Education Sustainability Initiative (HESI) has launched a Sustainability Literacy Test (SULITEST). The Sustainability Literacy Test (SuLitest) defines "Sustainability Literacy as the knowledge, skills, and mindsets that help compel an individual to become deeply committed to building a sustainable future and allow him or her to make informed and effective decisions to this end."[92]

SULITEST is an online multiple choice question assessment. It assesses, in 30 minutes, the minimum level of knowledge in economic, social, and environmental responsibility, applicable all over the world, in any kind of higher education institution (HEI), in any country, for students from any kind of tertiary-level course (bachelors, masters, MBAs, and PhD).

BOX 6.28 ADVANCING SUSTAINABILITY IN HIGHER EDUCATION[93]

Advancing Sustainability in Higher Education (AASHE) was formed in the United States in 2005. AASHE empowers higher education faculty, administrators, staff, and students to be effective change agents and drivers of sustainability innovation.

AASHE has developed Sustainability Tracking, Assessment & Rating System™ (STARS) that is a self-reporting framework for colleges and universities to measure their sustainability performance. It operates a Campus Sustainability Hub that is a one-stop shop for AASHE members to access toolkits and resource collections about all aspects of sustainability in higher education, from academics to operations to governance. AASHE's membership is mostly from the universities of North America and Canada. Members include a full spectrum of higher education institutions and organizations, from community colleges to research universities, and from institutions just starting their sustainability programs to long-time campus sustainability leaders. Over 900 higher education institutions, businesses, and nonprofit organizations comprise AASHE's membership base.

BOX 6.29 GREEN SCHOOLS

ECO-SCHOOLS PROGRAM, WORLD WILDLIFE FUND[94]

The Eco-Schools Program by WWF is an initiative offered by the Foundation for Environmental Education (FEE). The idea is to promote sustainable development through action-oriented environmental education. The emphasis is to integrate environmental sustainability in the school curriculum and school life; providing them with a simple framework for them to adopt. The program is structured in a 7-step framework. Once these steps are completed, the schools are awarded with the internationally recognized, Eco-Schools Green Flag award. The award lapses over two years and, hence, needs to be renewed.

The benefits to schools of participating in this program are[95]: An improved school environment; holistic learning through the integration of positive environmental aspects; better integration with the local community through joint efforts to tackle environmental issues; greater student empowerment; and an opportunity to tap into a vast global network of schools involved in the program.

U.S. DEPARTMENT OF EDUCATION GREEN RIBBON SCHOOLS (ED-GRS)

In 2011, the U.S Department of Education created the Green Ribbon Award program, with the aim to inspire districts, private and public elementary, middle, and high schools, and institutions of higher education to employ green school practices. There are three main components or pillars, based on which green schools are awarded[96]:

- Reduction of environmental impacts and costs
- Improvement of occupants' health & performance
- Increase environmental and sustainability literacy civic skills and green career pathways

Awarding the schools is a part of the U.S. Department of Education's effort to identify and communicate good practices that result in improved student engagement, academic achievement, graduation rates, and workforce preparedness; and reinforces federal efforts to increase energy independence and economic security.

 **DISCUSSION QUESTIONS**

- *Why should parents consider sending their children to Green Schools?*
- *Can the school alumni play a role?*
- *Is it necessary that the infrastructure of the school is green—say by the LEED standard?*
- *Is it necessary that the teachers in the school receive a formal training on greening?*

All the questions in this assessment will ensure that future graduates have basic knowledge on sustainable development and both individual and organizational sustainability and responsibility. For this purpose, the scope of this assessment covers two types of questions: Questions on challenges facing society and the planet, that is, general knowledge on social, environmental, and economic issues, basic understanding of the earth, for example, water and carbon cycles, greenhouse effect, etc. Questions on the organization's responsibility in general and on corporate responsibility, that is, questions on practices for integrating social responsibility throughout an organization and questions on the responsibility of individuals as employees and citizens are also asked.

There is now a *Handbook of Sustainability Literacy*: a multimedia version edited by Poppy Villiers-Stuart and Arran Stibbe. This two-part handbook has several interesting topics. You can browse this handbook's online resource by chapter from the paperback and additional chapters, as well as video interviews.[97] This book covers a wide range of skills and attributes from technology appraisal to ecological intelligence, and includes active learning exercises to help develop those skills.

6.6.1.3 Learning Cities

Enabling people to continue learning throughout their lives has become a priority for communities the world over. This is largely due to a growing awareness that lifelong learning, a holistic, inclusive and sector-wide approach to learning, is crucial not just for individuals' well-being, but indeed for the future of society. With their high population densities and complex infrastructures, cities offer particularly favorable conditions for making lifelong learning opportunities available to all of their citizens. Enhancing and expanding such opportunities is at the heart of the learning city approach.

The UNESCO Institute for Lifelong Learning (UIL)[98] defines a learning city as: a city which effectively mobilizes its resources in every sector to promote inclusive learning from basic to higher education; revitalize learning in families and communities; facilitate learning for and in the workplace; extend the use of modern learning technologies, enhance quality and excellence in learning; and nurture a culture of learning throughout life. The benefits of building a learning city include more empowered citizens, improved social cohesion, increased economic and cultural prosperity, and more sustainable development. Thus, more and more urban communities are adopting the learning city approach as a means of unlocking their potential.

The concept of learning cities is defined where there are city level efforts made to improve access to and improve quality of learning to all ages, abilities, ethnicities of people, to all the citizens irrespective of their social standing. The idea is that in providing such learning, the citizens become empowered to improve their city and their lives. Learning is to be lifelong. Many positive impacts are felt as a ripple effect among the community when a city adopts the approach towards becoming a learning city; the impact is on a global level. In this process, concerns regarding urban sustainability get addressed.

UNESCO has launched a learning cities program in 12 cities across the world that includes Melton (Australia), Sorocaba (Brazil), Beijing (China), Bahir Dar (Ethiopia), Espoo (Finland), Cork (Ireland), Amman (Jordan), Mexico City (Mexico), Ybycuí (Paraguay), Balanga (Philippines), Namyangju (Republic of Korea), and Swansea (United Kingdom of Great Britain and Northern Ireland).[99]

We describe in Box 6.30 the Learning Lighthouse in Namyangju in Seoul.

6.6.2 Traditional (Ecological) Knowledge

Traditional ecological knowledge (TEK) refers to knowledge that has been known and passed down for generations. These practices have evolved through trial and error and have been proved flexible enough to adapt to change.[100]

BOX 6.30 LEARNING LIGHTHOUSE IN NAMYANGJU IN SEOUL[101]

Many of Namyangju's residents work in Seoul, which has resulted in a low sense of belonging and community. As in many of the Republic of Korea's cities, a sense of loneliness and alienation is widely reported among Namyangju's citizens. Building social cohesion and community and promoting communication among citizens was therefore a major issue for the city. Namyangju's economy is largely dependent on outside investment. In order to ensure sustainable growth, the city is seeking to foster talent, stimulate innovation, and promote entrepreneurship within the city.

Namyangju's 1-2-3 Lifelong Learning Infrastructure project is improving access to learning for citizens of all ages. Initiated by the citizens themselves, Learning Lighthouses turn unused spaces around the city into community learning spaces. The city government ultimately intends to ensure that no resident is more than a ten-minute walk away from the nearest Learning Lighthouse. Learning Lighthouses not only offer opportunities to learn; they also promote community, cooperation, and active citizenship.

In 2011, 835 citizens participated in Learning Lighthouse programs. In 2014, this number had increased to 10,402. In a survey conducted in 2011, 68% of participants said that the convenient location was the most attractive aspect of Learning Lighthouse programs. Furthermore, 96% stated that such programs are needed. As well as benefitting learners, Learning Lighthouses create jobs by employing Citizen Lecturers. Six hundred Citizen Lecturers were registered with the program in 2014, and of these, 373 taught at 640 Learning Lighthouse programs. This demonstrates that lifelong learning is providing paid employment as well as enabling citizens to contribute their knowledge and skills to their community.

An important feature of Learning Lighthouses is that, while the city administration provided support and guidance, citizens themselves were encouraged to take a proactive role in deciding how their Learning Lighthouses will develop. This ensured that Learning Lighthouses cater to specific local needs, but it also encourages active citizenship. Another important impact of Learning Lighthouses was that they promote communication and cooperation between citizens and the building of networks between Namyangju's various villages. The Learning Lighthouse began as Namyangju's innovative idea, but it is now having a national impact, as many other local governments throughout the Republic of Korea have since been inspired by this project. Fifty different local governments have visited Namyangju since 2011 to see how the project operates.

DISCUSSION QUESTIONS

- *Is the objective of the Lighthouse program only networking and learning and not so much towards actions?*
- *The lighthouse program is not explicitly linked to fostering sustainability. How can this be done? Study other Lighthouse projects under the UNESCO program.*

TEK is also known as *local knowledge, folk knowledge, people's knowledge, traditional wisdom or traditional science*. Sustainable development underpins TEK, and the strong links between the two have been recognized. Traditions often tell about practices carried out in the past on sustainability by various cultures and over generations. Unfortunately, connection between the generations is rapidly reducing especially in the urbanized societies.

We are now realizing that there is a lot to learn from communities who possess the traditional knowledge, especially communities of indigenous people. We present in Box 6.31 a case of the West Arnhem Land Fire Abatement (WALFA) Project in Australia. This knowledge sharing partnership between the Government and indigenous community won the first Eureka Prize for Innovative Solutions to Climate Change in 2007 and was also awarded the Banksia Foundation's Caring for Country Indigenous Award in 2011.

Innovations often come from grass root levels. These innovations need to be tapped, documented, and nurtured to up-scale and replicate while building entrepreneurships. Honey Bee Network is a network of innovators, farmers, scholars, academicians, policy makers, entrepreneurs, and NGOs. The network operates in more than seventy-five countries. The network recognizes the importance of knowledge and recognition of the knowledge provider and a mechanism of sharing the proceeds that are accrued from the value addition of local traditional knowledge and innovation in a fair and just manner. The Honey Bee Network focuses on the traditional knowledge holders and grassroots innovators. These have been the guiding principles of the Network. We present in Box 6.32 the Honey Bee Network in India.

BOX 6.31 THE WEST ARNHEM LAND FIRE ABATEMENT (WALFA) PROJECT, AUSTRALIA

Northern Australia is dominated by a savannah ecosystem that is prone to fires. The indigenous peoples that have lived in this area have historically managed these ecosystems from the fires through customary burning. Savannah fires are a major source of greenhouse gas (GHG) emissions. These emissions contribute to approximately 3% of the annual GHG emissions in Australia.[102]

The WALFA project was the first of its kind fire management project in collaboration with government agencies, scientists, and the carbon markets that used traditional fire management techniques used by the indigenous peoples to control the extent and severity of the fires and GHG emissions. It also helped in conserving the environmental and cultural values of the region.

The project set an annual target of reducing GHG emissions by 100,000 tons of CO_2, in the five years to 2010; however, it abated 707,000 tons.[103]

DISCUSSION QUESTIONS

- *How are indigenous peoples defined? Look for examples where efforts have been made to capture the knowledge resident with indigenous peoples on sustainable living.*
- *Do you think the traditions followed by indigenous peoples are possible to practice or replicate in our modern life? Are they still relevant?*

BOX 6.32 HONEY BEE NETWORK IN INDIA[104]

Over the last twenty years the Honey Bee network in India has documented more than 1,000,000 ideas, innovations and traditional knowledge practices—many of these help in practicing sustainability. Honey Bee, true to its metaphor, has been the source of pollination and cross-pollination of ideas, creativity and grassroots genius, without taking away the nectar from the flower forever.

Acknowledging the very source of the traditional knowledge, the Honey Bee network releases a newsletter in seven Indian languages (Hindi, Gujarati, Tamil, Kannada, Telugu, Malayalam, and Oriya) in addition to English. The regional language versions of the newsletter reach out to the thousands of grassroots knowledge holders, who otherwise would have been alienated from the benefits of knowledge.

The Honey Bee Network in India organizes *Shodh Yatra* (Journey of Discoveries) each year over a duration of a week. A Shodh Yatra is carried out to search for knowledge, creativity and innovations at the grassroots. The Yatra reaches to remote parts of the country and involves mutual sharing of knowledge and practices. The discussions establish a dialogue between the old and future generations.

 **DISCUSSION QUESTIONS**

- *Visit the Honey Bee Network website and look up some interesting innovations in the areas of agriculture and health care (http://www.sristi.org/hbnew/index.php).*
- *Look for examples where the innovations compiled in the Honey Bee Network have been scaled up and commercialized protecting the interest of the grass root innovator.*
- *Similar to the Shodh Yatra, Yatras are conducted on various other themes in India. Examples are Pani (Water) Yatra by the Centre for Science and Environment, New Delhi, and Urja (Energy) Yatra, supported by the Ministry of New Resources of Energy (MNRE). Read about them and make a comparison.*

6.7 Behavioral Change

The real savior to the sustainability of this planet is the change in the way we live. The regulations by the government, leadership taken by the businesses, or investments made by the financing institutions have limitations. The individual and his/her perception for practicing sustainability make a difference. Here, as discussed earlier, education, culture, and traditions matter. What we should be looking for is behavior change.

Consumptive lifestyles have perhaps been one of the major culprits. There have been examples of curbing consumption. Box 6.33 illustrates the initiative *Buy Nothing Day*. It is debatable, however, whether observing such days will make a difference on consumption patterns. We may need to complement such initiatives by raising awareness at the local level on concepts such as Ecological Footprints (see Chapter 1) and by inviting participation of the shopping malls with programs such as Green Public Procurement (discussed in

BOX 6.33 BUY NOTHING DAY[105]

Buy Nothing Day (BND) is an international day to take action against increasing consumerism. It originated in Canada in 1992, and in the United States it is celebrated after Black Friday, one of the world's biggest days of sales of consumer products. It is promoted to change behavioral habits not for that one day but for a lifetime. BND campaigns are held in the United States, the United Kingdom, Israel, Austria, Germany, New Zealand, Japan, the Netherlands, France, Norway, and Sweden. Participation now includes more than 65 nations.

 **DISCUSSION QUESTION**

- *Do you support the idea of Buy Nothing Day? If yes, why, and if not, why not?*

Chapter 2) that will greatly help. To bring in a behavior change, a multitude of simultaneous interventions are needed and on a sustained basis.

In some of the cities in India, people have started banning vehicles on certain streets and zones not just to curb air and noise pollution but open avenues for community engagements. *Raahgiri Day* in Delhi, India,[106] is a weekly street event which provides citizens with the opportunity to reclaim their streets, connect with their community, and celebrate their city. *Raahgiri* is a Hindi word signifying the temporary closure of a network of streets to cars so that they become open to people. The event takes place on all Sundays starting from November 17, 2013, in Gurgaon, New Delhi.

On June 5, 2015, the theme for UNEP's World Environment Day was on sustainable consumption through low carbon living as in *Seven Billion Dreams. One Planet. Consume with Care*. Low carbon living entails reducing the carbon footprint on the planet taking environmentally responsible actions. Some basic steps that can be taken on an individual level are[107]: switching to a car with better fuel economy; making your house better insulated; eating less meat; upgrading your refrigerator and air conditioners; using LED light bulbs; washing clothes in cold water; and buying fewer things and reducing, re-using, and recycling instead.

There have been some futuristic innovations that can help us lead a low carbon lifestyle as well. Take for example AMPY,[108] the world's first wearable motion-charger. It is a portable smartphone battery that charges as you move, by transforming the kinetic energy from your movement for power for your smartphones and wearables. Moreover, it charges just as fast as a wall outlet, while not adding to your carbon footprint! On average, up to one hour of exercise can produce one hour of smartphone battery life for normal use, five hours of battery life in standby use, and up to 24 hours of smartwatch battery life. It contains a 1,800 mAh battery, which will last the whole day. AMPY has also created a free app, which lets you track how much power is generated based on your phone usage and activity level.

Products are now designed to take advantage of densely populated areas. Pavegen, a flooring tile company,[109] converts kinetic energy from people's footsteps into renewable energy. Pavegen has set up streets with such tiles next to places of convention and sports where the commuters walk and generate electricity for street lighting and thus reduce the carbon emissions. In Chapter 5, we have illustrated several such examples.

BOX 6.34 SHARING OF RESOURCES AND COLLABORATIVE CONSUMPTION[110]

Fon attempts to solve the challenge of Internet accessibility for consumers while they are away from their home, office, or other readymade Internet networks. Fon allows home WiFi users to safely share a signal with others through a Fon Spot and in return use others' networks while away from home. All the Fon Spots together create a crowd sourced network where everyone who contributes connects for free.

AirBnB provides a platform for those with an empty room or apartment to rent it out on a short-term basis. ParkatmyHouse enables those with an available parking space, garage, or driveway to rent it out to others in the community.

RelayRides allows private car-owners to rent out their vehicles via an online interface. Zipcar, bought by Avis last year, is a member-based car-sharing company. Zilok provides a platform for owners of things like cameras, cars, or drills to rent them to others.

Sustainable Product Design is sometimes more than the design. It can even include a strategy. On an average 30% of garments in our closets have not been worn for almost a year. Lease A Jeans is a guilt-free solution from a company, Mud Jeans, based in the United States for people that have a desire for newness and are conscious about the environment. This company leases jeans instead of selling them, encouraging customers to return them after use. The old jeans are recycled to make new wonderful items. This leads to a considerable saving of water, energy, and leads to reduced generation of waste.

DISCUSSION QUESTION

- *Does collaborative consumption make a business case? Debate by researching examples such as Mud Jeans.*

Apart from minimizing consumption, concepts of sharing of resources and collaborative consumption have emerged. Box 6.34 describes a few inspirational examples.

The most significant source of food wastage in India is weddings. Indian Weddings are filled with rituals and celebrations for several days where Indians spend lavishly on weddings. A recent survey showed that annually, the Indian city Bangalore alone, wastes 943 tons of quality food during weddings. The team surveyed 75 of Bangalore's 531 marriage halls over a period of six months. As per a study by a team of ten professors from the University of Agricultural Sciences (UAS), Bangalore, this was enough to feed 26 million people a normal Indian meal.[111] Clearly, there are opportunities for recycling the wasted food to those who need it.

Feeding India,[112] which was founded with the object of eliminating hunger, aims to connect hunger and food waste as solutions for each other. They collect the food waste from individuals, weddings, canteens, and other events and redistribute it to the needy at no cost. Started in 2014, it now operates in more than 30 cities across India and has served more than 1.4 million meals. There have been similar such initiatives in India. Box 6.35 illustrates the work carried out by the Annakshetra Foundation Trust in Jaipur, India.

BOX 6.35 ANNAKSHETRA FOUNDATION TRUST IN JAIPUR, INDIA[113]

The Annakshetra Foundation Trust is an organization established in Jaipur that combats food wastage at Indian weddings and other socio-religious occasions like weddings, anniversaries, and birthday parties. Established in November 2010 by the Center for Development Communication (CDC), Jaipur, this non-profit organization collects unutilized surplus food and gives it to the underprivileged section of society through a network of volunteers. Between August 2012 and April 2013, Annakhetra served 64,100 people which is ~237 per day. The recipients of the surplus food were women and children between the ages of 4–15, waste sector workers, and construction laborers. Annakshetra operates in the cities of Jaipur and Allahabad in India.

DISCUSSION QUESTIONS

- *Like a wedding, describe another common event or venue that is a major producer of food waste where the concepts of recycling and sharing of resources have been applied.*
- *Are there any risks associated with donating surplus food? How can these risks be mitigated?*

Endnotes

1. Inclusive growth is economic growth that creates opportunity for all segments of the population and distributes the dividends of increased prosperity, both in monetary and non-monetary terms, fairly across society. See: Inclusive Growth, OECD, online source: http://www.oecd.org/inclusive-growth/
2. For further reading: Low Carbon Green Growth Roadmap for Asia and the Pacific, Case Study: Charging ahead into green lifestyles: Republic of Korea's green credit card, online source: http://www.unescap.org/sites/default/files/32.%20CS-Republic-of-Korea-green-credit-card.pdf
3. Based on: Employee Involvement—Framework on Information and Consultation, European Commission: Employment, Social Affairs & Inclusion, online source: http://ec.europa.eu/social/main.jsp?catId=707&langId=en&intPageId=210
4. You may like to read Country Partnership Frameworks developed by the World Bank http://www.worldbank.org/en/projects-operations/country-strategies and Country Partnership Strategies prepared by the Asian Development Bank at https://www.adb.org/documents/series/country-partnership-strategies
5. Based on: Tidal Thames Pilot Project: Detailed information on engagement approach, Your Tidal Thames, Thames21 and Thames Estuary Partnership, 2012, online source: http://www.thames21.org.uk/wp-content/uploads/2013/07/Detailed-information-on-Engagement-process.pdf
6. Based on: About Us, Your Tidal Thames, 2012, online source: http://www.thames21.org.uk/project/your-tidal-thames/
7. Based on: Change by US, 2017, online source: http://newyork.thecityatlas.org/change/
8. Based on: Digital Sustainability Conversations—How Local Governments can Engage Residents Online, pp. II-22, 2012, online source: http://sustainablecommunitiesleadershipacademy.org/resource_files/documents/community-social-engagement-guidebook-and-case-studies.pdf

9. Based on: City of Vancouver, 2014, online source: http://vancouver.ca/green-vancouver.aspx
10. Based on: How can we reach our 2020 Greenest City Targets?, City of Vancouver, 2010, online source: https://vancouver.uservoice.com/forums/56390-gc-2020/suggestions/857509-drinking-water-fountains-bottled-water-ban#comments
11. Taken from: More than street art—Fighting air pollution through graffiti, *Georgian Journal*, 2015, online source: http://www.georgianjournal.ge/arts-a-culture/31599-more-than-street-art-fighting-air-pollution-through-graffiti.html
12. Based on: River Listening, Leah Barclay, 2015, http://www.australianmusiccentre.com.au/article/river-listening
13. Based on: Hard Rain Project, online source: http://www.hardrainproject.com/d
14. Most of the text is taken from the source: River Listening, Leah Barclay, 2015, http://www.australianmusiccentre.com.au/article/river-listening
15. Most of the text has been taken from the source: Malhotra, K., Integrated Community Sustainability Plan—Town of Antigonish, Nova Scotia, 2010, online source: https://www.townofantigonish.ca/departments/planning-building-services/42-plans-and-strategies-integrated-community-sustainability-plan/file.html
16. Based on: Redkar, S., *Enhancing Solid Waste Management Capacity of Local Government Authority & Community Participation*, Advanced Locality Management, online source: http://emcentre.com/wasterecycling/presentation/MCGM_%20Seema.pdf
17. For further reading: Menon, S., Participatory Budgeting, Education for Change, online source: http://www.desd.org/efc/Participatory%20Budgeting.htm
18. Based on: Keruwala, N. et al., Presentation on Participatory Budgeting in Pune, online source: http://spa.ac.in/writereaddata/Session6bMrNaimKeruwalaArKomalPotdarandMsMayaRoy.pdf
19. Taken from: Involving waste-pickers to improve door-to-door collection, online source: http://www.swachcoop.com/pdf/wastepickerstoimprovedoor-to-doorcollection.pdf
20. For further reading: Saving Kenya's Black Rhinos, WWF Global, online source: http://wwf.panda.org/?210210/Saving--Kenyas-Black-Rhinos
21. For further reading: Rhino Return to Samburu, LEWA Wildlife Conservancy, 2015, online source: http://www.lewa.org/nc/stay-connected/news-on-lewa/news-on-lewa/article/rhino-return-to-samburu/?tx_ttnews%5BbackPid%5D=75
22. For further reading: United Nations Development Programme, Monks Community Forest, Cambodia, Equator Initiative Case Study Series, New York, NY, 2012, online source: http://www.kh.undp.org/content/dam/cambodia/docs/EnvEnergy/Case%20study_Monks%20Community%20Forest_Eng.pdf
23. Based on: Chi Phat Community-Based Ecotourism in the Cardamom Mountain Range, UIAA, 2016, online source: https://theuiaa.org/conservation-of-biodiversity/chi-phat-community-based-ecotourism-in-the-cardamom-mountain-range-cambodia/
24. Based on: Talwar, A., Saving the Whale Shark—No Room for Ahabs, Sustainability Zero, 2017, online source: http://sustainabilityzero.com/interview_alkatalwar/
25. Based on: Tata Chemicals' Whale Shark Conservation Campaign, CSR Mandate, 2016, online source: http://csrmandate.org/tata-chemicals-whale-shark-conservation-campaign/
26. Based on: The Chipko Movement, online source: http://edugreen.teri.res.in/explore/forestry/chipko.htm
27. For further reading: Chipko: an unfinished mission, 1993, Down to Earth, online source: http://www.downtoearth.org.in/coverage/chipko-an-unfinished-mission-30883
28. Based on: SunderlalBahuguna, the defender of Himalayas, turns 89: Some facts On Chipko Movement, 2016, http://indiatoday.intoday.in/education/story/sunderlal-bahuguna/1/566324.html
29. Based on: Forest Heroes, Our Planet, online source: http://www.ourplanet.com/tunza/issue0403en/pages/forest_8.html
30. Taken from: Smart City Air Challenge, online source: https://www.challenge.gov/challenge/smart-city-air-challenge/

31. Based on: Citizen Science projects, EPA Environment Protection Authority Victoria, online source: http://www.epa.vic.gov.au/our-work/programs/citizen-science-program/citizen-science-projects
32. Taken from: Air Quality Egg: community-led sensing network, online source: http://airqualityegg.com/
33. Visit: Smart Citizen v2 BETA—Open source technology for citizens political participation in smarter cities, online source: https://smartcitizen.me/
34. Visit: Kids making sense, 2016, online source: http://kidsmakingsense.org/
35. Visit: EnviroDIY: An initiative of Stroud Water Research Center, 2017, online source: https://envirodiy.org/
36. Taken from: EnviroDIY: An initiative of Stroud Water Research Center, 2017, https://envirodiy.org/about/
37. Based on: Hale, R., Environmental monitoring to surge via potential super sensors, Science Network Western Australia, 2016, online source: https://phys.org/news/2016-07-environmental-surge-potential-super-sensors.html
38. Taken from: PLEASED—Plants Employed As SEnsing Devices, online source: http://pleased-fp7.eu/
39. Taken from: DENR Administrative Order No. 96-37 December 02, 1996, online source: http://oneocean.org/download/db_files/denr_ao_96-37.pdf
40. Taken from: Chapter 8, *Environmental Compliance Monitoring*, online source: http://www.geocities.ws/denrcaraga/CHAP8.html
41. Based on: DENR Administrative Order No. 2003-30 on Implementing Rules and Regulations (IRR) for the Philippine Environmental Impact Statement (EIS) System
42. For further reading: Niyati, M., Judicial Activism for Environment Protection in India, *International Research Journal of Social Sciences*, Vol. 4(4), pp. 7–14, 2015, online source: http://www.isca.in/IJSS/Archive/v4/i4/2.ISCA-IRJSS-2014-327.pdf
43. For further reading: Bangladesh Environmental Lawyers Association, Climate Action Network International, online source: http://www.climatenetwork.org/profile/member/bangladesh-environmental-lawyers-association-bela
44. For further reading: Public Interest Litigation, Bangladesh Environmental Lawyers Association, online source: http://www.belabangla.org/public-interest-litigation/
45. Based on: Dirty and Dangerous shipbreaking in Chittagong, Bangladesh, EJAtlas, online source: https://ejatlas.org/print/dirty-and-dangerous-shipbreaking-in-chittagong
46. For further reading: A pioneer in waste management, *The Hindu*, online source: http://www.thehindu.com/lf/2003/12/14/stories/2003121408150200.htm
47. For further reading: A woman's battle to keep waste from ending up in landfills, *The Hindu*, online source: http://www.thehindu.com/news/cities/chennai/a-womans-battle-to-keep-waste-from-ending-up-in-landfills/article5023257.ece
48. Based on: Important High Court/Supreme Court Orders, (7) Public Interest Litigation No. 888/1996 Almitra H. Patel & Ars. v/s Union of India, Maharashtra Pollution Control Board, online source: http://www.mpcb.gov.in/legal/imphighcourt7.php
49. Visit: https://www.facebook.com/SwachhWarangal/ to learn about the Warangal experiment
50. You may like to watch the video on 'Alang, Gujarat: World's biggest ship breaking yard turns killer' on Youtube at: https://www.youtube.com/watch?v=iHlvAE1me5Y
51. Based on: About Us, The World Bank, online source: http://ewebapps.worldbank.org/apps/ip/Pages/AboutUs.aspx
52. Based on: Compliance Review Panel, Asian Development Bank, online source: https://lnadbg4.adb.org/dir0035p.nsf/alldocs/BDAO-7XGAGP?OpenDocument&expandable=
53. Based On: Learning from Narmada, Independent Evaluation Group (IEG), http://lnweb90.worldbank.org/oed/oeddoclib.nsf/DocUNIDViewForJavaSearch/12A795722EA20F6E852567F5005D8933
54. Based on: Investigations Report Released, The World Bank, online source: http://ewebapps.worldbank.org/apps/ip/Pages/2017%20January%20Newsletter.html

55. For further reading: 2016 Learning Report on the Implementation of the Accountability Mechanism Policy, Accountability Mechanism Learning Report, ADB, 2016, https://www.adb.org/sites/default/files/institutional-document/193411/am-learning-report-2016.pdf
56. Based on: *Economic and Community Development Project Strategies*, Rotary, p. 4, online source: https://www.rotary.org/myrotary/en/learning-reference/about-rotary/economic-and-community-development
57. Based on: Rotarians taking action to empower communities, 2015, online source: https://rotaryservice.wordpress.com/tag/microcredit/
58. Based on: Making an impact in Honduras through economic and community development, 2016, online source: https://rotaryservice.wordpress.com/2016/10/27/making-an-impact-in-honduras-through-economic-and-community-development/
59. Based on: MICROCAPITAL STORY: Participatory Sustainable Waste Management Project Extends Microfinance to Informal Recyclers in Brazil, 2009, online source: http://www.microcapital.org/microcapital-story-participatory-sustainable-waste-management-project-extends-microfinance-to-informal-recyclers-in-brazil/
60. Based on: Microfinance for solar power, One World South Asia, online source: http://southasia.oneworld.net/features/microfinance-for-solar-power#.WOIdJTuGPb1
61. Based on: http://www.millionsoul.iitb.ac.in/home_about_the_project.html
62. Based on: A new concept from Gramalaya Smart Toilet, Annual Report 2015–2016, online source: http://guardianmfi.org/pdf/annual-report-2016.pdf
63. Based on: Townsend, C.J., A Better Way to 'Buy One, Give One', 2014, online source: https://www.forbes.com/sites/ashoka/2014/10/08/a-better-way-to-buy-one-give-one/#5a0f5945485e
64. Based on: Mealshare, 2017, online source: http://www.mealshare.ca/en/home/
65. Based on: 3M, *Sustaining our future-* Pollution, Prevention, Pays (3P) moving 3M towards sustainability, 2008, [http://solutions9.3m.com/3MContentRetrievalAPI/BlobServlet?locale=en_US&lmd=1240969645000&assetId=1180581674144&assetType=MMM_Image&blobAttribute=ImageFile], [http://www.ideasaccelerator.com/wp-content/uploads/2011/06/3M-3P-Presentation.pdf]
66. Based on online source: https://9to5google.com/2013/08/16/googles-20-percent-time-birthplace-of-gmail-google-maps-adsense-now-effectively-dead/
67. Based on: Solving for sustainability, Google, online source: https://www.google.com/green/efficiency/oncampus/
68. For further reading: https://www.starbucks.com/responsibility/sourcing/coffee
69. For further reading: https://www.scsglobalservices.com/starbucks-cafe-practices
70. Based on: https://www.google.co.in/url?sa=t&rct=j&q=&esrc=s&source=web&cd=3&cad=rja&uact=8&ved=0ahUKEwip5fmyZPRAhWKs48KHX9uDeYQFgglMAI&url=https%3A%2F%2Fwww.mahindralifespaces.com%2Fdownload-file%2Fmldlcsrpol-48c78307a44cca6.pdf&usg=AFQjCNGlxxVceOKO3bBTprrOUACFVBNMpw&bvm=bv.142059868,d.c2I
71. Taken from: http://www.businesscalltoaction.org/
72. Taken from: Natura: Multi-level sales for multi-level impact, Business Call to Action, online source: http://www.businesscalltoaction.org/sites/default/files/resources/bcta_casestudy_natura_web.pdf
73. Taken from: http://www.undp.org/content/undp/en/home/presscenter/pressreleases/2013/03/08/natura-cosmeticos-empowers-women-in-mexico.html
74. Taken from: http://www.businesscalltoaction.org/member/access-afya-
75. Taken from: Eureka Forbes is promoting water entrepreneurship among India's slum communities, 2016, online source: https://yourstory.com/2016/02/eureka-forbes-water-shops/
76. Based on: Eureka Forbes Purifies Water for Rural India, *Forbes India*, 2013, online source: http://www.forbesindia.com/article/real-issue/eureka-forbes-purifies-water-for-rural-india/35737/1
77. Taken from: Greenstar Solar Community Center, Greenstar Foundation, 2008, online source: http://www.greenstar.org/

78. Taken from: Introduction to Greenstar, 2004, online source: http://www.greenstar.org/introduction.htm
79. Taken from: Oracle buys Opower, 2016, https://www.oracle.com/corporate/pressrelease/oracle-buys-opower-050216.html
80. See details at https://www.oracle.com/corporate/acquisitions/opower/index.html
81. Based on: Environmental Entrepreneurship & Women Waste Pickers—Commitment by Stree Mukti Sanghatana, Clinton Foundation, online source: https://www.clintonfoundation.org/clinton-global-initiative/commitments/environmental-entrepreneurship-women-waste-pickers
82. Based on: All Women Recycling, 2011, online source: http://www.allwomenrecycling.com/recycling-products.php
83. Based on: https://www.adelphi.de/en/publication/waste-food
84. Access the Toolkit at: https://wedocs.unep.org/bitstream/handle/20.500.11822/11964/Greening%20University%20Toolkit%20V2.0.pdf?sequence=1&isAllowed=y
85. Taken from: UNEP Green Universities Toolkit, online source: http://www.sustainabilityexchange.ac.uk/unep-greening-universities-toolkit
86. Taken from: Empowering Rural India the RE Way: Inspiring Success Stories, Ministry of New and Renewable Energy (MNRE), Government of India (GoI), and UNDP, 2012, online source: http://mnre.gov.in/file-manager/UserFiles/compendium.pdf
87. Taken from: Business Model, Husk Power Systems, http://www.huskpowersystems.com/innerpagedata.php?pageT=Business%20Model&page_id=77&pagesub_id=114
88. Taken from: Husk Power Systems Electrify India, online source: http://acumen.org/wp-content/uploads/2013/03/Husk-1024x500.jpg
89. Taken from: GramPower, 2015, online source: https://www.grampower.com/
90. Taken from: Sanghvi, N., How a Start-up with UC Berkeley Roots is setting up smart microgrids in India's Remotest Regions, BERC- Berkeley Energy & Resources Collaborative, 2015, online source: http://berc.berkeley.edu/start-uc-berkeley-roots-setting-smart-microg rids-indias-remotest-regions/
91. Taken from: Sustainability Literacy Test (SULITEST) of the Higher Education Sustainability Initiative (HESI), https://sustainabledevelopment.un.org/partnership/?p=9551
92. For further reading: http://sulitest.org/en/index.html for more details on SuLitest
93. Visit: http://www.aashe.org/ for more details
94. Taken from: http://www.wwf.sg/for_schools/eco_schools_programme/
95. Taken from: http://www.wwf.sg/for_schools/eco_schools_programme/
96. Based on: The Center for Green Schools, U.S. Green Building Council, 2017, online source: http://www.centerforgreenschools.org/green-school
97. Access the handbook at http://arts.brighton.ac.uk/stibbe-handbook-of-sustainability
98. Based on: http://unesdoc.unesco.org/images/0023/002345/234536E.pdf
99. Taken from: http://learningcities.uil.unesco.org/key-features/purpose
100. For further reading: Melchias, G., *Biodiversity and Conservation*, Science Publishers, Inc, Enfield, 2001.
101. Based on: http://www.uil.unesco.org/lifelong-learning/project/namyangjus-learning-light-houses. See Namyangu's video at https://youtu.be/AsOhZMq_xvs
102. For further reading: Narayan, M., Reardon-Smith, K, Griffinths, G., Apan, A., Savanna burning methodology for fire management and emissions reduction: A critical review of influencing factors, *Carbon Balance and Management*, 2016, online source: https://cbmjournal.springeropen.com/articles/10.1186/s13021-016-0067-4
103. Based on: WALFA—West Arnhem Land Fire Abatement Project, NAILSMA: North Australian Indigenous Land and Sea Management Alliance, 2012, online source: https://www.nailsma.org.au/walfa-west-arnhem-land-fire-abatement-projecthtml.html
104. Taken from: *Honey Bee*, 2016, online source http://www.sristi.org/hbnew/aboutus.php
105. Based on: https://www.adbusters.org/bnd/

106. Taken from: http://www.sutp.org/en/news-reader/raahgiri-day-the-first-weekly-car-free-event-in-india.html
107. Taken from: http://www.ucsusa.org/sites/default/files/legacy/assets/documents/global_warming/Cooler-Smarter-Top-Ten-List.pdf
108. Taken from: http://www.getampy.com/ampy-move.html
109. For further reading: http://www.pavegen.com/ to know more
110. Taken from: Model Behavior—20 Business Model Innovations for Sustainability, SustainAbility, pp. 46, 2014, online source: http://sun-connect-news.org/fileadmin/DATEIEN/Dateien/New/model_behavior_20_business_model_innovations_for_sustainability.pdf
111. Read the post by Sahithi Andoju in Society, Volunteerism at https://www.youthkiawaaz.com/2017/03/what-you-can-do-about-food-wastage-in-india/
112. For further reading: https://www.feedingindia.org/ for more details
113. For further reading: http://www.annakshetra.org/ for more details

Index

D

E

F

G

H

I

T

U

V

W

X

Y

Z

PGSTL 12/09/2017